T0091525

An Introduction to Complex Systems

Paul Fieguth

An Introduction to Complex Systems

Society, Ecology, and Nonlinear Dynamics

Paul Fieguth
Faculty of Engineering
University of Waterloo
Waterloo
Ontario, Canada

ISBN 978-3-319-83093-3 ISBN 978-3-319-44606-6 (eBook)
DOI 10.1007/978-3-319-44606-6

Softcover reprint of the hardcover 1st edition 2016

Printed on acid-free paper

This Springer imprint is published by Springer Nature
The registered company is Springer International Publishing AG
The registered company address is: Gewerbestrasse 11, 6330 Cham, Switzerland

Preface

Although I had studied nonlinear dynamic systems, bifurcations, and bistable systems as a graduate student, I had clearly never really *absorbed* the ideas at a conceptual level, since it was ten years later that I was struck by a simple figure of catastrophic nonlinear state transitions in the context of ecological systems. I was immediately hooked: the concept was so clear, so elegant, and so easy to understand. How was it possible that I had never really encountered this?

Over time I became convinced that not just I, but indeed most of the undergraduate students with whom I interact, fundamentally do not grasp the big picture of the issues surrounding them, even though the underlying mathematical concepts are well within reach. The underlying problem is clear: for pedagogical reasons nearly all of the courses which my students take focus on mathematics and systems which are small scale, linear, and Gaussian. Unfortunately, there is not a single large-scale ecological or social phenomenon which is scalar, linear, and Gaussian.

This is, very simply, the rationale for this text: to explore a variety of large issues, global warming, ice ages, water, and poverty, and to motivate and teach the material of the course, nonlinear systems, non-Gaussian statistics, spatial systems, and complex systems — motivated by these case studies.

The large-scale problems challenging the world are complex and multifaceted and will not be solved by a single strategy, academic field, or perspective. This book cannot claim to teach how to solve such enormous problems; however, the intent is very much to draw explicit parallels and connections between the mathematical theory, on the one hand, and world issues/case studies on the other.

The specific topics being taught are nonlinear dynamic systems, spatial systems, power-law systems, complex systems, and inverse problems. To be sure, these fields have been around for some time, and many books have been written on the subjects; however, the fields are, at best, only weakly present in most undergraduate programs.

This book is intended for readers having a technical background, such as engineering, computer science, mathematics, or environmental studies. The associated course which I have taught is open to third and fourth year undergraduate students; however, this book should, I hope, be of interest and mostly accessible to a significantly wider audience. The only actual prerequisites are some background in algebra and in probability and statistics, both of which are summarized in the appendices. The reader who would prefer to get a perspective of the text might prefer to first read the two overview chapters, on *global warming* in Chapter 2 and on *water* in Chapter 12.

There are many online resources related to nonlinear dynamics and complex systems; however, online links can frequently change or become outdated, so I am reluctant to list such links here in the text. Instead, I am maintaining a web page associated with this book, at

http://complex.uwaterloo.ca/text

to which the reader is referred for further reading and other material.

A number of people were of significant support in the undertaking of this textbook. Most significantly, I would like to thank my wife, Betty Pries, who was tireless in her enthusiasm and support for this project and regularly articulated the value which she perceived in it. My thanks to Professor Andrea Scott and Doctor Werner Fieguth, both of whom read every page of the book from beginning to end and provided detailed feedback. Appreciation to Doctor Christoph Garbe, my host and research collaborator at the University of Heidelberg, where much of this text was written.

Teaching this material to students at the University of Waterloo has allowed me to benefit from their creative ideas. Here, I particularly need to recognize the contribution of the project reports of Maria Rodriguez Anton (discount function), Victor Gan (cities), Kirsten Robinson (resilience), Douglas Swanson (SOC control), and Patrick Tardif (Zipf's law).

Finally, my thanks to the contributions of my children:

- Anya, for allowing her artwork to appear in print in Figure A.2
- Thomas, for posing at Versailles and appearing in Example 3.1
- Stefan, for demonstrating an inverted pendulum in Figure 5.8

Waterloo, Ontario, Canada Paul Fieguth

Contents

List of Examples

Chapter 1
Introduction

For every complex problem there is an answer that is clear, simple, and wrong

Paraphrased from H. L. Mencken

The world is a complex place, and simple strategies based on simple assumptions are just not sufficient.

For very good reasons of pedagogy, the vast majority of systems concepts to which undergraduate students are exposed are

Linear: Since there exist elegant analytical and algorithmic solutions to allow linear problems to be easily solved;
Gaussian: Since Gaussian statistics emerge from the Central Limit Theorem, are well behaved, and lead to linear estimation problems;
Small: Since high-dimensional problems are impractical to illustrate or solve on the blackboard, and large matrices too complex to invert.

Unfortunately, nearly all major environmental, ecological, and social problems facing humankind are *non*-linear, *non*-Gaussian, and *large*.

To be sure, significant research has been undertaken in the fields of large-scale, nonlinear and non-Gaussian problems, so a great deal *is* in fact known, however the analysis of such systems is really very challenging, so textbooks discussing these concepts are mostly at the level of graduate texts and research monographs.

However there is a huge difference between *analyzing* a nonlinear system or deriving its behaviour, as opposed to understanding the *consequences* of a system being nonlinear, which is much simpler. It is the latter consequences which are the scope and aim of this book.

That is, although a detailed analysis of a complex system is, in most cases, too difficult to consider teaching, the consequences of such systems are quite easily understood:

A Nonlinear System is subject to irreversibility, such that given some change in the inputs to the system, undoing the change does *not* necessarily return the system to its start, whereas all linear systems are reversible. Furthermore nonlinear systems can be subject to discontinuous or catastrophic state changes, which is not possible in linear systems.

P. Fieguth, *An Introduction to Complex Systems*,
DOI 10.1007/978-3-319-44606-6_1

Non-Gaussian/Power-Law Systems may be characterized by extreme behaviours which would appear to be unpredictable or unprecedented based on historical data. In contrast, Gaussian statistics converge reliably and effectively assign a probability of zero to extreme events, giving a false sense of security in those circumstances where the underlying behaviour is, in fact, power-law.

Large Coupled Nonlinear Spatial Systems, also known as complex systems, can give rise to highly surprising macroscopic behaviour that would scarcely be recognizable from the microscopic model, a phenomenon known as emergence. Managing such systems, particularly in response to some sort of failure, is very difficult. In contrast, the nature of linear systems is unaffected by scale.

The sorts of systems that we are talking about are not arcane or abstract, rather there are well-known systems, such as geysers and toy blocks, or indeed everyday systems such as stock markets or weather. It is precisely because such systems are so common and everyday that engineers or other technical professionals are likely to encounter them and need to be informed of the possible consequences of interacting with and influencing such systems.

The context of this text is sketched, necessarily oversimplified, at a high level in Figure 1.1. Essentially we have two interacting classes of systems, the human/societal/economic and natural/ecological/environmental systems, both of which will exhibit one or more elements of non-linearity and spatial interaction

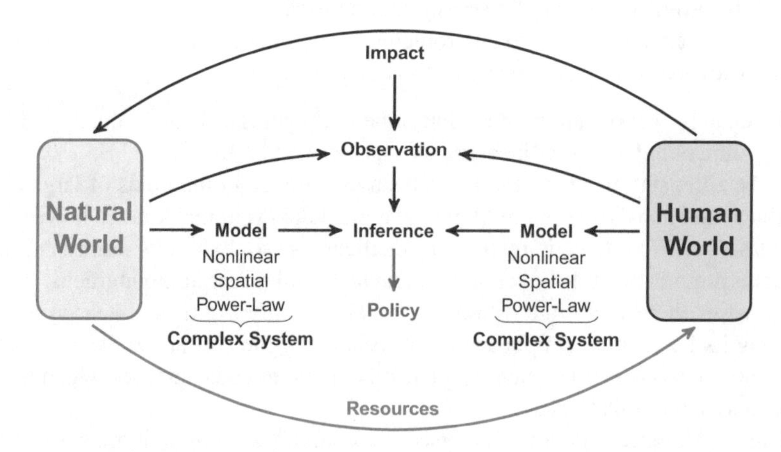

Fig. 1.1 An overview: human/societal/economic systems (right) draw resources from and have an impact on natural/ecological/environmental systems (left). Both domains contain many examples of systems which exhibit one or more elements of non-linearity and spatial interaction, leading to complex-systems and power-law behaviours. Observations of one or more systems are combined with models in an inference process, leading to deeper insights and (ideally) better policy. The red portions are the focus of this text, with the blue, green, and grey components illustrating the broader, motivating context.

which lead to complex-systems and power-law behaviours. Since a model alone is of limited utility, we are interested in performing inference by combining models with measurements, particularly global-scale remotely-sensed measurements from satellites.

1.1 How to Read This Text

This text is aimed at a target audience of undergraduate engineering students, but is intended to be more broadly of interest and accessible.

Those readers unfamiliar with this text, or with complex systems in general, may wish to begin with the overview in Chapter 2, followed by a survey of the case studies which are presented at the end of every chapter, and which are listed on page xi.

An explicit goal of this text is not to focus attention on the mathematics behind complex systems, but to develop an understanding of the *interaction* between complex systems theory and the real world, how complex systems properties actually manifest themselves. For this reason there are, in addition to the end-of-chapter case studies, a large number of examples, listed on page xi, and those readers more interested in a high level or qualitative understanding of the book may want to start by focusing on these.

For those readers interested in the mathematics and technical details, the chapters of the book are designed to be read in sequence, from beginning to end, although the spatial and power law chapters can be read somewhat independently from the preceding material, Chapter 4 through Chapter 7, on dynamics and nonlinear systems.

Complex systems can be studied and understood from a variety of angles and levels of technical depth, and the suggested problems at the end of every chapter are intended to reflect this variety, in that there are problems which are mathematical/analytical, computational/numeric, reading/essay, and policy related.

This book is, to be sure, only an introduction, and there is a great deal more to explore. Directions for further reading are proposed at the end of every chapter, and the bibliography is organized, topically, by chapter.

References

1. C. Martenson. *The Crash Course: The Unsustainable Future of our Economy, Energy, and Environment*. Wiley, 2011.
2. M. Scheffer. *Critical transitions in nature and society*. Princeton University Press, 2009.
3. A. Weisman. *The World Without Us*. Picador, 2007.
4. R. Wright. *A Short History of Progress*. House of Anansi Press, 2004.

Chapter 2
Global Warming and Climate Change

There are few global challenges as widespread, as complex, as significant to our future, and politically as controversial as that of global warming.

The goal of this chapter is, in essence, to motivate this textbook; to convince you, the reader, that very nearly *all* of the topics in this text need to be understood in order to grasp the subtleties of a subject such as global warming.

However the specific problem of global warming is not at all unique, in this regard. That is, after all, the premise of this book: that there is a wide variety of ecological and social challenges, all of which are highly interdisciplinary, and for which a person unfamiliar with one or more of systems theory, nonlinear dynamics, non-Gaussian statistics, and inverse problems is simply ill-equipped to understand.

P. Fieguth, *An Introduction to Complex Systems*,
DOI 10.1007/978-3-319-44606-6_2

Other similarly broad problems would include

- Ecological pressures and extinction,
- Human Poverty,
- Energy, and
- Water, to which we shall return in Chapter 12.

Allow me now to take you on a tour of the entire book, through the lens of global warming.

Chapter 3: Systems Theory

Global warming is the warming of the earth's climate caused by an increase in the concentrations of carbon-dioxide and methane gases in the atmosphere, due to human industry, fossil fuel consumption, and land use.

Suppose we begin with a rather naïve model of atmospheric carbon:

Human society, as a system, interacts with the rest of the world through the boundaries of the system. The naïve model has simplistically assumed fixed boundary conditions, that the sources (energy) and sinks (carbon pollution) are infinite, implying that humans do not influence the global climate.

A more realistic model, but also more complicated, understands energy sources and carbon-dioxide sinks to be finite, and therefore subject to influence by human activity:

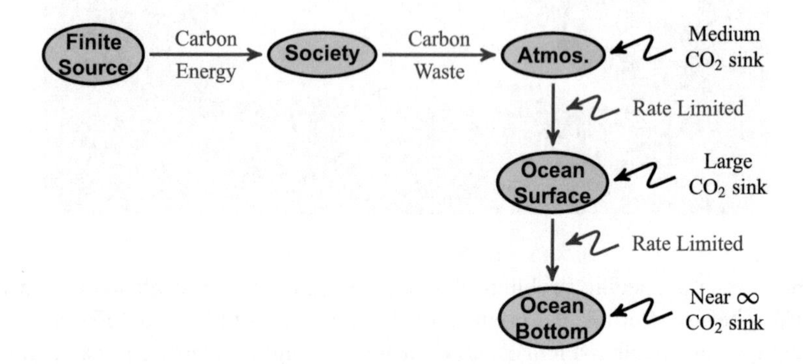

So we cannot understand a system, such as a human society, in the absence of the *interactions* that the system has with others around it, such as how rapidly or easily carbon can flow from one place to another.

Chapter 4: Dynamic Systems

An additional subtlety is that the latter model is implicitly time-dynamic: there are *rates* of carbon transfer, and a continual transfer of carbon into a finite system implies a system which changes over time.

Two time-dynamic changes associated with carbon are indisputable:

1. CO_2 concentrations in the atmosphere are increasing over time, from 315 ppm in 1960 to around 400 ppm today;
2. Since 1800 the world's upper oceans have increased in acidity by 0.1 pH point, corresponding to a 30 % increase in acidity, due to the uptake of CO_2 into the ocean and the formation of carbonic acid [2].

On the other hand, when we try to articulate questions of global warming, or any other sort of climate change over time, the problem becomes much more slippery:

What Is Our Baseline? We know that the climate has always been changing, switching between epochs with and without ice ages, for example. Indeed, the earth's climate changes over time periods of all scales:

Tens of millions of years . . .	Tropical Jurassic (dinosaur) era
Tens of thousands of years . . .	Ice ages
Hundreds of years . . .	Medieval warming period (950–1250 AD), So-called little ice age (1550–1850 AD)
Years . . .	1930's dust bowl, El-Niña / El-Niño

So at what point in time do we actually *start* measuring, in order to know whether the earth is warming or not?

What Is Actually Warming? The atmosphere is well-mixed and relatively easy to measure, whereas the heat uptake patterns in the oceans and the ground are far more variable, depending on the presence of subsident currents in the ocean, or geothermal activity underground.

On the other hand, although land and ocean have a far greater mass than the atmosphere, it is primarily the upper surfaces of land and ocean which would be warming, only a small fraction of the total:

Atmosphere	$6 \cdot 10^{14}$ kg per metre at sea level,	$5 \cdot 10^{18}$ kg in total
Ocean	$3 \cdot 10^{17}$ kg per metre of depth,	$1 \cdot 10^{21}$ kg in total
Land	$4 \cdot 10^{17}$ kg per metre of depth,	Ill-defined total mass

Furthermore it is primarily the upper surfaces of land and ocean which are biologically active, and thus a warming of only this top sliver may produce a disproportionate ecological impact. So is *global warming* . . .

1. a *physical concept*, a warming of the land-oceans-atmosphere in total, which can be measured as increases in global average temperature?, or
2. an *environmental concept*, a disturbance to ecological balance caused by the warming of some *part* of the land-oceans-atmosphere, where the localized warming may produce almost no impact on the global average temperature, and so must be assessed indirectly via some other measurement?

What Is Causing the Warming? The increase, over time, in human fossil-fuel consumption is well documented. Similarly the increase, over a similar period of time, in atmospheric CO_2 concentration has been accurately measured. However a *correlation*, a statistical relationship, between two events is not the same as *causation*, where one event can be said to cause another.

Measuring a correlation is an exceptionally simple statistical test, whereas causation requires a much deeper understanding. In the case of global warming we need to understand the carbon cycle: the industrial and natural global sources and sinks of carbon.

Chapter 7: Coupled Nonlinear Systems

On top of this, in trying to model the presence or flow of heat and energy present in the ocean-atmosphere system, over half of all kinetic energy is not in atmospheric storms or in ocean currents, rather in ocean eddies.

However an eddy, like other similar forms of oceanic and atmospheric turbulence, is one of the hallmarks of coupled nonlinear systems, a nonlinear relationship between multiple elements. No linear system, regardless how complex or large, will exhibit turbulent behaviour, so we cannot fully model the ocean-atmosphere system without studying nonlinear systems.

Chapter 8: Spatial Systems

The atmosphere and oceans are not just nonlinear, or coupled, but indeed very *large* spatially. The governing equation for both air and water flow is the Navier–Stokes nonlinear partial differential equation:

Water/Ocean	Navier–Stokes	*Small* spatial details	Challenging Misfit
	Incompressible	*Slow* changes over time	
Air/Atmosphere	Navier-Stokes	*Large* spatial scales	
	Compressible	*Fast* changes over time	

with interesting challenges due to the different time/space scales of water and air. Really the modelling challenge is much worse, since the range of temporal and spatial scales is actually quite tremendous, as shown in Table 2.1, particularly when the near-fractal[1] behaviour of ocean ice is included.

Executing such a model has been and continues to be a huge numerical challenge: dense spatial grids, with layers in depth (ocean) or height (atmosphere) or thickness (ice) and over time.

Chapter 11: System Inversion

Now what would we do with such a model? Ideally we would initialize the model to the earth's current state, and then run it forward to get an idea of future climatic behaviour:

[1] Meaning that there is a self-similar behaviour on a wide range of length scales, such as cracks in ice on all scales from nanometre to kilometres.

Table 2.1 Spatial and temporal scales: typical scales in space and time for structures in ice, water, and air.

System	Typical structure	Spatial scales	Temporal scales
Ocean surface	Eddies	10–100 km	Days–months
Ocean mid-depth	Southern oscillation	100–1000 km	Years
Ocean deep	Thermohaline	1000 km	1000 years
Atmos. local	Storms	1–100 km	Hours
Atmos. nonlocal	Pressure systems	1000 km	Weeks
Ice local	Cracks	cm–km	Seconds–years
Ice nonlocal	Sheets	1–100 km	1–100 years

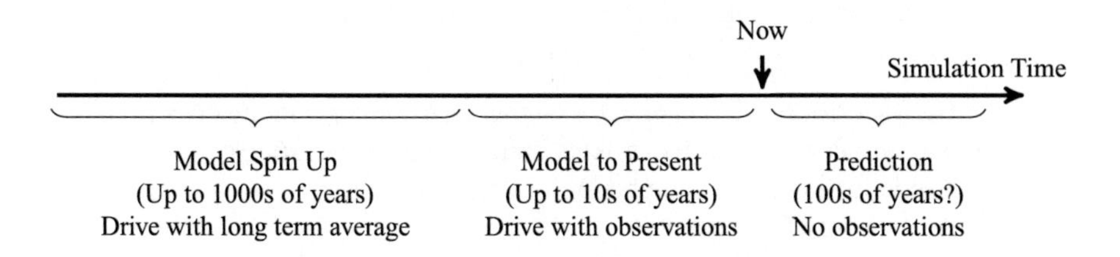

Such a process of coupling a model with real data is known as data assimilation or as an inverse problem. There is a the true earth state $\underline{z}$, which is unknown, but which can be measured via a forward model $C()$:

True state $\underline{z}(t)$ $\longrightarrow$ Observations $\underline{m}(t) = C\big(\underline{z}(t)\big) + \text{noise}$

In principle we want to solve the inverse problem, to estimate the unknown state of the earth:

Inverse Problem: find $\hat{\underline{z}}$, an estimate of $\underline{z}$, by inverting $C()$:

$$\hat{\underline{z}} = C^{-1}(\underline{m}) \tag{2.1}$$

However almost certainly we cannot possibly obtain enough observations, especially of the deep oceans, to actually allow the inverse problem to be solved, analogous to an underconstrained linear system of equations.

Instead, we perform data assimilation, integrating the observations into the simulated model:

We want to push the simulated state $\tilde{\underline{z}}(t) \longrightarrow \underline{z}(t)$
The idea is to iteratively nudge $\tilde{\underline{z}}$ in a direction to reduce

$$\left\| \underline{m}(t) - C\big(\tilde{\underline{z}}(t)\big) \right\| \tag{2.2}$$

where $\|\cdot\|$ measures inconsistency. Thus we are trying to push the simulation to be consistent with real-world measurements, and therefore hopefully towards the real-world state.

Chapter 11: System Sensing

So measurement is key to successfully modelling and predicting climate. What are the things we can actually measure:

Atmospheric Temperature:
- Weather stations on the earth's surface
- Weather balloons
- Commercial aircraft
- Satellite radiometers

Oceanic Temperature:
- Satellite infrared measurements of ocean surface temperature
- Ocean surface height measurements (thermal expansion)
- Ocean sound speed measurements
- Buoys, drifters, gliders directly taking measurements in the ocean

Temperature Proxies (indirect effects indicative of temperature):
- Arctic ice extent and number of ice-free days
- Arctic permafrost extent
- Date of tree budding/leaf-out/insect appearance/bird migrations

If we consider global remote sensing via satellite, really the *only* possible measurement is of electromagnetic signals. Therefore a satellite is limited to measuring signal *strength* (brightness) and signal *time* (range or distance).

The key idea, however, is that there are a *great* many phenomena $\underline{z}$ which affect an electromagnetic signal via a forward model $C()$, meaning that from the measured electromagnetic signals we can infer all manners of things, from soil moisture to tree species types to urban sprawl to ocean salinity, temperature, and currents.

Chapter 9: Non-Gaussian Statistics

Given adequate observations to allow a model to be simulated and run, what do we do with the results?

Understanding climate models and validating the simulated results are huge topics which could certainly fill another textbook. But even something much simpler, such as a time series of historical temperature data, can lead to challenges.

All students are familiar with the Gaussian distribution,[2] a distribution that very accurately describes the number of heads you might expect in tossing a handful of 100 coins. Phenomena which follow a Gaussian distribution, such as human height, are convenient to work with because you can take the average and obtain a meaningful number. This seems obvious.

[2]Or bell curve or normal distribution; all refer to the same thing.

However most climate phenomena, and also a great many social systems, do *not* follow a Gaussian distribution, and are instead characterized by what is called a *power law*. Examples of power laws include meteor impacts, earthquake sizes, and book popularity. Given power law data (say of meteor impacts over the last 10 years), the average is *not* representative of the possible extremes (think dinosaur extinction . . .). In fact, taking an average over longer and longer periods of time still fails to converge. That seems strange.

We humans only barely learn from history at the best of times; learning from historical power law data is even worse, because it is difficult to know how much data we need to reach a conclusion.

Chapters 6 and 7: Linear and Nonlinear Systems

Lastly, it is important to understand the macro-behaviour of climate as a nonlinear system:

> **Linear systems** are subject to *superposition* and have no *hysteresis*.
> **Nonlinear systems** are subject to *catastrophes* and *hysteresis*.

These concepts are most effectively explained in the context of a plot of stable climate states, shown in Figure 2.1.

The principle of superposition says that if increasing CO_2 by some amount leads to a reduction in ice, then twice the CO_2 leads to twice the reduction.

Superposition is highly intuitive, very simple, and usually wrong. We are presently on the lower curve in Figure 2.1, a planet with a mixture of ice and water. As CO_2 is increased, the amount of ice indeed slowly decreases, until point "A", at which point an *infinitesimal* increase in CO_2 leads to the *complete* disappearance of *all* ice. We have here a bi-stable nonlinear system with a bifurcation at point "A", leading to a discontinuous state transition known as a "catastrophe."

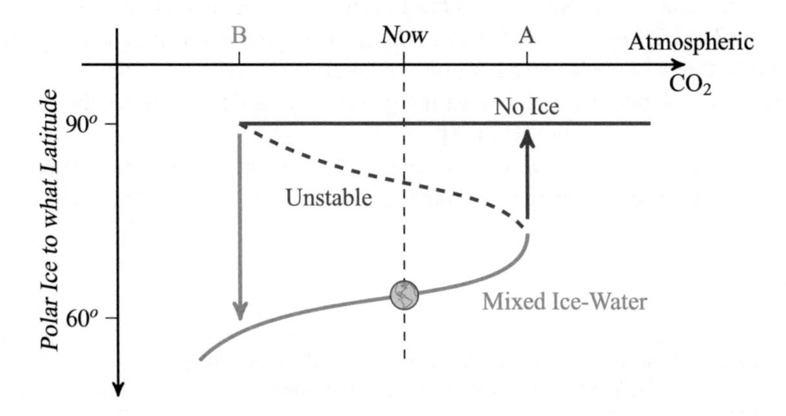

Fig. 2.1 Stable climate states as a function of atmospheric carbon-dioxide: the earth may have no ice (top) or may have polar ice to some latitude (middle); the arrows indicate the discontinuous climatic jumps ("catastrophes"). Although obviously an oversimplification of global climate, the effects illustrated here are nevertheless real.

In a linear system, to undo the climate damage we would need to reduce the CO_2 level back to below "A". A nonlinear system, in contrast, has memory, what is called hysteresis. Reducing CO_2 to just below "A" has no effect, as we are stuck on the ice-free stable state; to return to the mixed water-ice stable state we need to reduce CO_2 much, *much* further, to "B".

Chapter 10: Complex Systems

Climate, ecology, human wealth and poverty, energy, water are all complex systems: nonlinear, non-Gaussian, coupled, poorly-measured, spatial dynamic problems.

Viewing any of these as systems with isolated or fixed boundaries, subject to superposition and Gaussian statistics, is to misrepresent the problem to such a degree as to render useless any proposed engineering, social, economic, or political solution.

This book cannot pretend to solve global warming or any of many other complex problems, but perhaps we can take one or two steps towards understanding the subtleties of complex systems, as a first step in identifying meaningful solutions.

Further Reading

The references may be found at the end of each chapter. Also note that the textbook further reading page maintains updated references and links.

Wikipedia Links: Global Warming, Climate Change

A challenge, particularly with global warming, is that there is an enormous number of books, most polarized to the extremes of either denial or despair. Books such as *What We Know About Climate Change* [1], *Global Warming Reader* [5], or the personable *Walden Warming* [6] offer a broad spectrum on the subject.

Good starting points for global warming science would be the respective chapters in earth systems books, such as [3, 4].

Regarding the role of nonlinear systems in climate, the reader is referred to Case Study 6. More information may be found in the book by Scheffer [7].

References

1. K. Emanual. *What We Know About Climate Change*. MIT Press, 2012.
2. A. Johnson and N. White. Ocean acidification: The other climate change issue. *American Scientist*, 102(1), 2014.
3. L. Kump et al. *The Earth System*. Prentice Hall, 2010.
4. F. Mackenzie. *Our Changing Planet*. Prentice Hall, 2011.
5. B. McKibben. *The Global Warming Reader*. OR Books, 2011.
6. R. Primack. *Walden Warming*. University of Chicago Press, 2014.
7. M. Scheffer. *Critical transitions in nature and society*. Princeton University Press, 2009.

Chapter 3
Systems Theory

There are many very concrete topics of study in engineering and computer science, such as Ohm's law in circuit theory or Newton's laws in basic dynamics, which allow the subject to be first examined from a small, narrow context.

We do not have this luxury. A topic such as global warming, from Chapter 2, is an inherently large, interconnected problem, as are topics such as urban sprawl, poverty, habitat destruction, and energy. For example the latter issue of fossil fuels and global energy limitations is truly interdisciplinary, requiring the understanding, in my opinion, of many fields:

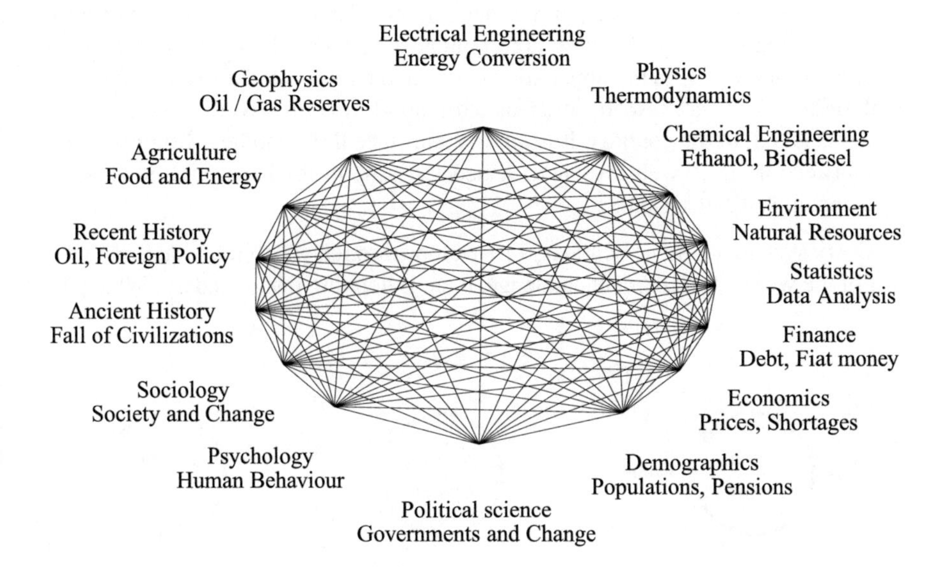

P. Fieguth, *An Introduction to Complex Systems*,
DOI 10.1007/978-3-319-44606-6_3

This cannot be studied as a set of small, isolated problems, since the interconnections *are* what make the problem. In order to understand interconnected systems, we have to understand something regarding systems theory.

3.1 Systems and Boundaries

A *system* is most fundamentally characterized by the manner in which it interacts with its surroundings:

System type	Mass transfer	Energy transfer
Open system	✓	✓
Closed system	✗	✓
Isolated system	✗	✗

These basic systems are shown graphically in Figure 3.1. A quick illustration should make the definitions clear:

Open: Most human systems are open, since virtually all human societies and companies are based around trade and the exchange of goods.

Closed: The earth as a whole is very nearly closed, since the inputs to the earth are dominated by solar energy (with a tiny amount of mass transfer from meteors), and the outputs from the earth are dominated by thermal energy (with a small amount of mass from the upper atmosphere and the occasional space probe).

Isolated: There are few, if any, isolated natural systems. Even a black hole is not isolated; to the contrary, it is quite open, since there can be substantial mass transfer into it. A well insulated, sealed test-tube in the lab would be close to being an isolated system.

Every system has some boundary or envelope through which it interacts with the rest of the world. A significant challenge, then, is defining this boundary: What part

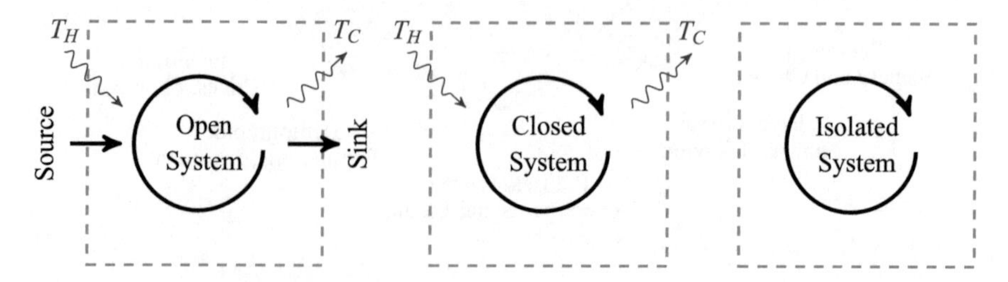

Fig. 3.1 Basic system categorization: systems are distinguished on the basis of whether energy and/or mass can cross their boundaries.

of the greater system is placed inside the boundary, and modelled *explicitly*, and what is outside, ignored or modelled *implicitly* via the boundary?

Interestingly, there is a tradeoff, illustrated in Figure 3.2, between the complexity of the represented system and the complexity of the boundary. This makes sense:

- The more of the system I move *outside* of the boundary, the simpler the remaining system, but the more complex the interactions through the boundary with the other parts of the system.

Example 3.1: Three systems and their envelopes

The earth is very nearly a closed system, primarily sunlight coming in and thermal energy going out, with only tiny changes in mass:

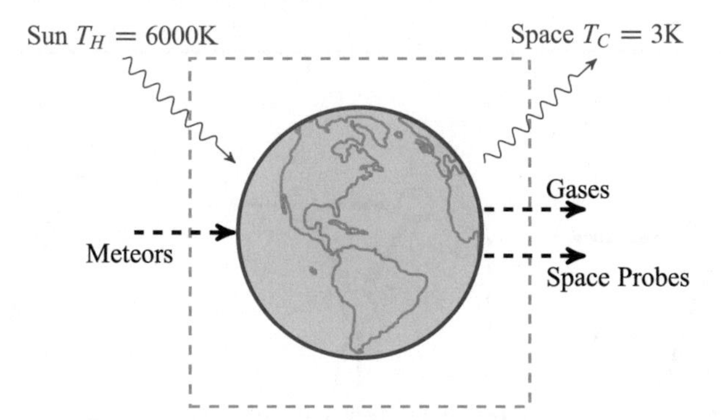

A single human is very much an open system, since food and water come into our bodies from outside of us. The goal in such a system diagram is to annotate the predominant interactions between a system and the outside world, not necessarily to be exhaustive and complete. Here, for example, we consider only movement of mass to/from a body, and not other interactions such as senses of touch, smell, sight, and hearing:

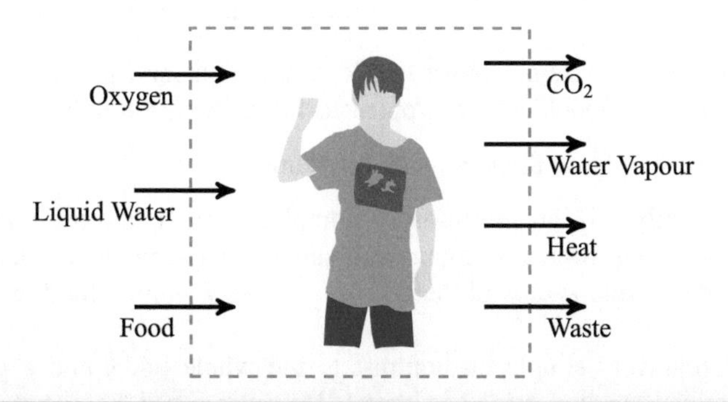

Example continues . . .

Example 3.1: Three systems and their envelopes (continued)

Human society is very open; indeed, the exchange of goods and trade is the hallmark of advanced societies. This diagram is terribly incomplete, as the range of materials coming into and out of a city is vast:

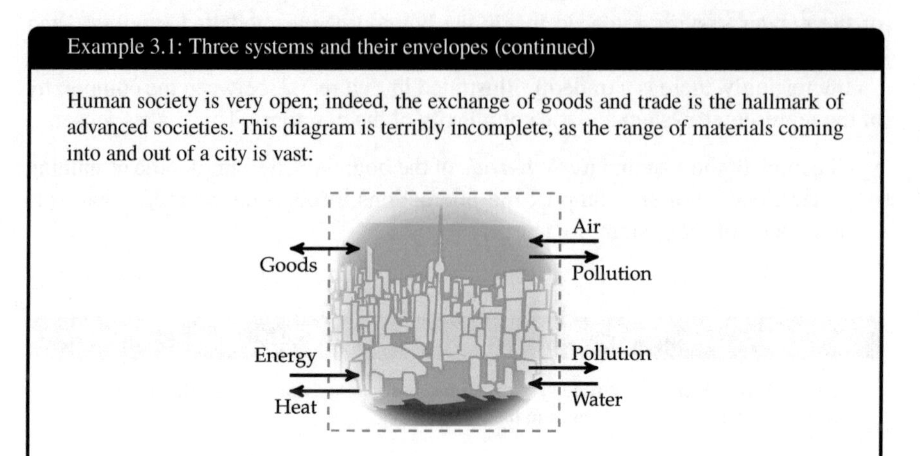

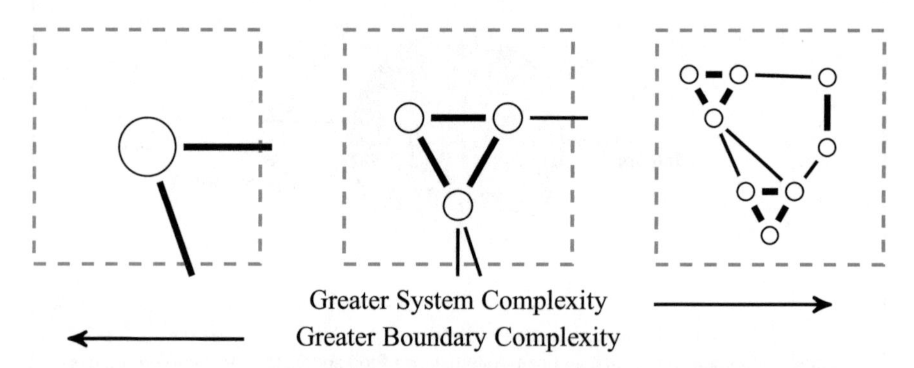

Fig. 3.2 System versus boundary complexity: a self-contained system, right, may be relatively complex but has a simple boundary. Representing only a small part of a system, left, leaves us with a simple dynamic model, but the strong dependencies (thick lines) on unmodelled aspects of the system may leave us with a more difficult boundary representation than in the middle, where the dependencies are more modest (thin lines).

- The more of the system I move *inside* of the boundary, the more complex a system I have to model, but the simpler the interactions through the boundary.

This is perhaps simpler to illustrate with an example:

The Human Body is an enormously complex system, almost impossible to understand given the many organs and subsystems (circulatory, nervous etc.), however the envelope around the body is relatively simple: food, water, waste, air etc.

A Cell is relatively simple, in contrast to the whole body, and a great many cell functions are understood in detail. However a cell has, at its boundary, a tremendously complex envelope of proteins, hormones, other molecules and physical forces induced by the body within which it lives.

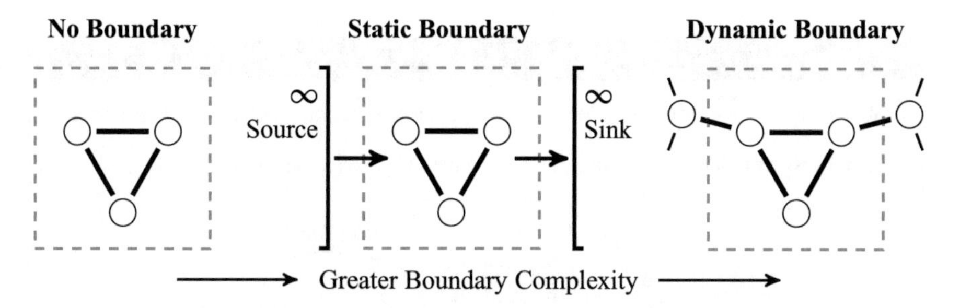

Fig. 3.3 Boundary conditions: an isolated system has no boundary dependencies. If the sources and sinks are infinite, then their behaviour is unaffected by the modelled system and the boundary is static. The most difficult circumstances are boundary interactions, right, with finite systems.

Although there are clearly many possible ways of representing and modelling boundary conditions, very roughly there are three groupings, illustrated in Figure 3.3:

No Boundary: Isolated systems, with no interactions modelled outside of the boundary.

Static Boundary: Systems having infinite sources and sinks, meaning that the boundary conditions can simply be held fixed and are not influenced by the system in any way.

Dynamic Boundary: Systems having finite sources and sinks, external systems which can be affected by the modelled system.

In most circumstances the boundary between societal and environmental systems should fall into the third category, interacting with and affecting each other. However, as discussed in Example 3.2, given a societal system we often *represent* external environmental systems as being in one of the first two categories: either dismissed entirely (a so-called *externality*), or assumed to be fixed and immune from human influence, such as assuming the ocean to be an infinite carbon sink.

3.2 Systems and Thermodynamics

The ability of a system to do something, to be dynamic, relies on some sort of input–output gradient or difference: temperature, pressure, height, chemical etc. A turbine relies on a high–low pressure gradient, the human body relies on a food–waste chemical gradient, even a photovoltaic solar panel relies on an thermal-like gradient between the effective temperature of incoming sunlight (6000 K) and that of the local environment (300 K).

The basic thermodynamic system in Figure 3.4 shows input energy Q_H, waste heat Q_C, and work W. By conservation of energy, we know that

$$Q_H = W + Q_C. \tag{3.1}$$

Example 3.2: Complete Accounting

Where do we draw system envelopes when it comes to economics and GDP calculations?

In most current GDP calculations, the natural world is outside our system envelope:

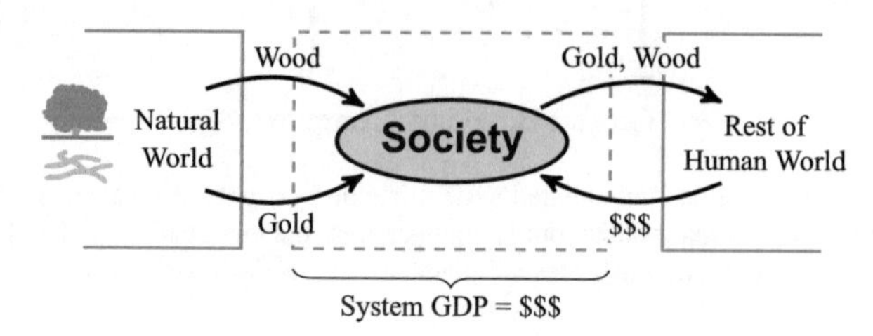

The society has traded the mined gold for money. But after selling the gold, it is *no longer in the ground!* The gold in the ground *was* worth something, but that resource loss, or the fact that the forest takes time to regrow, is not accounted for.

The gold and the forest are examples of *externalities*, aspects or components of a problem left outside of the system envelope. We may choose to limit the system envelope this way for a variety of plausible reasons: to limit model complexity, because of significant uncertainties in the external component, or because of standardized accounting practices. However just as often it may be for reasons of political expediency or accountability — if we don't model or measure something (humanitarian issues, overfishing, clearcutting), then we can claim innocence or ignorance of the problem: out of sight, out of mind.

Environmental or full-cost accounting extends the system boundaries to include the natural world, to specifically account for removal losses and liabilities such as strip mining, clearcutting, and overfishing:

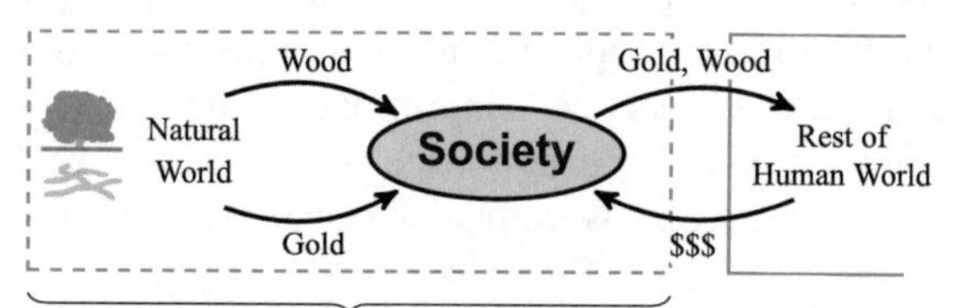

System GDP = $$$ - Loss of Trees + Regrowth of Previously Cut Forest
- Loss of Gold Ore - Pollution Due to Mining

There are, of course, significant challenges and uncertainties associated with assigning "costs" to buried gold, water pollution, and extinct species. However such challenges should not stand in the way of including the natural world within our system envelope.

Further Reading: There are many books on environmental accounting, such as [10, 13] or [14]. Also see Full-cost accounting or Environmental economics in *Wikipedia*.

What interests us is to maximize W for a given amount of input energy Q_H. Thermodynamics places an upper limit on the efficiency of any system:

$$W \leq W_{opt} = Q_H \left(1 - \frac{T_C}{T_H}\right) \tag{3.2}$$

however this is an upper limit; it is certainly possible for W to be significantly below W_{opt}. We observe from (3.2) that a steeper gradient T_H/T_C leads to a more efficient system. It is for this reason that one needs to be careful in proposing schemes to harness "waste" energy: any additional system introduced on the "Cold" side requires a temperature gradient of its own, raising $\bar{T}_C$ and *reducing* the efficiency of the original system, as shown in the right half of Figure 3.4. That being said, there are certainly many examples of successful waste heat recovery and co-generation where the waste heat is used directly, to heat a building, rather than to run a machine.

Because of the limit to efficiency, it follows that

$$\text{Total input energy } Q_H \neq \text{Total usable energy } W$$

leading us to a definition for *entropy*:

Entropy: A measure of the energy not available for doing work.

In general, entropy is also a measure of energy dispersal: greater dispersal corresponds to lower gradients, corresponding to lower efficiency and less work.

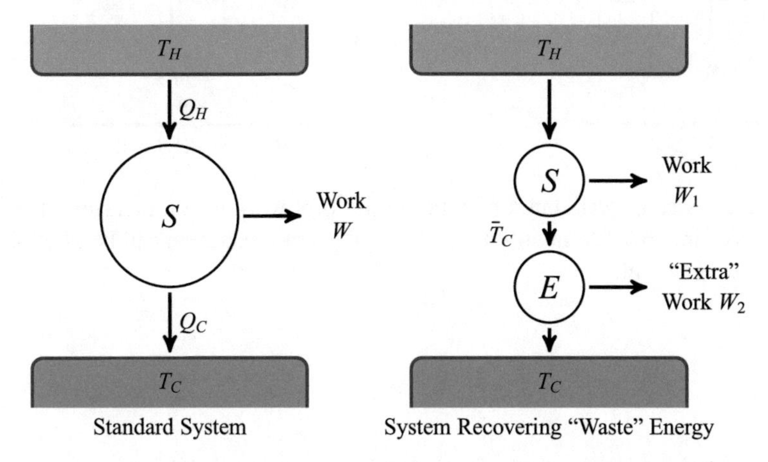

Fig. 3.4 The heat flow Q across a temperature gradient from hot T_H to cold T_C allows a system S to do work W. The output heat flow Q_C is commonly seen as waste energy, however attempting to do work on the basis of this waste, right, leads to a temperature drop across introduced system E, leading to a reduced gradient across S and a correspondingly reduced output $W_1 < W$.

It is essential to understand this dispersal effect; for example, it is widely quoted that

Solar Energy Arriving on Earth	10^{17}W,
Total Human Energy Use	10^{13}W.

It is claimed, therefore, that there is no energy shortage, only a shortage of political will to invest in solar energy. However the 10^{17}W are exceptionally dilute (effectively a very low T_H), spread over the entire planet (including the oceans). To be useful, this energy needs to be collected/concentrated (high T_H) in some form:

Solar-Thermal energy	Mirrors
Solar-Electric energy	Conversion, transmission
Solar-Agricultural energy	Harvest, transport, processing

All of these require infrastructure and the investment of material and energy resources, a topic further discussed in Example 3.5. A dilute energy source is simply not equivalent to a concentrated one, even though both represent the *same* amount of energy.

Although energy is always conserved, clearly *useful* energy is *not*.[1] So two cups of water, one hot and one cold, present a temperature gradient that could perform work, an ability which is lost if the cups are mixed:

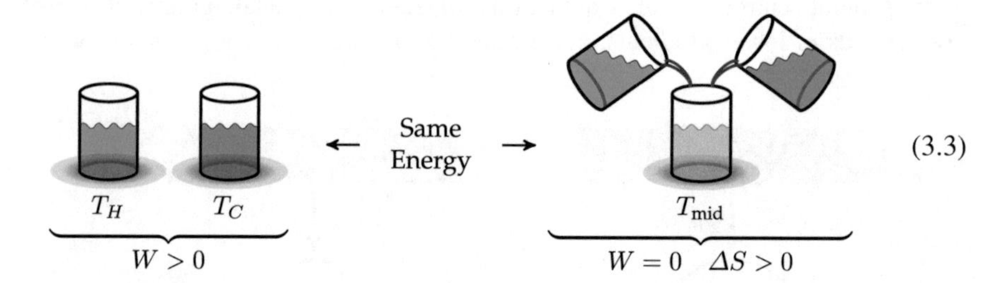

(3.3)

Energy is conserved, but the ability to perform work is not, therefore there is a corresponding increase in entropy S. An equivalent illustration can be offered with a pressure gradient:

[1] The ambiguity in definitions of *useful* or *available* energy has led to a proliferation of concepts: energy, free-energy, embodied-energy, entropy, emergy, exergy etc. I find the discussion clearer when limited to energy and entropy, however the reader should be aware that these other terms are widely used in the systems/energy literature.

Example 3.3: Entropy Reduction

The second law of thermodynamics tells us that entropy cannot decrease.

Consider, then, the process of crystallization:

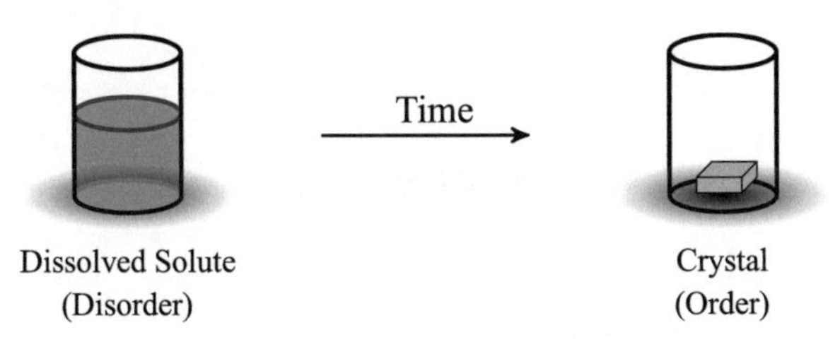

We appear to have a spontaneous reduction in entropy, from disordered (high entropy) to ordered (low entropy)!

However, considering a system envelope drawn around the beaker of solute, this is *not* an isolated system. Water has evaporated ($S\uparrow\uparrow$) and crossed through the envelope, taking a great deal of entropy with it.

So, indeed, the entropy *inside* the system envelope *has* decreased, however global entropy has increased, as expected.

System boundaries matter!

P_H P_L ← Same Energy → P_{Mid} P_{Mid} (3.4)

$W > 0$ $\quad$ $W = 0 \quad \Delta S > 0$

It is clear that gradients can spontaneously disappear, by mixing, but cannot spontaneously appear.[2] From this observation follows the second law of thermodynamics, that entropy cannot decrease in an isolated system.

However, as is made clear in Example 3.3, although entropy cannot decrease in an isolated system, it can most certainly decrease in an open system, since open systems allow energy (and entropy!) to cross their boundaries.

[2]There are phenomena such as Rogue Waves in the ocean and Shock Waves from supersonic flow, in which gradients do appear, however these are *driven* effects, induced by an energy input.

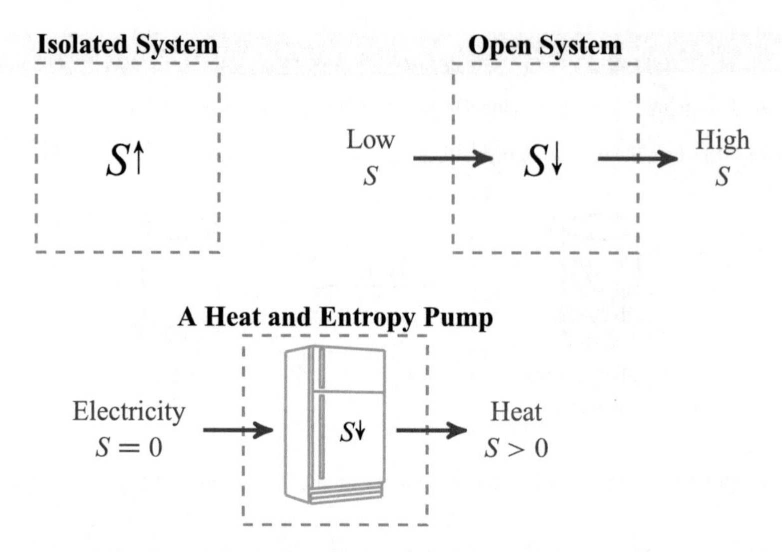

Fig. 3.5 In an isolated system entropy S cannot decrease, however entropy can most certainly decrease in *open* systems, provided that sufficient entropy is pumped across the system boundary, the principle under which all refrigerators, freezers, and air conditioners work.

As shown in Figure 3.5, an open system can import *low*-entropy inputs ...

high gradients (very hot and very cold), fuel, mechanical work, electricity

and subsequently export *high*-entropy outputs ...

low gradients (warm thermal energy), burnt fuel, smoke, water vapour.

So although a gradient cannot spontaneously appear, a refrigerator, for example, *does* indeed create a thermal gradient (cool fridge in a warm room). However to create this gradient requires it to consume an even *greater* gradient (energy) from somewhere else, and that energy (electricity) must have consumed an ever greater gradient (burning wood or coal) somewhere else, and so on.

Furthermore the gradient metaphor generalizes to non-energetic contexts: to clean your hands (establishing a cleanliness gradient) requires you to consume an existing gradient, such as purified tap-water or a clean towel; much more abstractly, entropy also applies to information, such as the computation or re-writing of a binary digit, known as Landauer's principle.

Finally, not only do gradients not spontaneously appear, but nature exerts a force towards equilibrium and randomness: wood rots, metal rusts, stone weathers, animals die, things break. This tendency towards equilibrium consumes any kind of gradient, whether temperature (coffee cools, ice cream melts), pressure, chemical, cleanliness etc.

However in order to do something, any form of work, a system cannot be in equilibrium. Life is therefore a striving towards maintaining gradients, holding *off* equilibrium, where

$$\dot{S}_{\text{System}}(t) = \dot{S}_{\text{Input}}(t) - \dot{S}_{\text{Output}}(t) + \dot{S}_{\text{Nature}}(t) \tag{3.5}$$

So we need sufficiently low-entropy inputs (food, liquid water) and sufficiently high-entropy outputs (heat, water vapour, waste) to counter the natural push towards equilibrium (death).

We are finally led to the following summary perspective on entropy:

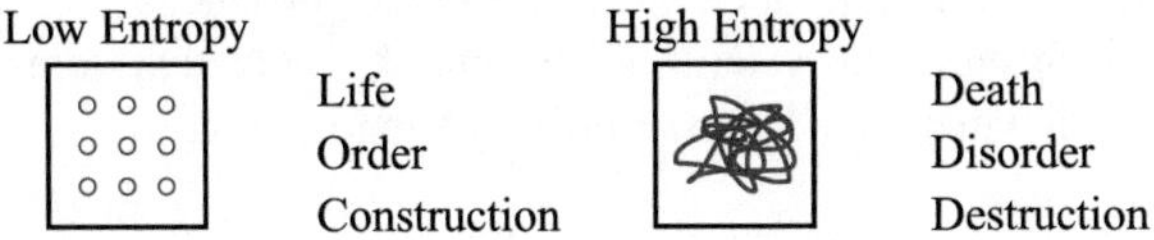

Low Entropy		High Entropy	
	Life Order Construction		Death Disorder Destruction

3.3 Systems of Systems

In principle a single system could contain *many* internal pieces, and therefore could be arbitrarily complex.

However it is usually simpler to think of a "system" as being uniform or self-contained, with a natural boundary or envelope. In this case several separate, but interacting, systems would be referred to as a "system of systems," as summarized in Table 3.1.

Table 3.1 Systems of systems: an interacting collection of systems is not just a larger system. Heterogeneous collections of systems possess unique attributes and pose challenges, yet are common in ecological and social contexts.

	System	System of systems
Similarity	Homogeneous	Heterogeneous
Operation	Autonomous	Interacting
Behaviour	Straightforward	Emergent
Geography	Local	Distributed
Concepts	Simple	Interdisciplinary
	↓	↓
Methods	Linear systems Nonlinear systems Partial diff. equations	Complex systems Agent-based models Cellular automata
Strategy	Analysis	Simulation

The heterogeneity of Systems of Systems make them a significant challenge, so we need to maintain a certain level of humility in attempting to model or analyze such systems. Nevertheless Systems of Systems are all around us,

Creek or Stream:	plankton, fish, plants, crustaceans, birds ...
Meadow:	perennials, annuals, worms, insects, microbes, ...
Human Body:	organs, blood, immune system, bacteria, ...
Automobile:	drive train, suspension, climate control, lighting, ...
Major City:	buildings, transportation, businesses, schools, ...

so at the very least we want to open our eyes to their presence.

In most cases, Systems of Systems are nonlinear, complex systems which will be further examined in Chapter 10, particularly in Section 10.4.

Case Study 3: Nutrient Flows, Irrigation, and Desertification

Suppose I have an apple tree. Thinking about the system envelope around the tree, can I keep taking apples indefinitely? At some point, don't I have to put something back?

The question of nutrient flows is an important one, and one which can be examined through the context of systems concepts.

There are a great many nutrients present in the natural world, and a certain balance is required to ensure optimum plant (and later human) health. We will simplify the discussion by focusing on those six nutrients which overwhelmingly dominate plant mass. Of these six, three are readily available from the air, and three present a challenge:

Mineral	Source	Mineral	Natural source	Modern source
C	Air	N	Lighting, bacteria, animals	Natural gas
H	Air/water	P	Soil weathering, volcanoes	Mining
O	Air/water	K	Soil weathering, volcanoes	Mining (potash)
Easy minerals		Difficult minerals		

Example 3.4: Society, Civilization, and Complexity

Formally, thermodynamics applies to large collections of particles (fluids and gases), to heat flow, and to issues of energy and energy conversion. To the extent that the internal combustion engine, electricity production, and agricultural fertilizers are critical components of modern civilization, thermodynamics is certainly highly relevant to world issues.

We should not really be applying principles of thermodynamics informally to collections of people and treating them as "particles"; nevertheless thermodynamics *can* give insights into large-scale complex systems, which includes human society.

Our study of thermodynamics concluded with two facts:

1. Nature exerts a force towards increased entropy or, equivalently, towards increased randomness, on every part of every system.
2. Opposing an increase in entropy, possible only in an open system, requires an energy flow.

Therefore the greater the complexity of a system,
The greater the force towards randomness,
The greater the energy flow required to maintain the integrity of the system.

This phenomenon will be clear to anyone who has done wiring or plumbing: It is easy to cut copper wire or pipe to length, but you cannot uncut a pipe, and at some point you are left with lengths which simply do not reach from A to B. The copper must then be recycled and melted (energy flow) to form a new material.

More buildings, more computers, more information to maintain means more things that break, wear out, and require energy to fix or replace.

If human civilizations invariably move towards greater complexity[17],
And if energy flows are bounded or limited,
Then all civilizations must eventually collapse.

That is, as formulated by Greer [4] and discussed in other texts [2, 17, 20], waste production (entropy) is proportional to capital, so a certain rate of production is required to maintain the productive capital.

If production is insufficient to maintain the capital base, then some capital is forcibly turned to waste, leading to further reductions in capital, and further reductions in production. How far a society collapses depends on whether the resource base is intact (a so-called maintenance collapse), such that a supply of resources can be used at some stage to rebuild capital; or whether resources have been overexploited, preventing or greatly limiting the rebuilding of productive capital (a catabolic collapse).

Example continues ...

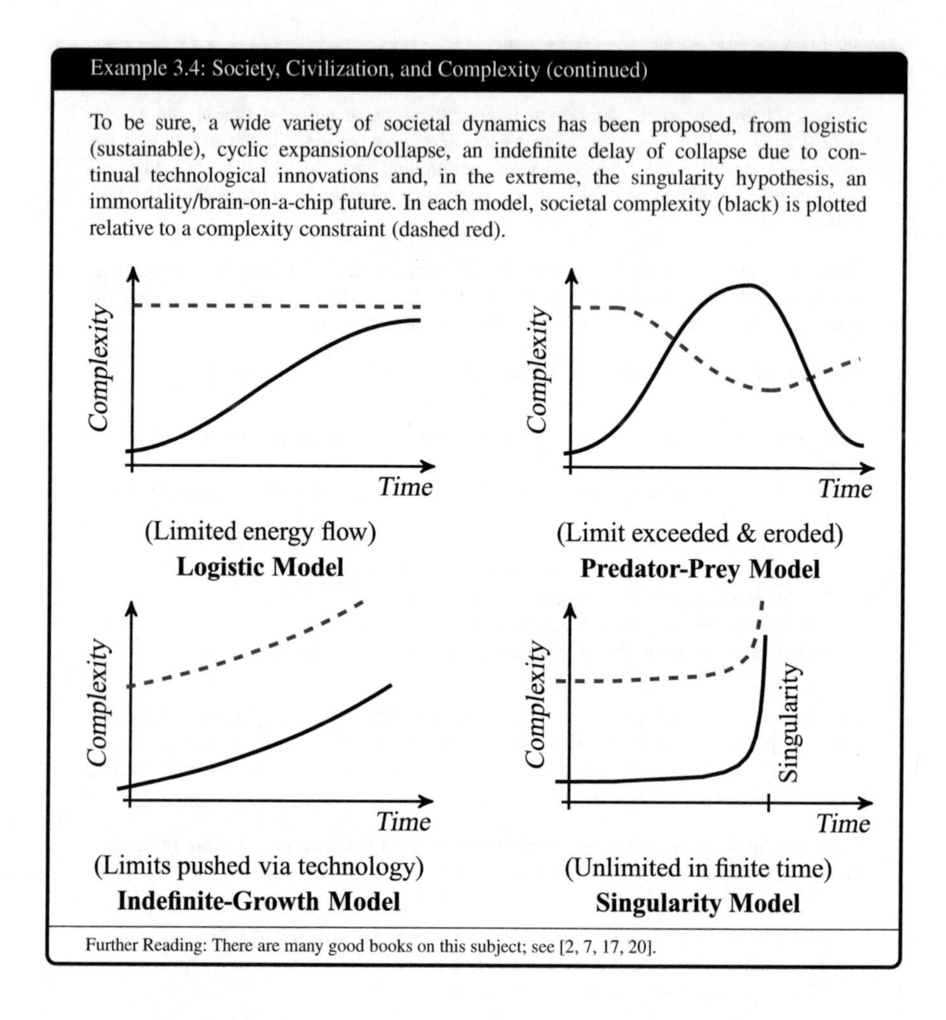

Example 3.4: Society, Civilization, and Complexity (continued)

To be sure, a wide variety of societal dynamics has been proposed, from logistic (sustainable), cyclic expansion/collapse, an indefinite delay of collapse due to continual technological innovations and, in the extreme, the singularity hypothesis, an immortality/brain-on-a-chip future. In each model, societal complexity (black) is plotted relative to a complexity constraint (dashed red).

Further Reading: There are many good books on this subject; see [2, 7, 17, 20].

The minerals NPK[3] are frequently growth-limiting, dominate many aspects of agricultural policy, and play a central role in questions of the sustainability of global food production.

To be sure, the behaviour of nutrient flows in/out of a natural system will depend on time scale:

Short Term (< 1000 Years):	Most natural systems are nutrient–closed,
Long Term (to 10^9 Years):	Most natural systems are nutrient–open.

[3]Nitrogen, Phosphorus, and Potassium. Nitrogen is, of course, highly plentiful in the atmosphere, but in very stable molecular form. When we talk about agricultural nitrogen we mean *fixed* or reactive nitrogen, normally in the form of ammonium.

Indeed, in the long term, geological processes such as ice ages, volcanoes, erosion, and continental drift lead to large-scale reallocations of minerals.

Here we are considering shorter, human–societal time scales. Most human influences on the natural world take a *closed* nutrient cycle and *open* it, as shown in Figure 3.6. Arguably, historical human influences did not so much open a given system as widen its envelope to that of a farm or town, however the advent of modern transport technology has allowed the bulk transport of materials world-wide.

Because a European/Western perspective or attitude dominates much of modern agriculture, it is illuminating to observe the system-level differences in nutrients between temperate and tropical soils, discussed in Table 3.2. Tropical forests, with prodigious growth rates, gave Europeans the impression of a robust ecosystem that would support heavy agricultural harvests. However the robustness is deceiving: the majority of the system nutrients are locked up in the living matter, so opening the system and exporting a harvest led to a significant loss of nutrients which are only slowly replenished. Tropical nutrient replenishment is primarily via atmospheric deposition, such as from volcanoes or, in the case of the Amazon rainforest, primarily as dust transport from the Sahara desert (page 169). Those tropical areas which do not experience significant atmospheric nutrient deposition, such as Easter Island [2, 17], are therefore at the highest risk of system disturbance and nutrient loss.

Example 3.5: Energy Returned on Energy Invested (EROEI)

The world runs on energy. As we saw in Example 3.4, the energy crossing the boundary into a complex system matters a great deal.

World oil and gas production figures quote *primary* production of energy, in terms of the oil and gas coming out of the ground. However all of that energy is not actually available to society, since some of that energy got expended to get the oil and gas out of the ground in the first place:

$$\text{Net Energy} = \text{Primary Energy} - \text{Energy Expended to obtain Primary Energy}$$

What really matters to a society is the ratio

$$\text{EROEI} = \text{Energy return on investment} = \frac{\text{Primary Energy}}{\text{Expended Energy}}$$

Example continues ...

Example 3.5: Energy Returned on Energy Invested (EROEI) (continued)

There are, of course, other possible ratios. Most significantly, historically, would have been the Food Returned on Food Invested (FROFI), where the 90 % fraction of society necessarily involved in agriculture pretty much limited what fraction remained to do everything else — literature, philosophy, government, engineering etc.

With less than 2 % of the population in western countries now involved in agriculture, the food ratio is no longer a limiting factor on maintaining and growing society, in contrast to energy.

Raw energy is easily quantified, and widely reported, but the assessment of expended energy is much harder. The case of ethanol is particularly striking, since both oil and ethanol production are reported, however there is absolutely no way that

$$\text{Oil Production} + \text{Ethanol Production}$$

is actually available to society as energy, since there are very significant amounts of energy *used* in the agricultural production of corn and later distillation of ethanol.

EROEI assessments suggest numbers on the order of

Classic Texas gusher	EROEI $\approx$ 100
Saudi Arabian oil	20
Ocean deepwater oil	3
Canadian tar sands	1–3
Wind	10–20
Solar Photovoltaics	6–10
Ethanol	< 1 ?

Meaning that in the case of a Texas gusher, for only a small amount of energy invested in the form of drilling, vast amounts of energy became available for others to use.

In significant contrast, in the case of the Canadian tar sands a large amount of energy is *expended* before the resulting oil is available to society ...

Natural gas for extraction
Energy for bitumen cracking/upgrading
Energy for piping or transport
Energy to build the vehicles involved in transport?
Energy to make the lunches for the workers involved??

... but where to stop? The problem is that nearly all natural and human social systems are coupled, with no natural boundaries, with a few exceptions like deep sea vents, remote islands etc.

But it matters: this is not a theoretical or philosophical question. The energy expended to acquire energy, no matter how remote or indirect, is no longer available to society for other purposes.

Further Reading: A very good overview is provided by Heinberg in [5], plus EROEI in *Wikipedia*.

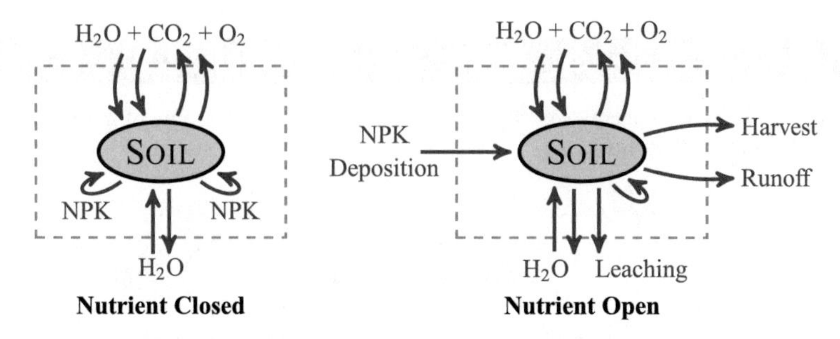

Fig. 3.6 Natural systems are nutrient-closed on human time scales, left. However most human influences, particularly logging, farming, and fishing, take a closed nutrient cycle and open it, right.

Table 3.2 Temperate versus tropical soils: temperate soils, left, are highly robust and nutrient-rich compared to their tropical counterparts, right.

Temperate soils	Tropical soils
Fresh minerals exposed by ice ages ⟶ Soil is young	Soils essentially untouched for millennia ⟶ Soil is old
Soil is relatively rich in minerals, Weathering releases minerals ⟶ Robust to disturbance	Soil is essentially depleted of minerals, Heavy rainfall leaches minerals ⟶ Sensitive to disturbance
Most of the system minerals are located in the ground ⟶ Slow to degrade after clearcut	Most of the system minerals are tied up in living matter ⟶ Fast to degrade after clearcut

Figure 3.7 shows the progression of a temperate ecosystem, from a forest state, through historical agriculture, to modern agriculture. In particular, it is interesting to observe that modern agriculture is dominated by *annuals* (plants which live only for one season), such as corn, which put all of their energy into seed production and therefore maximize agricultural production, but with shallow roots and high nutrient needs. In contrast, historical agriculture involved many more perennials (which live for years), which put more energy into root development and therefore are more limited in yield but with reduced nutrient requirements.

Indeed, compare the root depths [19] of perennials like the Wild Rose (6 m), Alfalfa (5 m) and Horseradish (4 m) to annuals such as Corn (1 m), Tomato (1 m), or Wheat (1–2 m).

Example 3.6: Maximum Power Principle

Suppose we have an energy source with flow restrictions, which is true of most energy sources in natural systems ...

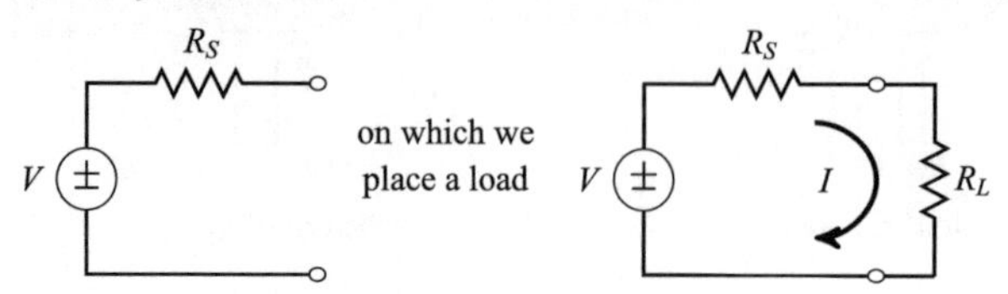

The load extremes are easily analyzed:

$R_L \to 0$ maximizes current I, thus maximizing the flow;
$R_L \to \infty$ has waste power = $I^2 R_S \to 0$, thus maximizing efficiency.

Thermodynamically, a system operating at the efficiency limit is reversible, implying tiny energy flows. However the *useful* energy flow, the energy flow that actually lets us get something done, is the power in the load:

$$P_L = I^2 R_L = \left(\frac{V}{R_L + R_S}\right)^2 R_L \tag{3.6}$$

We observe the output power behaviour

$$\begin{aligned} & P_L \to 0 \quad \text{as} \quad R_L \to 0 \quad \text{(Max. Current)} \\ \text{and} \quad & P_L \to 0 \quad \text{as} \quad R_L \to \infty \quad \text{(Max. Efficiency)} \end{aligned} \tag{3.7}$$

so maximizing efficiency in this context is not all that useful. What we actually want is to maximize the usable load power, thus

$$\frac{\partial P_L}{\partial R_L} = 0 \quad \longrightarrow \quad R_L = R_S$$

Similarly, the maximum power principle argues that biological systems evolve to maximize power, and not efficiency. That is, biological systems compete with each other on the basis of their *usable* power, their ability to do work, thus maximizing the processed energy flow (power) maximizes the evolutionary potential. The maximum power principle has its roots in the work of Lotka (of predator-prey model fame, discussed in Chapter 7) and particularly the later work of Odum. The motivation was that thermodynamics tells you what does *not* happen, but not so much what *does*. The maximum power principle was formulated to try to answer why ecological and economic systems tended to organize themselves into certain patterns or structures.

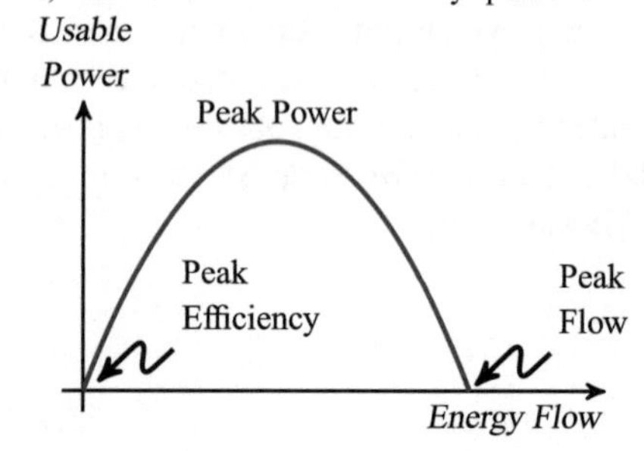

Further Reading: Any of a number of books by Odum, such as [11].
Maximum power principle in *Wikipedia*.

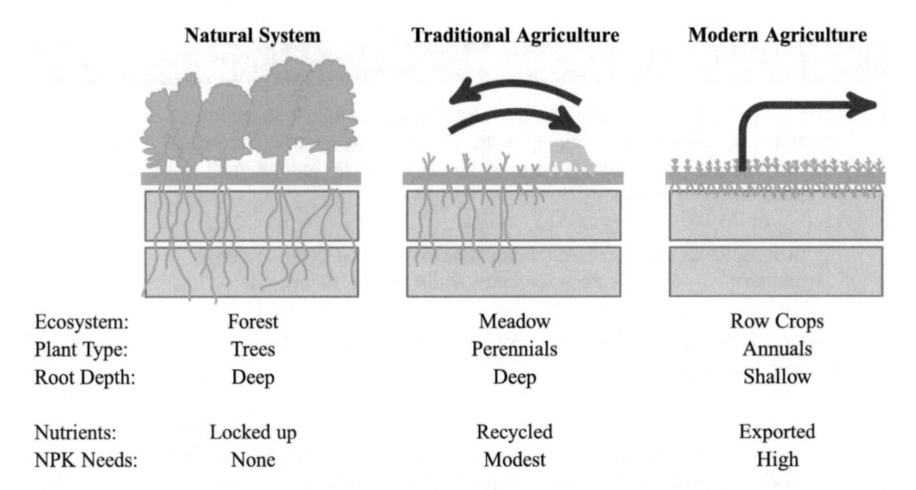

Fig. 3.7 Three ecosystems: the nutrient cycling, in a nutshell, for a forest, subsistence agriculture, and modern industrial agriculture. Historically, most cultures with sloppy nutrient practices would have failed, so successful historical farming practices include innovations such as legumes, meadows, humanure, and fallow cycles which all aim to close the nutrient cycle. Modern agriculture, illustrated further in Figures 3.8 and 3.9, is no longer nutrient-closed.

A more careful examination of the system diagrams for soil reveal two further insights.

First, at a small scale, Figure 3.8 shows the envelope for a parcel of soil. The actual pattern of flows will of course depend on whether the soil is irrigated, the precipitation patterns, the nature of the soil, and the type of crop. Nevertheless we observe inflows and outflows of a system in balance ... with one exception. In many arid parts of the world it is common to irrigate with well water. However unlike rain, groundwater contains dissolved minerals and salts. When groundwater is applied to the soil, particularly in arid regions, most of the water is lost through the plant (transpiration) or directly to the air (evaporation), however the salts and minerals do not evaporate. Minerals such as NPK will be taken up by the plant and harvested, however the salts remain in the soil. Thus the soil has become a *sink* for salts, an unsustainable situation whereby increased salt concentrations ultimately kill the soil.

Next, at a very large, societal scale, Figure 3.9 shows the worldwide agricultural nutrient flow. The open-ness of the system is striking: much of our food production is not at all in steady-state, and it is no exaggeration to suggest that our food is largely mined output and fossil fuels, precious minerals which we proceed to flush down the toilet to the ocean. The challenges associated with nutrient flows are not unavoidable, but the systems illustrated here do raise a number of important policy issues:

- The growth of phytoplankton and algae in rivers, lakes, and oceans are normally P or N limited, and under normal circumstances most algae are eaten. However in circumstances of agricultural P or N runoff there is an explosion of plankton

Example 3.7: Global Flows I

There are a number of global cycles, global planet–wide flows taking place via the oceans or the atmosphere. Although there are countless micro-nutrients and pollutants also transported, the dominant and most significant cycles are the main constituents of the atmosphere and oceans:

Atmosphere: water, carbon (carbon dioxide), heat, nitrogen, aerosols
Oceans: water, carbon (carbonic acid), heat, nutrients, salt

Based on these, there are five major cycles worldwide, each of which involve the coupling of multiple systems:

Water: Oceans, Ice, Surface water, Groundwater, Evaporation, Clouds, Rain
Heat: Radiation in, Radiation out, Clouds, Ice, Albedo, Winds, Currents
Carbon: Atmosphere, Ocean, Soils, Plants, Fossil fuels
Nitrogen: Atmosphere, Bacteria, Fungi, Plants, Animals, Lightning
Aerosols: Air-Water exchange, Pollution, Soot, Dust, Salt, Nucleation

Each of these cycles can be further separated into their natural, historic behaviour and their present changes due to human influences. In some cases, such as the heat/energy cycle, the human component is relatively small,

$$\text{Sun input } 1 \cdot 10^{17}\text{W} \quad \text{Ocean/air flux } 7 \cdot 10^{15}\text{W} \quad \text{Human use } 2 \cdot 10^{13}\text{W}$$

and superficially so for the carbon cycle,

$$\text{Land } 120 \cdot 10^{15}\text{g/y} \quad \text{Ocean } 90 \cdot 10^{15}\text{g/y} \quad \text{Human input } 9 \cdot 10^{15}\text{g/y}$$

however keep in mind that the terrestrial and ocean carbon cycles are in steady-state (zero-mean), whereas the human carbon input is *not* zero mean, rather it integrates over time to a large carbon input. In contrast, historical nitrogen fixing

$$\text{Biological } 120\,\text{Tg/y} \quad \text{Lightning } 10\,\text{Tg/y}$$

has become completely overwhelmed by human activities:

$$\text{Industrial } 80\,\text{Tg/y} \quad \text{Crops } 40\,\text{Tg/y} \quad \text{Pollution / Clearing } 80\,\text{Tg/y}$$

Far from being independent, the global cycles are themselves interacting:

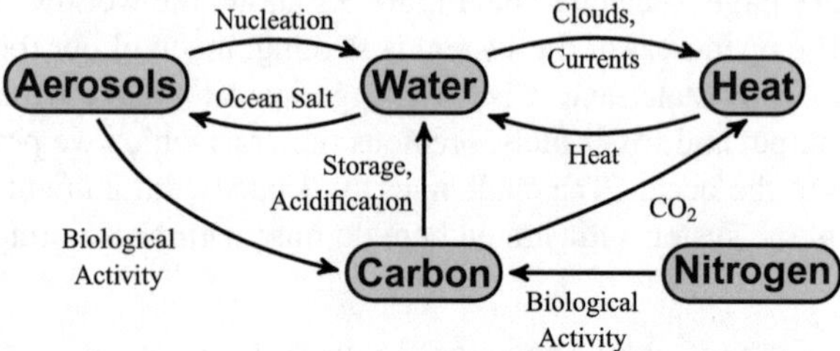

Spatial aspects of global cycles will be further considered in Example 8.4.

Further Reading: See Water cycle, Nitrogen cycle, Carbon cycle, Earth's Energy Budget in *Wikipedia*. Also Vitousek et al., "Human Alteration of the Global Nitrogen Cycle," *Issues in Ecology*, 1997.

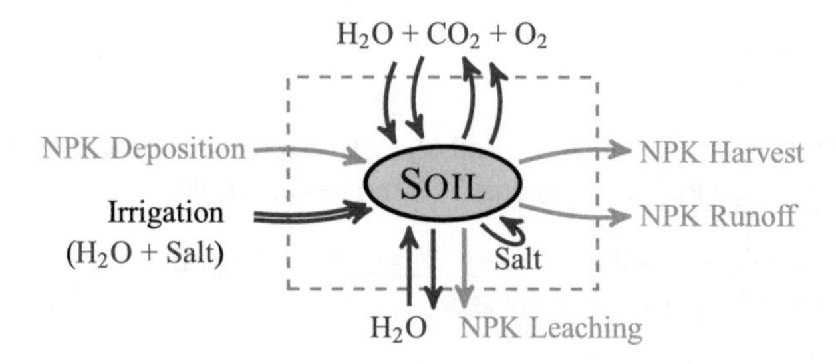

Fig. 3.8 System diagram for irrigated soil: following up on Figure 3.6, treating a parcel of soil as a system allows us to see, in details, the flows across the system envelope. One pernicious issue is that of irrigation: groundwater contains dissolved salts which concentrate in the soil as the water evaporates but the salts do not. Over time this concentration can, depending on precipitation and groundwater patterns, lead to soil salination and a loss of productivity.

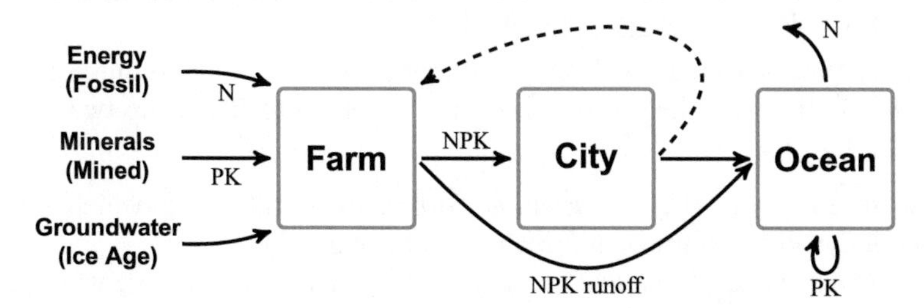

Fig. 3.9 Modern nutrient flows: nutrients flows in Western societies are very much one-way: from mining/fossil sources, left, ultimately to the ocean, right. To the extent that mined nutrients are a finite source, this open approach is clearly unsustainable. The current exception to the one-way flow is the dashed arc, middle, representing a recycling of sewage back to farms.

growth and the plankton then die rather than being eaten. The decomposing plankton absorb oxygen, leading to huge dead zones, most notably in Lake Erie, the Gulf of Mexico, and the Baltic Sea.

There has been significant progress in the reduction of phosphates in soaps and detergents, however the control of agricultural runoff remains a pressing issue, with related waterway and tilling policies.

- Based on Figure 3.9, there are clear policy issues surrounding food sustainability and nutrient recycling from sewage. Similarly, Figure 3.8 raises questions regarding the extent and sources of irrigation.
- Issues of nutrient leaching are not limited to tilled soil. Acid rain, in particular, increases nutrient leaching, thereby opening formerly-closed nutrient cycles even in forested/wilderness areas.

Further Reading

The references may be found at the end of each chapter. Also note that the textbook further reading page maintains updated references and links.

Wikipedia Links — Systems theory: System, Systems theory, Systems of Systems, Steady state

Wikipedia Links — Thermodynamics: Thermodynamics, Entropy, Energy, Landauer's principle

Wikipedia Links — Examples: EROEI, Environmental economics, Full-cost accounting, Environmental accounting, Externalities, Maximum power principle, Technological Singularity

This chapter has covered a wide range of topics, so it is difficult to suggest a single book which discusses the subject matter in further depth. There are two excellent places to start:

1. The recent book, *Why Information Grows*, by Hidalgo [6], which looks at thermodynamics, entropy, information, and complexity.
2. The recent book, *Dynamic Systems for Everyone*, by Ghosh [3], which offers a non-technical look at the role and appearance of systems in many aspects of everyday life.

In terms of books focused on individual topics, here are a variety of directions which the reader may wish to follow:

- Systems: There is a highly accessible book on systems theory by Meadows [8], the same person who co-authored the classic *Limits to Growth* book [9] which looked at applying systems theory to human and ecological systems. Another classic example of defining systems and boundaries is the concept of *ecological footprint* [18], which attempts to quantify the earth resources (land, water, energy) as a function of lifestyle.
- Thermodynamics: The reader will find an overwhelming selection of introductory and advanced textbooks, such as the book by Stowe [16].
- Energy: A very good overview of Energy Returned on Invested is offered by Heinberg in [5], plus many references therein.
- Environmental Accounting: There are several classic books, both that by Odum [10] and certainly *Small is Beautiful* by Schumacher [14]. The topic very much remains of current interest, for example in the more recent work of Schaltegger [13].
- Civilization & Complexity: Much has been written on this subject. Probably the best and most engaging book is that by Wright [20], which I would highly recommend. The classic texts are the large and thorough works of Tainter [17] and Diamond [2]. A more recent book written for the general public is that of

Homer-Dixon [7]. The extreme perspective of the technological singularity had quite balanced coverage in an *IEEE Spectrum* issue [1].
- Soil and Nutrients: A little outside the scope of this book, but a bit of a passion of mine. If you're interested, the book by Salatin [12] is quite inspiring, and the book by Weaver [19] from a century ago has beautiful drawings of underground root systems.

Sample Problems

Problem 3.1: Short Questions

Provide a brief explanation for each of the following:

(a) Why are there *no* truly linear systems in the natural world?
(b) Why are coal and nuclear power plants designed to run at a relatively high temperature?
(c) How does the second law of thermodynamics relate to systems theory:

  (i) Define "entropy"
  (ii) Why is it possible for entropy to *decrease* in open systems?
  (iii) Briefly describe how energy flows and entropy relate.
  (iv) Describe one way in which the concept of entropy relates to a complex society or civilization.

(d) State the "maximum power principle". What makes maximizing power more attractive than maximizing efficiency?
(e) How does a *System of Systems* differ from a regular System?
(f) Looking at agricultural nutrients from a systems level perspective, identify and discuss (in a few sentences) two policy issues related to one or more of nitrogen/potassium/phosphorus.

Problem 3.2: Conceptual — Thermodynamics

I take a bag of frozen peas out of my freezer (−25°C) and put them into the fridge (+4°C) to thaw. By putting them into the fridge I also keep the fridge cool and save energy. To my disappointment, a short while later I notice the fridge running, even though I observe that the peas are still frozen.

(a) Why is the fridge running when there are frozen peas in it?
(b) How is this phenomenon related to global energy issues?

Problem 3.3: Conceptual — Thermodynamics

In principle, any sort of gradient can allow work to be done:

- Height gradient
- Pressure gradient
- Temperature gradient
- Chemical gradient

Give an example of each of these and explain, very briefly (one or two sentences), how the gradient is used to do work.

Problem 3.4: Conceptual — Systems

Identify or briefly describe an example of an open system and an example of a (mostly) closed system for each of the following categories:

(a) A physical/non-living system,
(b) A natural/ecological system,
(c) A human/social system.

Problem 3.5: Conceptual — System Boundaries

Select some form of energy, preferably **one** of the following:

- Brazilian sugar-cane ethanol
- American corn ethanol
- Coal mining
- Canadian tar sands
- Wind power
- Photovoltaics (solar cells)
- Hydroelectric (water power)
- Tidal power
- Hydrogen

For your chosen energy context, develop three system envelopes, of different size or scope, which could be used to assess the *invested* energy, as discussed in Example 3.5. For each of these three systems, you should show an explicit diagram or a detailed list showing what is/is not contained in the system, and the predominant invested and output energy flows.

Do *not* attempt to actually calculate EROEI numbers. Rather, please discuss the pros and cons of each of your systems, where the pros and cons should relate to how complete the system model is, what is omitted, and the feasibility or practicality of computing EROEI based on such a system.

Problem 3.6: Conceptual — System Boundaries

Suppose we consider a university as a system, with a system envelope drawn around the physical campus. Draw a system diagram identifying the main energy, resource, and other physical flows (everything *except* the "flow" of people).

Problem 3.7: Analytical — Peak Power

For the given circuit, let

$$V = 1 \text{ Volt} \quad \text{and} \quad R_S = 1\,\Omega.$$

R_S V $\pm$ R_L

Define P_L to be the power in load resistor R_L.
Draw a sketch of P_L as a function of R_L.

Do *not* mathematically solve for the peak power. However on your plot, identify the approximate location of peak usable power, peak efficiency, and peak flow.

Problem 3.8: Numerical/Computational — Population Dynamics

Example 3.4 showed some simple illustrations of population dynamics, a topic to which we shall return in Example 6.1.

The simplest possible model is that of constant reproduction rate r, such that the population p evolves over time t as

$$\dot{p}(t) = rp(t) \tag{3.8}$$

To simulate this in a computer, we need to discretize time; for a time step δ_t, we can express the discrete-time analogue to (3.8) as

$$p(n+1) = p(n) + \delta_t \cdot r \cdot p(n) \tag{3.9}$$

where sequence p needs to be initialized at the first time step:

$$p(1) = p_o$$

We need to ensure that we interpret any such simulation in the original time t, so we do *not* want to plot p as a function of n. Instead, in Matlab/Octave we would issue a plot command such as

```
plot(d_t*(1:length(p)),p)
```

Unless otherwise specified, use $p_o = 1, \delta_t = 0.1, r = 0.1$:

(a) What curve do you obtain for p?
(b) How is the curve affected by discretization? If you use other time steps $\delta_t = 0.2, 0.3$, do you see much of a change in p?
(c) How is the curve affected by the reproduction rate? How does p change for $r = 0.2, 0.3$?

Problem 3.9: Numerical/Computational — Efficiency

From (3.2) we know the limiting thermodynamic efficiency, as a function of hot and cold temperatures T_H, T_C. Throughout this chapter we assumed T_H and T_C to be *fixed*, essentially like the *static* boundary condition of Figure 3.3. However what is the efficiency of *finite* temperature reservoirs, corresponding to the *dynamic* boundary in Figure 3.3.

Let $T_H(t), T_C(t)$ represent the hot and cold temperatures over time t. We will assume the heat flow at any time to be proportional to the temperature difference

$$Q(t) = T_H(t) - T_C(t)$$

Assuming perfect efficiency, the useful work that can be performed at that time is thus

$$W(t) = Q(t)\left(1 - \frac{T_C(t)}{T_H(t)}\right)$$

leading to the differential equations

$$\dot{T}_H(t) = -Q(t) \qquad \dot{T}_C(t) = Q(t) - W(t)$$

As in Problem 3.8, or looking ahead to Section 8.3, numerically simulate these equations, given

$$\text{Time step } \delta_t = 0.01 \qquad T_H(0) = 400 \qquad T_C(0) = 300$$

(a) If we had just mixed the two reservoirs the final temperature would have been $(400 + 300)/2$. Why is the final temperature now different?
(b) What would the *expected* efficiency have been, for infinite reservoirs at temperatures $T_H = 400,\ T_C = 300$?
(c) What is the *actual* efficiency in your simulation, computed as

$$\frac{\int W(t)\,dt}{\int Q(t)\,dt}$$

Problem 3.10: Reading — Energy Returned on Invested

Example 3.5 discussed the concept of EROEI (energy returned on energy invested). Read[4] *Chapter 3 — Net Energy (EROEI)* in *Searching for a Miracle* [5].

Comment briefly on the challenges of net energy evaluation and the challenges associated with ambiguities of where to assert system boundaries.

Problem 3.11: Policy — Accounting

Example 3.2 discussed the concept of economic accounting and system boundaries, in particular the role played by so-called *externalities* — air pollution, groundwater pollution, resource extraction, species loss.

Many ideas have been proposed, however very little has actually been accomplished, in terms of actually having these external costs recognized and acted upon.

What are the policy limitations or special interests that prevent western governments from changing their perspectives on the interpretation and calculation of GDP?

[4] The chapter can be found online at the Post Carbon Institute; a link is available from the textbook reading questions page.

References

1. Special issue on the singularity. *IEEE Spectrum*, 45(6), 2008.
2. J. Diamond. *Collapse: How Societies Choose to Fail or Succeed.* Penguin, 2011.
3. A. Ghosh. *Dynamic Systems for Everyone: Understanding How Our World Works.* Springer, 2015.
4. J. Greer. *How Civilizations Fall: A Theory of Catabolic Collapse.* Unpublished, 2005.
5. R. Heinberg. *Searching for a Miracle: Net Energy Limits & the Fate of Industrial Society.* Post Carbon Institute, 2009.
6. C. Hidalgo. *Why Information Grows: The Evolution of Order, from Atoms to Economies.* Basic Books, 2015.
7. T. Homer-Dixon. *The Upside of Down: Catastrophe, Creativity, and the Renewal of Civilization.* Knopf, 2006.
8. D. Meadows. *Thinking in Systems: A Primer.* Chelsea Green, 2008.
9. D. Meadows, J. Randers, and D. Meadows. *Limits to Growth: The 30-Year Update.* Chelsea Green, 2004.
10. H. Odum. *Environmental Accounting: Emergy and Environmental Decision Making.* Wiley, 1996.
11. H. Odum. *Environment, Power, and Society for the Twenty-First Century.* Columbia, 2007.
12. J. Salatin. *Salad Bar Beef.* Polyface, 1996.
13. S. Schaltegger and R. Burritt. *Contemporary Environmental Accounting: Issues, Concepts and Practice.* Greenleaf Publishing, 2000.
14. E. Schumacher. *Small is Beautiful: A Study of Economics as if People Mattered.* Vintage Digital, 2011.
15. J. Smillie and G. Gershuny. *The Soul of Soil.* Chelsea Green, 1999.
16. K. Stowe. *An Introduction to Thermodynamics and Statistical Mechanics.* Cambridge, 2007.
17. J. Tainter. *The Collapse of Complex Societies.* Cambridge, 1990.
18. M. Wackernagel and W. Rees. *Our Ecological Footprint.* New Society, 1995.
19. J. Weaver. *Root Development of Field Crops.* McGraw Hill, 1926.
20. R. Wright. *A Short History of Progress.* House of Anansi Press, 2004.

Chapter 4
Dynamic Systems

Consider nearly any major ecological or social phenomenon:

- Human driven climate change
- Fisheries collapse on the Grand Banks of the Atlantic ocean
- Perpetuated financial inequalities
- Periods of drought or flood, for example driven by El Niño

All of these are inherently *temporal* phenomena: a thing was once one way, and now it is different.

Are sea levels rising? Are temperatures changing? Are animal species going extinct? Are financial inequalities increasing?

Nearly all systems of interest to us, whether mechanical, ecological, or social, are systems which evolve over time; that is, they are *time dynamic* systems.

Consider, for example, the ecological catastrophe that is the Aral Sea, a huge body of water shrunk to less than *one tenth* its size due to water diversion for irrigation.

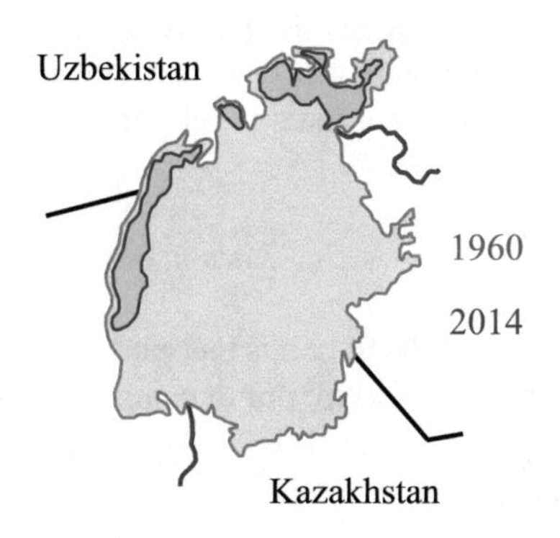

How do we analyze a time-dependent set of measurements, such as the area, depth, perimeter, or surface temperature of the sea?

What is our baseline: when did the sea start to shrink? Was its size historically constant, or variable? Are there correlations between the size of the sea, over time, and the construction of dams or other water diversion projects?

P. Fieguth, *An Introduction to Complex Systems*,
DOI 10.1007/978-3-319-44606-6_4

4.1 System State

We will characterize a system by its *state* $\underline{z}$, which captures everything we need to know, in principle, to describe the system at some point in time. Later, Chapter 11 will more clearly distinguish between what we observe about a system, as distinct from its underlying state.

The state $\underline{z}$ may be quite low dimensional, containing only two numbers,

$$\underline{z}(t) = \begin{bmatrix} \text{number of rabbits}(t) \\ \text{number of foxes}(t) \end{bmatrix} \tag{4.1}$$

in the case of a predator-prey sort of model, or possibly of exceptionally high dimension,

$$\underline{z}(t) = \boxed{}(t) \text{ or } \boxed{}(t) \text{ or } \begin{bmatrix} \text{Pressure} & \boxed{}(t) \\ \text{Temperature} & \boxed{}(t) \\ \text{Velocity} & \boxed{}(t) \end{bmatrix} \tag{4.2}$$

for spatial (2D) or volumetric (3D) climate models. In all cases, the state $\underline{z}$ evolves and therefore obeys some sort of dynamics, typically a form such as

$$\frac{\partial \underline{z}}{\partial t} = \dot{\underline{z}} = f(\underline{z}), \tag{4.3}$$

where the time-dynamic behaviour can be illustrated by means of a phase plot, illustrated in Figure 4.1. One of the most fundamental aspects of a dynamic system is the location of a fixed point $\bar{\underline{z}}$, a value of the state where the system does not change over time:

$$\bar{\underline{z}} \text{ is a fixed point of } f \text{ if } f(\bar{\underline{z}}) = \underline{0}. \tag{4.4}$$

Away from the fixed points the state certainly *does* change with time.

There are many types of dynamic systems:

Linear	$f(\underline{z}) = A \cdot \underline{z}$ for some matrix A
Linearized	$f(\underline{z}) = f(\underline{z}_o) + (\underline{z} - \underline{z}_o) \cdot {}^{\partial f}/_{\partial \underline{z}}$
Nonlinear	$f(\underline{z})$
Simple	Characterized by a single dynamic f
Complex	Characterized by an ensemble of dynamics $\{f_i\}$

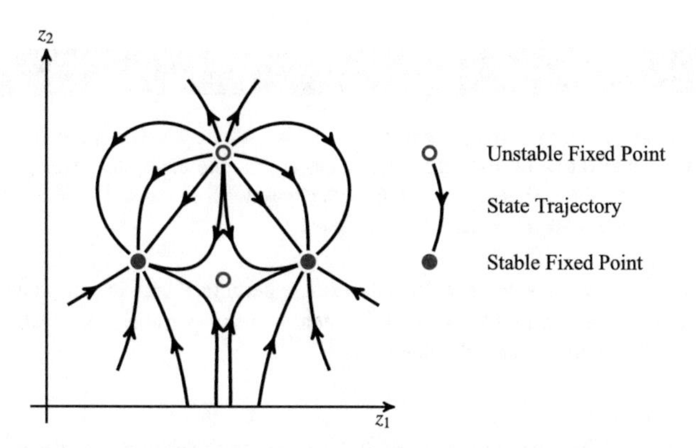

Fig. 4.1 A phase plot: an illustration of a two-dimensional dynamic system, showing the stable fixed points (filled circles), unstable fixed points (open circles), and arrows illustrating the dynamic behaviour of the state (z_1, z_2) over time. Be careful how you intuitively interpret such a diagram: unlike a magnet, the strength of attraction/repulsion is *not* greatest at the stable/unstable fixed points; in fact at the fixed points the strength is *zero*. The greatest rate of change of state occurs *between* the fixed points.

Time Stationary	The dynamics $f()$ are unchanging over time
Time Nonstationary	The dynamics $f(t)$ are time-dependent
Spatially Stationary	The dynamic kernel F is not a function of location
Spatially Nonstationary	The dynamics may vary from place to place
First Order	$\dot{\underline{z}}(t) = f\big(\underline{z}(t)\big)$
Second Order	$\ddot{\underline{z}}(t) = f\big(\underline{z}(t), \dot{\underline{z}}(t)\big)$
Third Order etc.	
Deterministic	The dynamics have no random term
Stochastic	There is some random aspect to the dynamics
Input Driven	The dynamic system $f(z, u)$ is subject to some input $u(t)$

A few simple examples of the preceding systems are listed in Example 4.1.

It is essential to understand that notions of system complexity or stationarity are a statement about $f()$, and *not* about the state $\underline{z}(t)$. That is, for a time-stationary system, clearly $\underline{z}(t)$ is changing over time, it is the function $f()$ *itself* which is not a function of time. Similarly for a spatially-stationary system, again clearly the elements of $\underline{z}$ can differ from one another, the stationarity asserts that $f()$ treats all elements consistently.

Building on (4.3), in most cases a given dynamic system will describe the time-dynamics of a state

$$\frac{\partial \underline{z}}{\partial t} = \dot{\underline{z}} = f(\underline{z}, \underline{\theta}, \underline{u}) \tag{4.5}$$

Example 4.1: System Examples

The list of dynamic system types starting on page 42 may strike some readers as a bit abstract, so the following list offers a few more concrete examples. Many systems will be studied throughout this text, and the reader may wish to consult entries under *Policy*, *System examples*, and *World issues* in the index.

Simple	A single system, such as a single particle, a human cell, or an ant.
Complex	An ensemble of systems, such as many particles, an organ of many cells, or an ant colony.

Time ...

Stationary	The nature of the system is unchanging over time, such as a grandfather clock: the hands (the system state) do change with time, however the actual *mechanism* of the clock does not.
Nonstationary	Systems in which the underlying behaviour changes over time, such as aging in the human body.

Time stationarity is often a question of time period. The human body, as a system, is essentially stationary from one day to the next, but *non*stationary from one decade to the next.

Space ...

Stationary	The laws of physics, for example, do not change from one laboratory to the next.
Nonstationary	Much of the earth's environment is location-dependent, with very different models governing the topical, desert, temperate, and arctic regions.

As with time stationarity, spatial stationarity is often a question of extent. Ecological system dynamics might vary within a few meters (microclimates), or be essentially stationary over many kilometres.

First Order	A model without momentum. The rate of growth of a population, for example, is primarily a function of the current population, not of past rates of growth.
Second Order	A model subject to inertia, momentum or friction, such as a rolling toy car.
Deterministic	Very rare in practice; much of idealized physics, Newtonian mechanics, simple collisions between particles are treated as deterministic.
Stochastic	Nearly any sort of system subject to outside disturbances, essentially most social, environmental, and thermodynamic systems.
Input Driven	Most human-designed systems deliberately include an input as a means of influence, such as a keyboard for a computer or the brake in a car. For an ecosystem, such as a river, there may be human influences such as pollution which can be modelled as an input. Very simple, contrived systems, such as a marble rolling in a bowl, have no input.

as a function of some vector of parameters $\underline{\theta}$ and possibly subject to an input $\underline{u}$. As we shall see in later chapters, there can be significant latitude in what is considered as state, and what is embedded in the parameters since, in principle, and as we shall in Example 6.3, the parameters $\underline{\theta}(t)$ could themselves be varying over time and/or subject to human influence. What we choose to capture in the state $\underline{z}(t)$ are those time-dynamic elements of interest or of relevance in studying a given system.

For example, the dynamic state of a car might be only its location and velocity, in the context of a mapping problem; or perhaps also engine RPM, fuel consumption, and oil temperature, in the context of studying engine performance; or perhaps details of airflow on every part of the car's surface, in designing a new car body.

Dynamic models can be characterized by their

Order:	The highest derivative in the differential equation,
Dimension:	The number of elements n in the state vector $\underline{z} \in \mathbb{R}^n$.

There is an interesting tradeoff between model order and state dimension. Specifically, it is possible to embed or re-cast a higher-order model as a first-order model via *state augmentation*. Suppose we have a simple spring with friction,

$$\ddot{x}(t) = -kx(t) - b\dot{x}(t) \qquad \begin{array}{l} k: \text{Spring constant} \\ b: \text{Coefficient of friction} \end{array} \tag{4.6}$$

This is a second-order equation, since we are expressing the second derivative of x. We can define an augmented state $\underline{z}(t)$,

$$\underline{z}(t) = \begin{bmatrix} \dot{x}(t) \\ x(t) \end{bmatrix} \tag{4.7}$$

allowing us to rewrite (4.6) as

$$\underbrace{\begin{bmatrix} \ddot{x}(t) \\ \dot{x}(t) \end{bmatrix}}_{\dot{\underline{z}}(t)} = \underbrace{\begin{bmatrix} -b & -k \\ 1 & 0 \end{bmatrix}}_{A} \underbrace{\begin{bmatrix} \dot{x}(t) \\ x(t) \end{bmatrix}}_{\underline{z}(t)} \tag{4.8}$$

from which we recognize

$$\dot{\underline{z}}(t) = A \cdot \underline{z}(t) \tag{4.9}$$

a *first* order equation.

This approach to state augmentation can also be applied to non-scalar states. Suppose we are given a second-order vector equation

$$\ddot{\underline{y}} = B\underline{y} \tag{4.10}$$

for some n dimensional $\underline{y}$ and $n \times n$ matrix B. Then with the same state augmentation as in (4.7),

$$\underline{z}(t) = \begin{bmatrix} \dot{\underline{y}}(t) \\ \underline{y}(t) \end{bmatrix} \tag{4.11}$$

we derive an equivalent first order augmented representation

$$\dot{\underline{z}}(t) = \underbrace{\begin{bmatrix} 0 & B \\ I & 0 \end{bmatrix}}_{2n \times 2n \text{ matrix}} \underline{z}(t) \tag{4.12}$$

Because it is possible to re-cast a higher-order system as a first-order one, it is reasonable for us to focus exclusively on the first-order case of (4.9).

4.2 Randomness

Most social/ecological dynamic models and time series will contain a random/stochastic component. Although a thorough discussion of stochastic processes is well outside the scope of this text, we at least want to be aware of how randomness creeps into a problem:

- Due to actual random errors in the measurements of a system, for example in measuring the surface temperature of the ocean.
- Due to influences, perturbations, and unmodelled inputs from outside of the system envelope: there are many such examples in finance, where the model envelope may include a single company, currency, or country, and will not be able to capture or represent the multitude of outside influences.
- Due to modelling errors from inside the system envelope: our system model can never be complete, down to the atomic or sub-atomic level, so there will always be effects which a given model cannot represent; the classic example is Brownian Motion, the constant random jitter of smoke or pollen particles when viewed under a microscope.
- Finally, related to the previous two points, stochastic elements may be deliberately introduced into a model to prevent the model (or the human observing the model results) from becoming overconfident.

Figure 4.2 compares the dynamics of

$$\dot{z}(t) = f\big(z(t)\big) \qquad \text{and} \qquad \dot{z}(t) = f\big(z(t)\big) + w(t) \tag{4.13}$$

for a Gaussian random noise process w. The deterministic model will be inclined to ignore measurements, as it rigidly believes its “exact” model; the stochastic

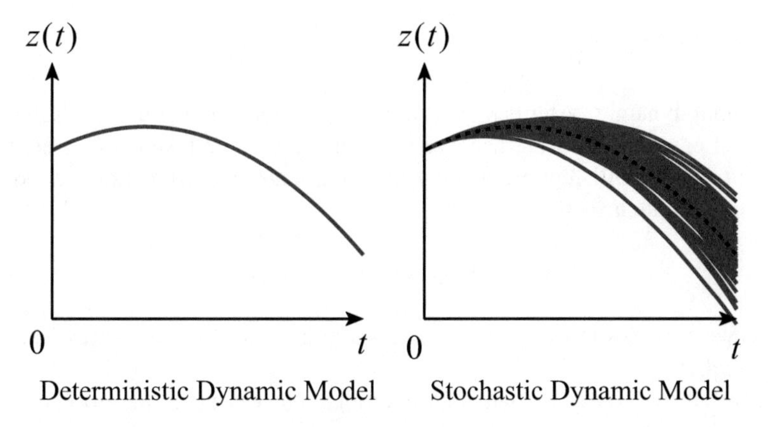

Fig. 4.2 Randomness: a deterministic dynamic model, left, is certain in its assertion of state, since its model gives no room for uncertainty. Nearly all models of dynamic systems contain some stochastic term, right, to reflect measurement noise, model uncertainty, and to prevent overconfidence.

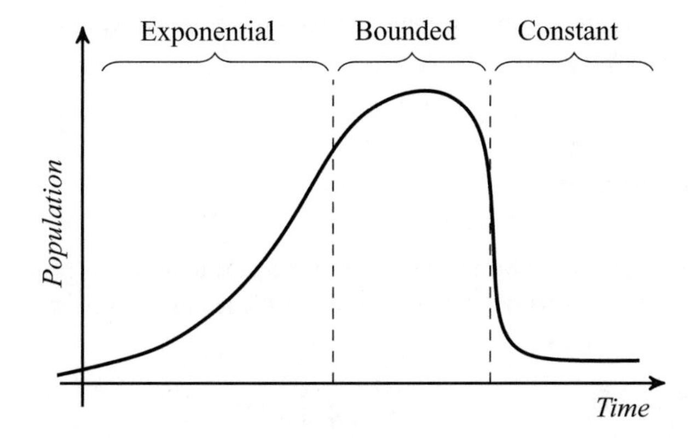

Fig. 4.3 Non-stationarity: the population of deer of St. Matthew island, off Alaska. When deer were introduced the island had previously been uninhabited and had built up extensive lichen beds. The population exploded until the slow-growing lichen had been consumed, at which point there was widespread starvation and a population crash. This is clearly a *non*-stationary time-dynamic signal, with roughly three stationary periods of behaviour: exponential, bounded, and constant.

model is much less sure of its state z, and will be much more easily influenced by measurements, a topic we will return to when examining *inverse problems* in Chapter 11.

4.3 Analysis

Although a dynamic system $\underline{z}(t)$ will normally evolve continuously with time t, in nearly all cases the measurements $m(t_i)$ of the system are taken at discrete points in time t_i. So we will suppose that our measurements, over some time period, have been stacked into a vector

$$\underline{m}^T = \begin{bmatrix} m(t_0) & m(t_1) & \cdots & m(t_{N-1}) \end{bmatrix} \tag{4.14}$$

Unless we need to know the specific points in time, it will be simpler just to refer to $m(i) \equiv m(t_i)$.

There are two fairly fundamental questions that we would like to know about measured signals:

Correlation: How are two or more signals related? For example, is there a relationship between atmospheric carbon dioxide and global temperature, or between regressive taxation and poverty?

Stationarity: What is the period of time or extent of space over which a signal has a consistent behaviour? If we wish to talk about a *change* over time (e.g., global warming), then we need to be able to define what it is from which we are changing.

4.3.1 *Correlation*

The correlation is the most basic concept regarding the connectedness of two signals or time series. We start with the concepts of sample mean and variance:

$$\text{Sample Mean} \qquad \mu_m = \frac{1}{N}\sum_{i=1}^{N} m(i) \tag{4.15}$$

$$\text{Sample Variance} \qquad \sigma_m^2 = \frac{1}{N-1}\sum_{i=1}^{N}\big(m(i) - \mu_m\big)^2 \tag{4.16}$$

Building on these, the correlation coefficient ρ_{xy} between signals $x(i)$ and $y(i)$ is defined as

$$\rho_{xy} = \frac{1}{N-1}\sum_{i=1}^{N}\left(\frac{x(i)-\mu_x}{\sigma_x}\right)\left(\frac{y(i)-\mu_y}{\sigma_y}\right) \tag{4.17}$$

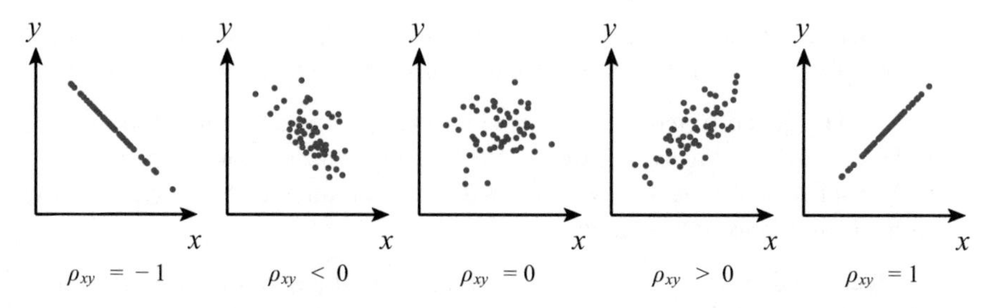

Fig. 4.4 Correlation: the correlation measures the linear relationship between two random variables x and y.

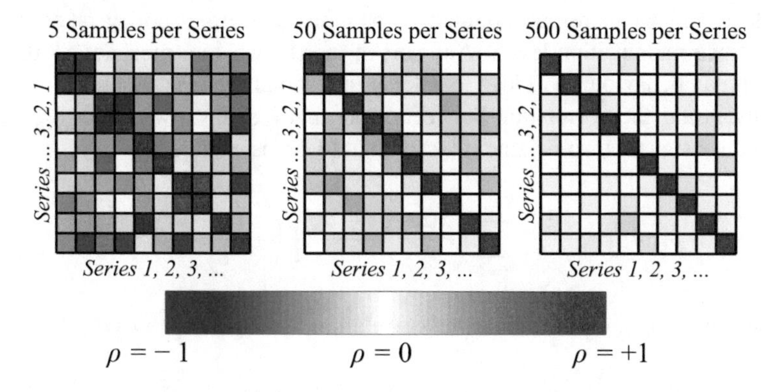

Fig. 4.5 Spurious correlations: the (i, j) element in each grid plots the correlation coefficient ρ_{ij} between time series i and j. In actual fact, *all* of the time series are *un*correlated; the apparent positive and negative correlations are coincident, and it takes a certain number of data points (right) to distinguish between real and spurious correlations.

As illustrated in Figure 4.4, the correlation coefficient is constrained to the range $-1 \leq \rho_{xy} \leq 1$, such that

$\rho_{xy} = -1$ x and y exactly fall on a line with negative slope
$\rho_{xy} < 0$ x and y negatively correlated: increasing x implies decreasing y
$\rho_{xy} = 0$ x and y uncorrelated: no linear relationship
$\rho_{xy} > 0$ x and y positively correlated: increasing x implies increasing y
$\rho_{xy} = 1$ x and y exactly fall on a line with positive slope

The correlation between two time series is a relatively simple concept, yet it is very easy to misinterpret. There are four common challenges or subtleties to keep in mind any time a correlation is measured:

Number of Samples: It takes data to establish a correlation: more data means greater confidence. With too few data points, even unrelated data sets can coincidentally appear to be correlated, an issue which is illustrated in Figure 4.5. There are formal statistical tests for this question (Pearson statistic, Chi-squared tests), and the reader is referred to further reading (page 61) for references.

At first glance the solution appears obvious, which is to acquire more data. However in practice the number of data points is very frequently limited:

- The data are collected over time, so more samples means more waiting;
- Data collection involves effort, so more samples means more money;
- Measuring instruments have a limited useful lifetime, and a given survey or field study runs only so long.

For these and many other reasons it is actually quite common to encounter attempts to establish correlations in marginally short data sets.

Time Lags: All real systems are causal, meaning that they cannot respond to an event until that event has taken place. So an abundance of food is only later reflected in an increased population, or a social welfare policy takes time to influence malnutrition levels. Any sort of inertia, whether mechanical, thermal, or human behaviour, will lead to a delay in responding to an input.

In assessing the correlation between input and response, we need to shift the response signal by some amount, such that (4.17) is changed to

$$\rho_{xy}(\delta) = \frac{1}{N-1}\sum_{i=1}^{N}\left(\frac{x(i-\delta)-\mu_x}{\sigma_x}\right)\left(\frac{y(i)-\mu_y}{\sigma_y}\right) \tag{4.18}$$

If, instead, we compute the lagged correlation of a signal with *itself*,

$$\rho_{x}(\delta) = \frac{1}{N-1}\sum_{i=1}^{N}\left(\frac{x(i-\delta)-\mu_x}{\sigma_x}\right)\left(\frac{x(i)-\mu_x}{\sigma_x}\right) \tag{4.19}$$

we obtain what is known as the autocorrelation, which measures how quickly or slowly a signal decorrelates over time.

Linear Versus Nonlinear: Correlation is only a measure of the *linear* relationship between two variables. However it is entirely plausible for two variables to be related, but *non*linearly so, such that there is no linear relationship, meaning that they are uncorrelated, even though the relationship could be exceptionally strong. One example of this effect is sketched in Figure 4.6.

Detecting a nonlinear relationship is more challenging than calculating a correlation, since the form of the nonlinearity (sine, parabolic, exponential) needs to be known ahead of time. The human visual system is very capable at detecting such patterns, so in many case the best choice of action is the visual inspection of plotted data.

Correlation Versus Causation: Ultimately we would like to understand cause and effect: Are humans causing global warming? Does free trade lead to increased employment?

However whereas there are definitive statistical tests to establish the statistical *significance* of a correlation, meaning that two time series are connected, it

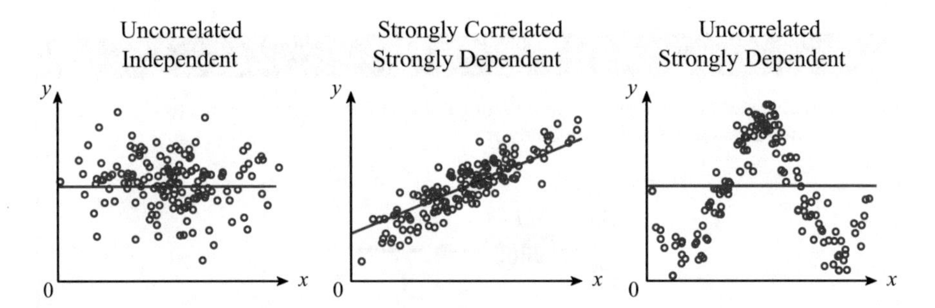

Fig. 4.6 Correlation and linearity: it is important to understand that correlation is a linear measure of dependence such that it is possible, right, to have random variables x and y strongly dependent but uncorrelated.

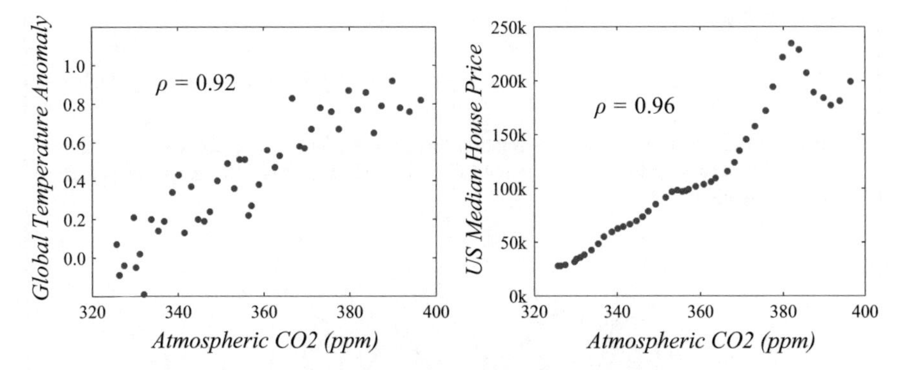

Fig. 4.7 One has to be very careful in attributing significance to correlations, even cases involving very strong correlation coefficients. Physics does tell us that global temperature, left, is related to the presence of CO_2, however house prices, right, are mostly driven by inflation and market prices, yet offer an even stronger correlation to atmospheric CO_2.

House Price data from the *S&P Case-Shiller National House Price Index*.

is not possible to establish *causation*, as illustrated in Figure 4.7. Causation can only be asserted on the basis of a model, such as a model for global carbon-dioxide sources, sinks, and flows, or an economic agent-based model representing businesses, consumers, and price signals.

4.3.2 *Stationarity*

For a deterministic signal, a transition from some behaviour A to B may be relatively clear. In contrast, for stochastic signals, subject to considerable variation and noise, a change may be much less obvious. A signal is *stationary* if its underlying model

Example 4.2: Correlation Lag — Mechanical System

Let us consider the example of an oscillating mass on a spring, an example to which we will return in some detail in Case Study 5:

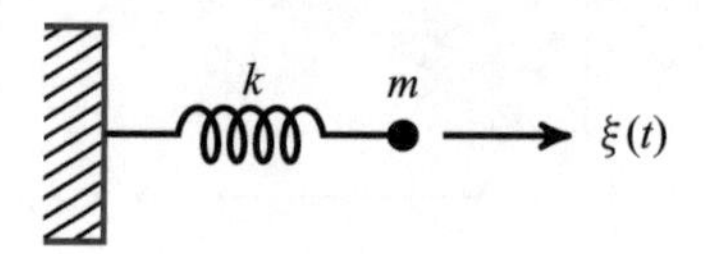

Suppose we had measurements of the applied force ξ, and we had an accelerometer and visual camera measuring the acceleration $\ddot{z}$, velocity $\dot{z}$, and position z of the mass. What might we expect to see?

We know that

$$\ddot{z} = \frac{\xi}{m} \qquad \dot{z} = \int \ddot{z} + c_1 \qquad z = \int \dot{z} + c_2 \tag{4.20}$$

Although there is an instantaneous relationship between force ξ and acceleration $\ddot{z}$, the inertia of the mass leads to a lag between force and velocity $\dot{z}$ or position z, as made clear in the simulation:

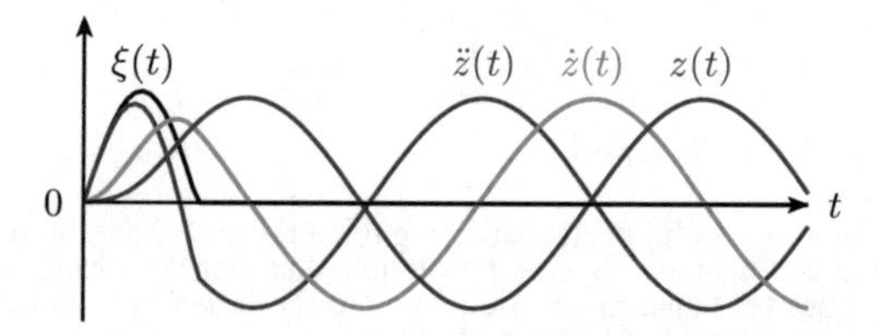

is constant, thus not that the signal itself is *constant*, rather that it its *behaviour* is constant with respect to some model:

Model	Stationarity assumption
Constant mean	Flat signal, possibly plus noise
Linear regression	Signal with constant slope, possibly plus noise
Fourier transform	Periodic signal with spectrum constant over time

Figure 4.3 shows one example of a non-stationary signal, having three time intervals of roughly stationary behaviour.

For the large–scale social or ecological systems of interest in this text, such as urban sprawl or global warming, the systems will possess both temporal and geographic extents. As a result, as was summarized on page 43, a given system

Example 4.3: Correlation Lag — Thermal System

Given a heat flow $h(t)$ into an object having a specific heat K, the temperature is given by

$$T(t) = \frac{1}{K} \int^{t} h(\tau)\, d\tau + c \tag{4.21}$$

so the thermal inertia of the object causes a lag between heat input and the resulting temperature. For example, although planet Earth is fed a nearly constant stream of input energy from the sun, a given location on the Earth has time-varying radiation input due to seasons and changing length of day:

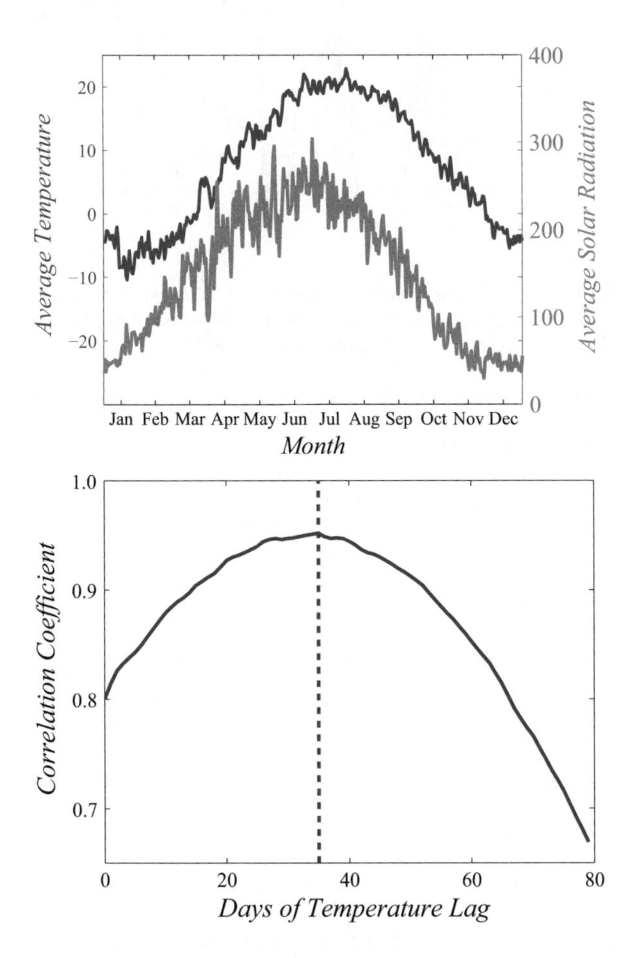

The left plot clearly shows the overall correlation between incoming radiation and the average temperature, however the peak in the lagged correlation, right, is at 35 days.

Example continues ...

Example 4.3: Correlation Lag — Thermal System (continued)

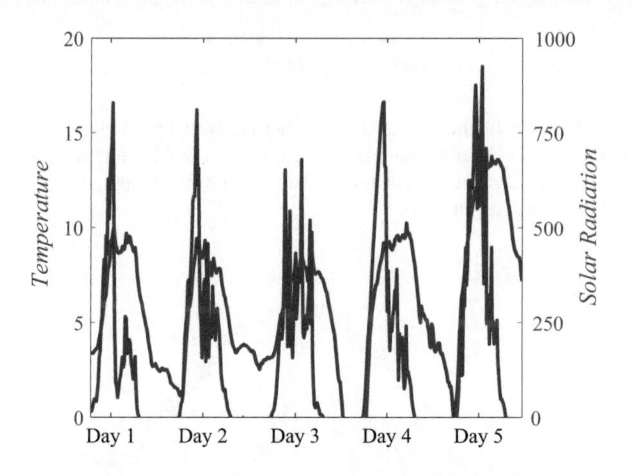

The thermal inertia can also be observed within single days. Day-time is clearly warmer than night-time, however the warmest part of the day is normally *not* right at noon (or at 1pm, with daylight savings time), when the sun is strongest.

Instead, due to the thermal inertia of the ground and the air, the peak in temperature is slightly later in the afternoon, as is very clearly revealed in plots correlating input–heat and lagged resulting–temperature:

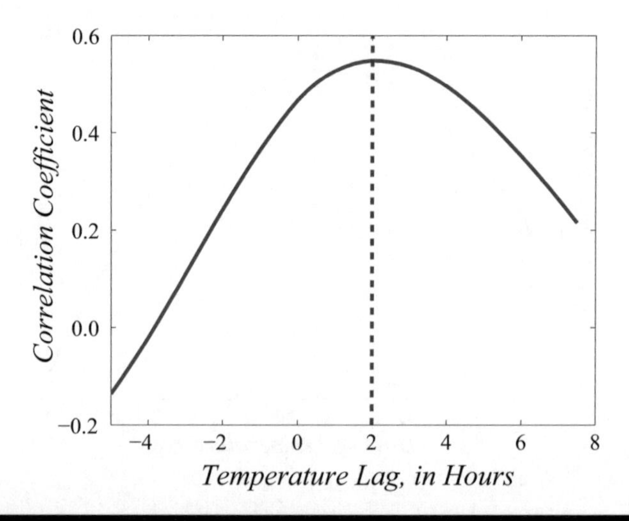

Example continues . . .

Example 4.3: Correlation Lag — Thermal System (continued)

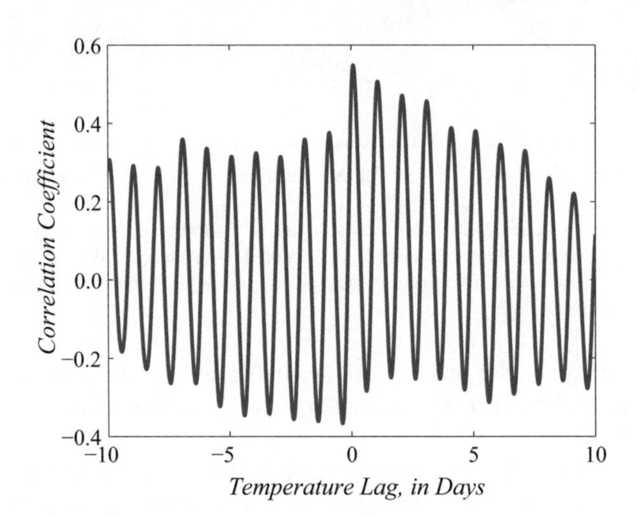

If we examine more distant lags, right, we see a sinusoidal pattern: the incoming sunlight today is more closely related to tomorrow's daytime temperature than last night's. The decrease over time is due to the fact today's sunlight is most closely related to today's temperature.

Further Reading: The data used in this example are available from the University of Waterloo weather station and from the textbook web site.

model may or may not be stationary over time, quite independent of whether it is stationary over space.

Suppose we are given data $\underline{m}$ and a *model* $M(\underline{\theta})$ predicting the behaviour of $\underline{m}$, parameterized by a vector of unknowns $\underline{\theta}$. To assess the stationarity of $\underline{m}$ against model M, we have two steps:

1. We need to fit the model to the data. That is, we need to learn the best choice of $\underline{\theta}$ from the given data $\underline{m}$:

$$\hat{\underline{\theta}} = \arg_{\theta} \min \|\underline{m} - M(\underline{\theta})\| \tag{4.22}$$

   such that $\hat{\underline{\theta}}$ is chosen to minimize some norm or penalty function $\|\cdot\|$.
2. We need to examine the stationarity of the *residuals*, the inconsistency between the data and the model:

$$\underline{r} = \underline{m} - M(\hat{\underline{\theta}}) \tag{4.23}$$

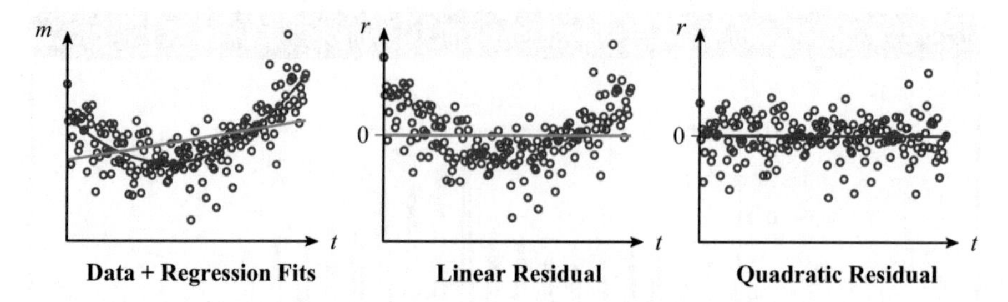

Fig. 4.8 Stationarity of residuals: stationarity is not an inherent property of a signal, rather a function of the *relationship* between a signal and a model. Given a set of points, left, the residuals are *not* stationary with respect to a linear model (green, middle), but *are* stationary with respect to a quadratic model (blue, right).

A visual illustration of this process is shown in Figure 4.8, where the data points are *non*-stationary with respect to a constant or linear fit, but are stationary with respect to a quadratic fit.

Given the residual signal $\underline{r}$, the remaining challenge is to identify the interval of time, the *baseline*, over which the residual is stationary. Ideally we would like to be able to assert a point in time t_o such that

$$\begin{aligned} &\{m(t), t < t_o\} \text{ is stationary with respect to } M(\underline{\theta}) \\ &\{m(t), t > t_o\} \text{ is statistically significantly } \textit{non}\text{-stationary to } M(\underline{\theta}). \end{aligned} \tag{4.24}$$

The baseline problem is then the combined determination of $M, \underline{\theta}$, and t_o. Such problems are actually quite widespread, typically in the form of *failure* or *anomaly* detection, such as the failure of a bearing in a motor or the insulation on an electrical conductor, with a interest in detecting the failure as quickly as possible to limit damage. However bearings and insulators, when working properly, do not exhibit significant fluctuations in behaviour, in contrast to social and environmental systems.

It is possible to design statistical tests for stationarity, however this is a relatively complicated question (see further reading on page 61). The human visual system is quite effective at assessing patterns and stationarity, so residual stationarity intervals are frequently determined by eye, even in research papers.

4.3.3 *Transformations*

Most signals are highly redundant; indeed, compare the hundreds of parameters needed to describe all of the data points in Figure 4.8, as opposed to the two parameters to describe a line or three parameters for a parabola. Because of this redundancy, most signals are analyzed by applying a transformation T to the signal or to a stationary portion of it, such that the transformed coefficients reveal more clearly the signal structure. We will briefly overview a few of the most common strategies:

Time Domain $T(\underline{m}) = \underline{m}$

Assumption: None

Here the transformation is just the identity $T = I$, so the signal is not actually changed. However the time domain may in fact be an effective domain for analyzing the signal, particularly for signals having abrupt discontinuities.

Mean $T(\underline{m}) = \frac{1}{N}\sum_{1}^{N} m(t_i)$

Assumption: Constant signal

The mean is very robust against noise, meaning that it is insensitive to the presence of noise in the signal. Constant signals will be relatively uncommon in measurements of time-dynamic systems, however they may be present for small periods of time.

Linear Regression $T(\underline{m}) = \begin{bmatrix} \sum_i t_i^2 & \sum_i t_i \\ \sum_i t_i & \sum_i 1 \end{bmatrix}^{-1} \begin{bmatrix} \sum_i t_i m(t_i) \\ \sum_i m(t_i) \end{bmatrix}$

Assumption: Signal with constant slope

A signal with constant slope over time is the most basic form of time dependency. Linear regression produces a mean and slope for a given set of data points; the method is easily generalized to fitting other functions, such as a parabola or an exponential.

Frequency Domain $T(\underline{m}) = \text{Fourier Transform}(\underline{m})$

Assumption: Signal is periodic with a frequency spectrum constant over time

The discrete Fourier Transform,

$$M_k = \sum_{j=0}^{N} m(j) \cdot e^{-2\pi ijk/N} \tag{4.25}$$

computes the complex sine/cosine component M_k from $\underline{m}$. It is possible to define the Fourier transform for irregularly spaced time points t_i, however (4.25) assumes regularly spaced points in time. The Fourier transform is particularly effective in identifying periodic behaviour — daily, weekly, yearly etc.

Time-Frequency Domain $T(\underline{m}) = \text{Wavelet Transform}(\underline{m})$

Assumption: None

We can view the time and frequency domains as two ends of a continuum:

- Time domain samples are highly local in time, but completely spread in frequency.
- Frequency domain samples are unlocalized in time, but are focused at a particular frequency.

The wavelet transform is intermediate between these two extremes, allowing for an assessment of time-varying frequency components.

Noise Reduction $$T(k) = \frac{\text{Power of Signal}(k)}{\text{Power of Signal}(k) + \text{Noise}(k)}$$

Assumption: Signal has frequency spectrum constant over time
All measurements have some degree of inaccuracy, what we refer to as measurement noise. As a result, a great many denoising strategies have been proposed, the simplest of which are the *mean* and *linear regression* cases discussed above, however most signals are not a constant mean or constant slope.
If the measurement noise is additive (added to the signal, as opposed to multiplied) and with a constant frequency spectrum, then the optimal denoising strategy is the Wiener Filter, the ratio $T(k)$ of power spectra, such that at some frequency k where the noise is strong the measurement is attenuated, and where the noise is weak the measurement is left intact.

Principal Components $$T(\underline{m}) = V^{-1}\underline{m}$$

Assumption: We need to know the statistics of $\underline{m}$
Principal components is one example of state dimensionality reduction, whereby we find an eigendecomposition and keep only the most significant modes. Most signals are dominated by relatively few modes, but the noise, being random and unrelated to the signal, is spread over all of the modes, therefore keeping only the first q modes preserves most of the signal but only q/N of the original noise. The key challenges are that we need to know the statistics of the signal in order to find its eigendecomposition (which gives us V), and that for large signals the eigendecomposition may be very hard or impossible to compute.

Since the number of transformations is modest, and are built into any modern numerical analysis program, selecting a transformation can be made on the basis of practice and experience, the assumptions associated with each transformation, or by trial and error.

Case Study 4: Water Levels of the Oceans and Great Lakes

Water levels are one of the most longstanding time series, since the timing and height of tides has been of importance into antiquity. Our goal in this case study will be to use water levels from three contexts as a vehicle for illustrating signal analysis.

We will begin with the Great Lakes of North America, focusing on the heights of Lakes Superior and Ontario, as shown in Figure 4.9.

The signals are clearly variable over time, however we see no long-term trend, and the variability appears to be more or less random, centred on a constant height. Therefore it is reasonable to apply the most basic transformation, that of the constant mean over time, shown dashed in the figure.

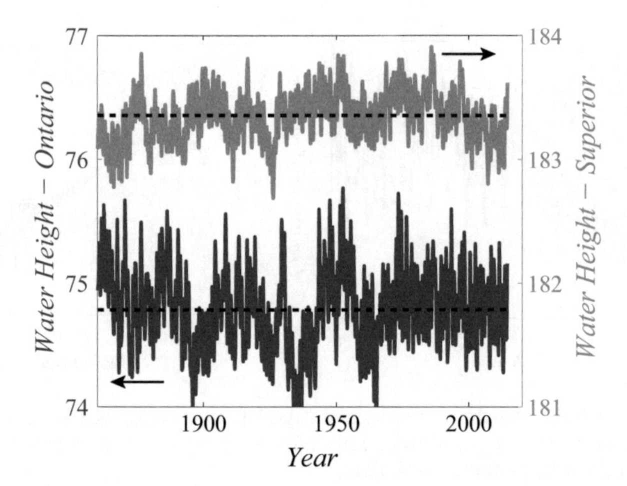

Fig. 4.9 The surface height of the Great Lakes: the time series are noisy, but appear to be mostly constant, so we begin with the most primitive transformation, finding the long-term means, dashed. Data from the NOAA Great Lakes Environmental Research Laboratory.

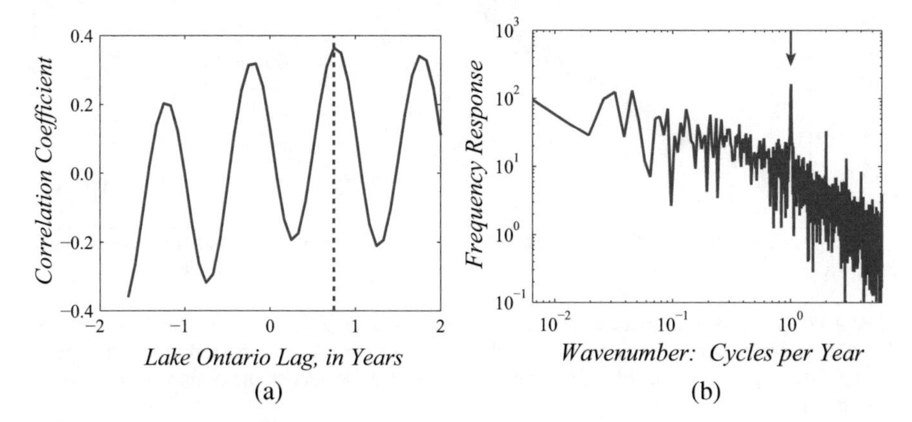

Fig. 4.10 Lake height analysis: from the data in Figure 4.9, we can compute the lagged correlation, left, between the time series. There is a modest correlation between the lake heights, with the Lake Ontario signal lagging that of Lake Superior. The annual cycle is most obvious (see arrow) in the frequency domain, right.

We expect a correlation between the heights of the two lakes, both because Lake Superior flows into Lake Ontario, but also because precipitation patterns will be similar for the two. Figure 4.10 (left) plots the correlation between the two time series in Figure 4.9, as a function of signal lag. The correlation is strongly oscillating, with a period of 1 year, suggesting that there is a yearly cycle present in the height of both lakes, however interestingly the peak correlation does correspond to a lag of approximately 9 months.

Figure 4.10 (right) plots the Lake Ontario data in the frequency domain. That a yearly cycle is present is quite striking, marked by the red arrow. You will observe

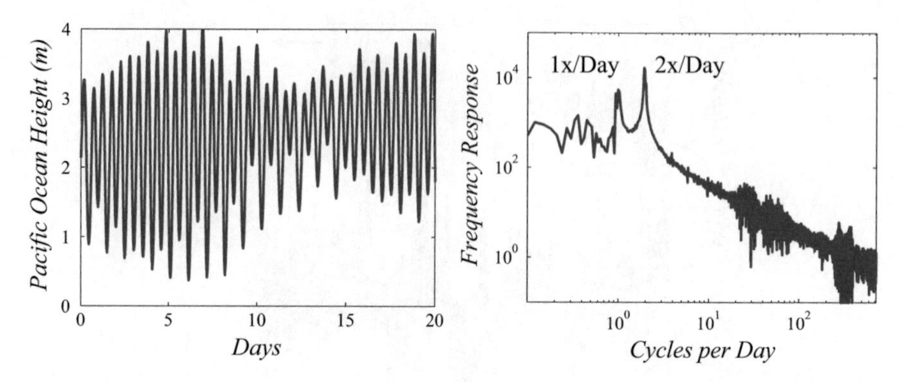

Fig. 4.11 Ocean surface height over a period of days, left, from which the regular tidal behaviour can clearly be seen. The daily and twice-daily tidal peaks are the most striking behaviour in a frequency domain analysis of the signal, right.

Data from Fisheries and Oceans Canada.

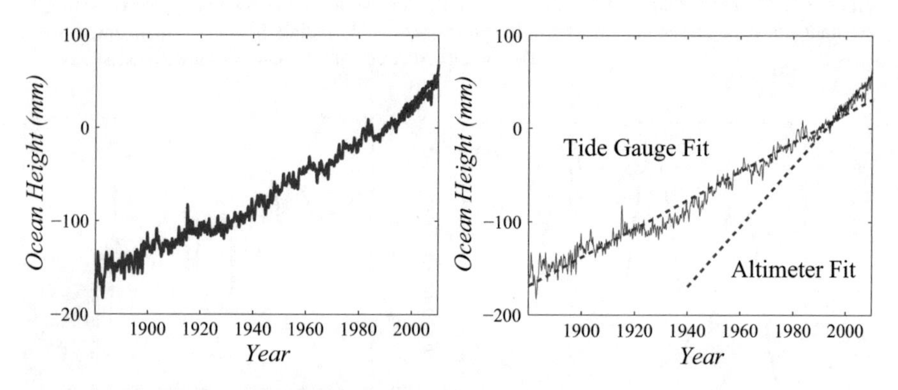

Fig. 4.12 Global ocean surface height over time based on tide gauges (blue) and, more recently, on satellite altimetry (red). The increase is relatively constant over time, so we can model the signals using linear regression (dashed), where we see that the satellite data implies a significantly steeper slope.

Data from CSIRO — Marine and Atmospheric Research.

further spectral lines at higher frequency harmonics (fractions) of 1 year; these are caused by the fact that the annual cycle is not sinusoidal, so the actual non-sinusoidal shape of the annual cycle introduces additional frequency components.

The Great Lakes are not subject to any meaningful tidal effect, whereas the world's oceans respond strongly to tidal effects of the moon. Figure 4.11 plots a tidal signal over time and in frequency, where the daily and twice-daily peaks emerge strongly.

The tidal signal of Figure 4.11 spans a period of only 1 month, making it impossible to observe yearly or longer term phenomena. In contrast, Figure 4.12 shows Sea-Surface-Height (SSH) data over a period of 130 years. There are two data sets shown: measurements from tide gauges and from satellite altimeters. The tide gauge series is far longer, spanning 130 years, but may be biased because the gauges

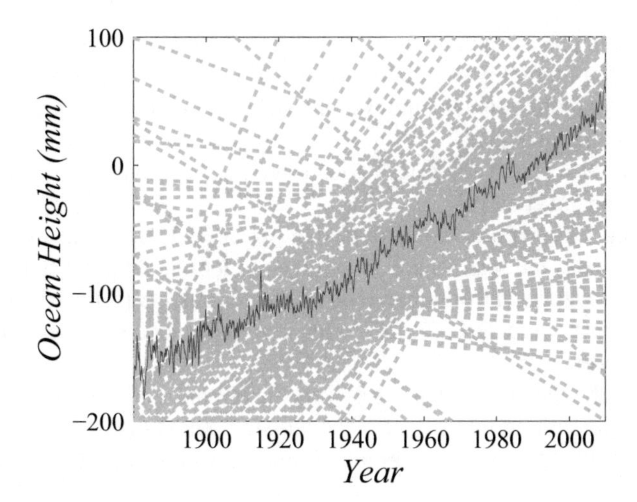

Fig. 4.13 The effect of baseline: if we compute a linear regression from the tidal data of Figure 4.12, but for a variety of different baselines, we can reach radically differing conclusions.

are mostly close to land and mostly in the northern hemisphere. Satellite altimeters, measuring ocean surface elevation based on radar echoes, are much more recent but are global in coverage.

The most striking aspect of the time series is the steady upwards increase of 150mm since 1880. The signal follows a relatively straight line, so a linear regression fit is appropriate, shown in Figure 4.12. Interesting the longer-term slope of the tide gauges and the more recent slope of the satellite altimetry are quite different. A longer baseline will be required to validate whether this difference in slope continues; the perils of baseline ambiguity are illustrated in Figure 4.13.

Further Reading

The references may be found at the end of each chapter. Also note that the textbook further reading page maintains updated references and links.

Wikipedia Links — Analysis: Dynamical system, Time series, Fourier transform, Wavelet transform, Principal components

Wikipedia Links — Statistics: Correlation and Dependence, Correlation coefficient, Correlation and Causality, Stationary Process, Pearson's Chi-Squared test

An excellent place to start is the recent *Dynamic Systems for Everyone* by Ghosh [6], which offers a highly accessible overview of dynamic systems. Further good introductory textbooks on dynamic systems include those by Devaney [4], Robinson [11], and Scheinerman [13].

With regards to signal analysis, a good reference book on signals and systems is that by Oppenheim et al. [9].

For statistics, a brief review is provided in Appendix B. There are endless choices for statistics texts, and readers are encouraged to look around the statistics section (Library of Congress call number HA29) of any university library. Some basic, introductory texts include [2, 7, 16], with more advanced treatments in Papoulis [10] and Duda et al. [5].

The field of time series analysis is very large, and only the most basic introduction was possible in this chapter. The classic book is that by Box and Jenkins, recently updated [1]. A very accessible introductory book is that of Brockwell and Davis [3].

The question of statistical stationarity is quite challenging. If the reader really wishes to explore the subject, there are tests built in to mathematical packages such as Matlab, PSPP, R, SPlus etc. The actual assessment of stationarity is via the Dickey–Fuller, Augmented Dickey–Fuller, and the Kwiatkowski–Phillips–Schmidt–Shin tests.

There are a number of books exploring the intersection of dynamic, ecological, and social systems. Recommended would be the book by Scheffer [12] and the series of books by Turchin, particularly those on the dynamics of social systems [15] and population [14]. A particularly influential and controversial work was that of *Limits to Growth* [8], originally published 40 years ago and since repeatedly updated.

Sample Problems

Problem 4.1: Short Questions

(a) Define state augmentation.
(b) What are a few reasons for including randomness in dynamic models?
(c) Specify three common signal transformations and their underlying assumptions.

Problem 4.2: Conceptual — Randomness

We would like to think about sources of noise and randomness in a variety of contexts. For example, suppose we have measurements of societal unemployment; two possible causes for randomness or uncertainty are

- Sampling errors — a survey asks only a few people, not everyone;
- Categorization errors — what is the definition of “employed”?

For each of the following measurements or data points, similarly propose two or three sources of randomness or uncertainty:

(a) North-Atlantic ocean surface temperature
(b) Atmospheric carbon-dioxide levels
(c) Clearcutting extent in the Amazon rainforest
(d) Poverty levels among children in North America

Problem 4.3: Conceptual — Stationarity

Give examples of systems or signals which are

(a) Time stationary
(b) Time nonstationary
(c) Spatially stationary
(d) Spatially nonstationary

Problem 4.4: Analytical — State Augmentation

An *autoregressive model*,

$$z(t) = w(t) + \sum_{i=1}^{n} \alpha_i z(t-i)$$

is very commonly used in the modelling of discrete-time signals and time-series. Although we did not discuss discrete-time series in this chapter, the principle of state augmentation is the same.

For the following third-order model,

$$z(t) = w(t) + \alpha_1 z(t-1) + \alpha_2 z(t-2) + \alpha_3 z(t-3)$$

perform state augmentation to convert it to first-order form.

Problem 4.5: Analytical — State Augmentation

Following up on Problem 4.4, a second widely-used model is the *moving average*,

$$z(t) = \sum_{i=0}^{n-1} \beta_i w(t-i),$$

driven by a random white-noise process $w(t)$.

For the following third-order model,

$$z(t) = \beta_0 w(t) + \beta_1 w(t-1) + \beta_2 w(t-2)$$

perform state augmentation to convert it to first-order form.

Problem 4.6: Analytical — Modelling

Define the state (not the dynamics) for the following systems:

(a) Pendulum
(b) Double pendulum
(c) A frictionless point mass free to move on a surface

Problem 4.7: Numerical/Computation — Time Series Analysis

We wish to do some basic data analysis of weather data, to reproduce parts of Example 4.3. You may acquire weather data for your local area, or you can download the data from the textbook web site.

(a) The first plot of Example 4.3 shows temperature and radiation as a function of time. Reproduce the plot, although clearly details of colour and formatting are not important.

    What transformation has been applied here? What assumptions does this transformation make, and how valid are they?

(b) The second plot shows the correlation coefficient as a function of lag. Because the signals are of finite length, when we shift them (due to the lag) they will no longer line up at the ends, so we need to assert a boundary behaviour:

    (i) We can truncate the ends so that the signals match
    (ii) We can extend the missing ends by padding with zeros
    (iii) We can extend the missing ends by reflecting the signals at the endpoints
    (iv) We can extend the missing ends by assuming the signals to be periodic

    You may select any of these four. Because the temperature and radiation signals really are, in fact, close to being periodic, the last of the four approaches is most likely, in this case, to give the best results.

(c) Compute the FFT (Fast Fourier Transform) and examine the magnitude (absolute value) of the frequency components:

```
plot( abs( fft( temp_signal ) ) );
```

    What do you observe?

    Because of aliasing, the frequency domain signal is periodic; you want to plot only the first half of the FFT samples.

    Next, the horizontal axis is hard to interpret. Think about the meaning of the x-axis, and scale the plot as was done in Figure 4.10 (cycles per year) or in Figure 4.11 (cycles per day).

Problem 4.8: Reading — Limits to Growth

This chapter looked at dynamic systems. One of the classic books in this area is *Limits to Growth*, a relatively controversial text using systems methods to model and simulate the interaction between human society, capital, pollution etc. Look over[1] *Chapter 3 — Growth in the World System* in *Limits to Growth*.

Comment briefly on the types of dynamic systems being modelled—deterministic vs. stochastic, linear vs. nonlinear. What are the modelling challenges identified by the authors?

Problem 4.9: Policy — Baselines

The question of *baseline* was raised at the outset of this chapter and it is probably one of the most contentious issues in the analysis of controversial data sets, such as climate change or global warming.

However the definition of a baseline is essential, since in the absence of a target it is impossible to formulate a policy goal.

Select a current environmental or social issue and discuss the challenges, the alternatives, or the controversies associated with defining an appropriate baseline.

References

1. G. Box, G. Jenkins, and G. Reinsel. *Time series analysis: forecasting and control*. Wiley, 4th edition, 2008.
2. C. Brase and C. Brase. *Understanding Basic Statistics*. Cengage Learning, 2012.
3. P. Brockwell and R. Davis. *Introduction to Time Series and Forecasting*. Springer, 2010.
4. R. Devaney. *A First Course in Dynamical Systems*. Westview Press, 1992.
5. R. Duda, R. Hart, and D. Stork. *Pattern Classification*. Wiley, 2000.
6. A. Ghosh. *Dynamic Systems for Everyone: Understanding How Our World Works*. Springer, 2015.
7. M. Kiemele. *Basic Statistics*. Air Academy, 1997.
8. D. Meadows, J. Randers, and D. Meadows. *Limits to Growth: The 30-Year Update*. Chelsea Green, 2004.
9. A. Oppenheim, A. Willsky, and H. Nawab. *Signals & Systems*. Prentice Hall, 1997.
10. A. Papoulis and S. Pillai. *Probability, Random Variables, and Stochastic Processes*. McGraw Hill, 2002.
11. R. Robinson. *An Introduction to Dynamical Systems*. Prentice Hall, 2004.
12. M. Scheffer. *Critical transitions in nature and society*. Princeton University Press, 2009.
13. E. Scheinerman. *Invitation to Dynamical Systems*. Dover, 2013.
14. P. Turchin. *Complex Population Dynamics*. Princeton, 2003.
15. P. Turchin. *Historical Dynamics: Why States Rise and Fall*. Princeton, 2003.
16. T. Urdan. *Statistics in Plain English*. Routledge, 2010.

[1]The text can be found online at the Donella Meadows Institute; a link is available from the textbook reading questions page.

Chapter 5
Linear Systems

One of the key issues in systems behaviour is that of *coupling*:

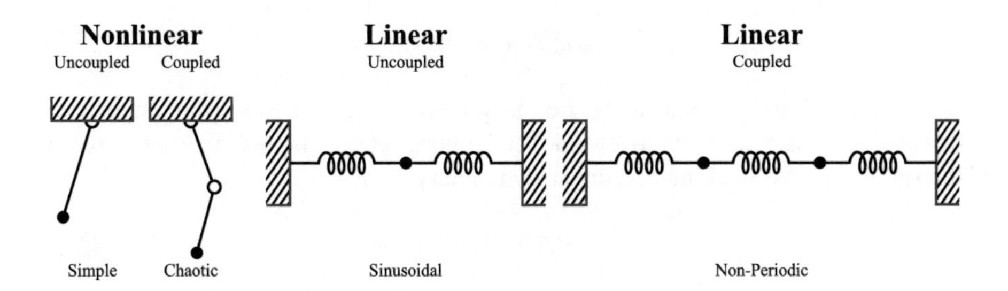

A single pendulum exhibits simple, uninteresting periodic motion, however coupling *two* pendula is a classic example of chaotic behaviour. Even in linear systems, a single mass attached to springs is periodic and sinusoidal, whereas two such masses, although still oscillating, can exhibit motion which is *non*-periodic.

Coupling is central to the study of ecological and social phenomena, since it appears in so many contexts:

- A town consists of many people, influencing each others' habits;
- A city consists of many businesses, who compete in buying and selling;
- A forest consists of many species, some parasitic, some symbiotic;
- An ecosystem consists of many ponds, rivers, lakes, streams, with water, nutrients, and pollution flowing between them.

On what basis might we be able to take a complex, interacting system and somehow break it into constituent parts? And under what circumstances might decoupling be possible for one type of system and not for another?

These are not easy questions to answer, so we need to start with those systems which are relatively simple and mathematically tractable — *linear* systems.

P. Fieguth, *An Introduction to Complex Systems*,
DOI 10.1007/978-3-319-44606-6_5

5.1 Linearity

We begin with an arbitrary nth order model

$$\underline{z}^{(n)}(t) = f\big(\underline{z}^{(n-1)}(t), \ldots, \underline{z}^{(1)}(t), \underline{z}^{(0)}(t), \underline{u}(t), t\big) \tag{5.1}$$

where $\underline{z}^{(i)}$ is the ith order time-derivative of $\underline{z}$, and $\underline{u}(t)$ is some external deterministic input. Recall the discussion of state augmentation, allowing higher-order dynamic models to be re-cast as first-order ones. On the basis of that idea, we can narrow our focus to the first-order case:

$$\dot{\underline{z}}(t) = f\big(\underline{z}(t), \underline{u}(t), t\big). \tag{5.2}$$

In this chapter we are considering *linear* dynamic models, and any linear relationship can be expressed as a matrix-vector product:

$$\dot{\underline{z}}(t) = A(t)\underline{z}(t) + B(t)\underline{u}(t) + \underline{w}(t) \tag{5.3}$$

for some matrices A, B and stochastic (noise) term $\underline{w}$. If we assume the model A to be time–stationary (constant over time) and ignore external inputs, then we arrive at the simplest, most fundamental linear-dynamic system:

$$\dot{\underline{z}}(t) = A\underline{z}(t). \tag{5.4}$$

To be clear, the concept of *linear system* can mean at least three different, but highly related ideas, illustrated in Figure 5.1. The distinction between these three will be clear from context, as we will not be mixing them. For the time being, we are focusing on linear dynamic systems.

In all cases, the key principle of *superposition* applies:

For a system obeying superposition, the system response to a sum of multiple inputs is just the sum of responses to the individual inputs.

Thus, for a linear system of equations $A\underline{x} = \underline{b}$,

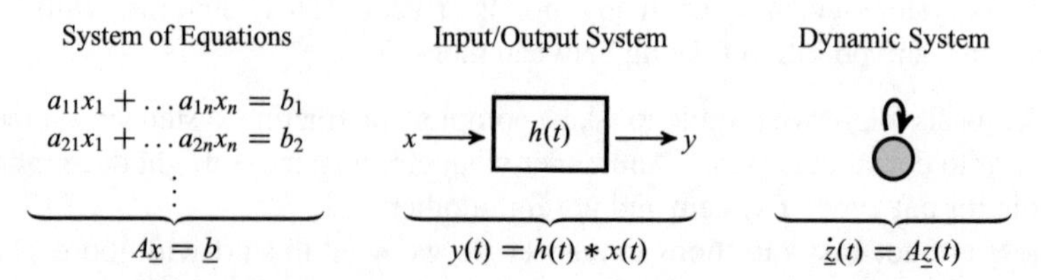

Fig. 5.1 Linear systems: a linear system can refer to several different concepts, depending upon the context. In this chapter we will focus primarily on linear dynamic systems.

If $\underline{b}_1$ leads to solution $\underline{x}_1$, and $\underline{b}_2$ leads to solution $\underline{x}_2$
Then $\alpha\underline{b}_1 + \beta\underline{b}_2$ leads to solution $\alpha\underline{x}_1 + \beta\underline{x}_2$

For a linear input–output[1] system $y(t) = h(t) * x(t)$,
If input $x_1(t)$ leads to output $y_1(t)$, and $x_2(t)$ leads to output $y_2(t)$
Then input $\alpha x_1(t) + \beta x_2(t)$ leads to output $\alpha y_1(t) + \beta y_2(t)$

For a linear dynamic system[1] $\dot{\underline{z}} = A\underline{z}(t)$,
If initial condition $\underline{z}(0) = \underline{i}_1$ leads to time behaviour $\underline{z}_1(t)$,
and $\underline{z}(0) = \underline{i}_2$ leads to time behaviour $\underline{z}_2(t)$,
Then initial condition $\alpha\underline{i}_1 + \beta\underline{i}_2$ leads to behaviour $\alpha\underline{z}_1(t) + \beta\underline{z}_2(t)$

The principle of superposition is one of the key points of this text and a deep understanding is essential, precisely because it seems so obvious and intuitive. In plain terms, for a linear system the average response is the same as the response to the average:

$$\text{Average}\Big(\text{System Response}\big(\text{Input}(t)\big)\Big) = \text{System Response}\Big(\text{Average}\big(\text{Input}(t)\big)\Big) \tag{5.5}$$

Yet virtually all social/ecological/natural systems do *not* satisfy superposition!, such that (5.5) does *not* hold. That is, in an ecological (*non*linear) system, two simultaneous disturbances (logging and pollution, for example) which, on their own might have had a modest effect, may together wound or break the system, such that the net response is, in fact, quite different from the sum of the individual ones. For the same reason it is inappropriate to report *average* pollution levels, such as in a river downstream from a mine, since a constant pollution level τ, versus a level of pollution fluctuating about τ and occasionally surpassing a toxicity threshold, can lead to very different consequences.

5.2 Modes

So, given a first-order linear dynamic system

$$\dot{\underline{z}} = A\underline{z} \tag{5.6}$$

how can we understand its behaviour?

[1]A dynamic system can have inputs and outputs, and a linear input–output system can be time-dynamic. The distinction is primarily one of representation; that is, whether the system is characterized by its impulse response $h(t)$, or by its time dynamics A. This text will focus almost entirely on the latter time-dynamic representation, since the concept of impulse response does not generalize well to nonlinear systems. Readers not familiar with convolution ($*$) and impulse response $h(t)$ may wish to consult a text on signals and systems, such as [8].

The most beautiful tool in linear algebra is the *eigendecomposition*. Sadly, most students recall (or were primarily taught) only the mechanics, that the eigendecomposition involves taking a determinant of some matrix and finding the roots of a polynomial, which leaves us with essentially no insight whatsoever.

The eigendecomposition takes a system and breaks it into decoupled *modes*, where the modes are the characteristic or unchanging behaviours of the system. Specifically, suppose we are given the dynamics matrix A in (5.6); if

$$A\underline{v} = \lambda \underline{v} \tag{5.7}$$

then we refer to λ as an eigenvalue of A, and $\underline{v}$ as its corresponding eigenvector or mode. Why does this matter? Observe that if the state at time $t = 0$ is initialized in the direction of mode $\underline{v}$,

$$\underline{z}(0) = \underline{v} \quad \longrightarrow \quad \dot{\underline{z}}(0) = A\underline{z}(0) = A\underline{v} = \lambda \underline{v}, \tag{5.8}$$

then the change $\dot{\underline{z}}$ points in the direction of $\underline{v}$, so that $\underline{z}$ *continues* to point in the direction of mode $\underline{v}$. Therefore we can write

$$\underline{z}(t) = \alpha(t)\,\underline{v} \quad \text{and} \quad \dot{\underline{z}}(t) = \dot{\alpha}(t)\,\underline{v}, \tag{5.9}$$

where $\alpha(t)$ is just a scalar, measuring how *far* $\underline{z}(t)$ points in the direction of $\underline{v}$. Then, combining (5.8) and (5.9),

$$\dot{\alpha}(t)\,\underline{v} = \dot{\underline{z}}(t) = A\underline{z}(t) = A\alpha(t)\underline{v} = \alpha(t)A\underline{v} = \alpha(t)\lambda\underline{v} \tag{5.10}$$

from which we find the dynamics of $\alpha(t)$ as

$$\dot{\alpha}(t) = \lambda\alpha(t), \tag{5.11}$$

a simple first-order differential equation, for which the solution follows as

$$\alpha(t) = \exp(\lambda t) \quad \longrightarrow \quad \underline{z}(t) = \exp(\lambda t)\,\underline{v}. \tag{5.12}$$

Thus if the state $\underline{z}$ points in the direction of an eigenvector, then it continues to point in that *same* direction over time; only the amplitude changes, with the rate of change measured by the *time constant* $\tau = {}^{1}/_{\lambda}$.

In general, however, an $n \times n$ matrix A will not have a *single* mode, rather n of them. If we assume, for the moment, that matrix A is diagonalizable, then we can collect the eigenvalues/eigenvectors

$$A\underline{v}_i = \lambda_i \underline{v}_i, \quad 1 \le i \le n \quad \longrightarrow \quad V = \begin{bmatrix} | & & | \\ \underline{v}_1 & \cdots & \underline{v}_n \\ | & & | \end{bmatrix} \quad \Lambda = \begin{bmatrix} \lambda_1 & & \\ & \ddots & \\ & & \lambda_n \end{bmatrix} \tag{5.13}$$

where V is invertible and Λ is diagonal, from which it follows that

$$A = V\Lambda V^{-1} \qquad \Lambda = V^{-1}A\,V. \tag{5.14}$$

The diagonalization of A in (5.14) refers to the ability of V to transform A to Λ.

The power of the eigendecomposition is that, whereas solving $\dot{\underline{z}} = A\underline{z}$ directly is quite challenging, the eigendecomposition transforms the problem into something simple. Let $\underline{y}(t)$ represent the amplitudes of the modes of the system, such that

$$\underline{y}(t) = V^{-1}\underline{z}(t) \qquad \underline{z}(t) = V\underline{y}(t). \tag{5.15}$$

Then

$$\dot{\underline{y}}(t) = V^{-1}\dot{\underline{z}}(t) = V^{-1}A\underline{z}(t) = V^{-1}A\,V\underline{y}(t) = \Lambda\underline{y}(t) \tag{5.16}$$

from which it follows that

$$\begin{array}{ccc} \dot{\underline{z}}(t) = A\underline{z}(t) & \xrightarrow[\text{Eig.}]{V^{-1}} & \dot{\underline{y}}(t) = \Lambda\underline{y}(t) \\ & & \therefore\ \dot{y}_i(t) = \lambda_i y_i(t) \\ & & \Big\downarrow \text{Easy} \\ \underline{z}(t) = V\underline{y}(t) & \xleftarrow[\text{Eig.}]{V} & \therefore\ y_i(t) = e^{\lambda_i t}\,y_i(0) \end{array} \tag{5.17}$$

We can immediately notice a few things:

- Because Λ is a diagonal matrix, the dynamics for $\underline{y}$ are decoupled into individual simple scalar dynamics for each y_i, the modes.
- We see that the general solution for $\underline{z}(t)$ is nothing more than a weighted combination of modes $\underline{y}$, where the weights are predetermined by V. However in a given problem, the actual amplitude of each mode is determined by the initial state $\underline{y}(0) = V^{-1}\underline{z}(0)$.
- Since $\underline{z}(t)$ is just a weighted sum of exponentials, broadly speaking there are only two possible behaviours:

$$\begin{array}{lll} \text{If Real}(\lambda_i) < 0 \text{ for all } i & |\underline{z}(t)| \longrightarrow 0 & \therefore \textit{Stable} \text{ Dynamics} \\ \text{If Real}(\lambda_i) > 0 \text{ for any } i & |\underline{z}(t)| \longrightarrow \infty & \therefore \textit{Unstable} \text{ Dynamics} \end{array} \tag{5.18}$$

Therefore *all* diagonalizable linear systems either decay to zero or grow exponentially.

Since all natural systems are finite, indefinite exponential growth is not possible, so for any real physical system a linear model is necessarily only an approximation to the actual, bounded, nonlinear system, such as the example in Figure 4.3.

Example 5.1: The Exponential Function

Anyone who believes exponential growth can go on forever in a finite world is either a madman or an economist.

– Kenneth Boulding

The characteristic behaviour (or modes or eigenfunctions) of *all* linear systems consists of the family of complex exponentials. As a result, with the exception of degenerate cases, all linear systems produce either exponentially shrinking or exponentially growing output over time. In general it is, of course, the exponentially growing systems which concern us.

Since people produce more people, and dollars (with interest) produce more dollars, in both cases

$$\dot{z}(t) \propto z(t) \tag{5.19}$$

the solution to which is exponential, which is why such behaviour appears so commonly in societal and economic time series, such as global debt, deforestation, or energy use.

So what makes an exponential behaviour so dangerous?

In a finite, closed system indefinite exponential growth is not possible, so at some point a limit or carrying capacity is reached, as illustrated in Figure 4.3 and discussed further in Example 6.1. An exponential function *must* eventually encounter a limit, and history suggests that such limits are encountered in difficult and painful ways.

Many people appear to be lulled or complacent of phrases like "modest growth" of 1 % or a "low inflation band" of 1–3 %. However a compounding growth of 1 % is *still* exponential and will ultimately encounter the same system limits, albeit more slowly than under rapid growth. Indeed, our economic system is premised on capital being paid, with interest, in the future; however for capital and interest able to be repaid necessarily and dangerously implies that the money supply must expand exponentially, and indefinitely.

The key challenge with exponentials is that they make it exceptionally difficult to learn from history. An exponential function has a finite doubling time:

$$e^{\alpha(t+T_{\text{Double}})} = 2e^{\alpha t} \quad\longrightarrow\quad \text{Doubling time } T_{\text{Double}} = \frac{\ln 2}{\alpha}, \tag{5.20}$$

meaning that the time from which a resource is only half used, to being completely exhausted, is just T_{Double}. Similarly, if the *rate* at which a resource is used grows exponentially as $e^{\alpha t}$, then since

$$\int_{-\infty}^{T} e^{\alpha t}\,dt = \frac{1}{\alpha}e^{\alpha T} \quad \text{and} \quad \int_{T}^{T+T_{\text{Double}}} e^{\alpha t}\,dt = \frac{1}{\alpha}e^{\alpha T} \tag{5.21}$$

it follows that the same amount of a resource was used, for *all* of recorded (and pre-recorded) history up to time T, as is used in the finite time interval from T to $T + T_{\text{Double}}$.

Further Reading: You may wish to compare the challenges raised here with those of Example 9.6. Problem 5.11 lists a number of follow-up readings and videos related to exponential functions.

5.3 System Coupling

We now know that the eigendecomposition of diagonalizable $n \times n$ dynamics A transforms a coupled system of n variables to n decoupled first-order systems:

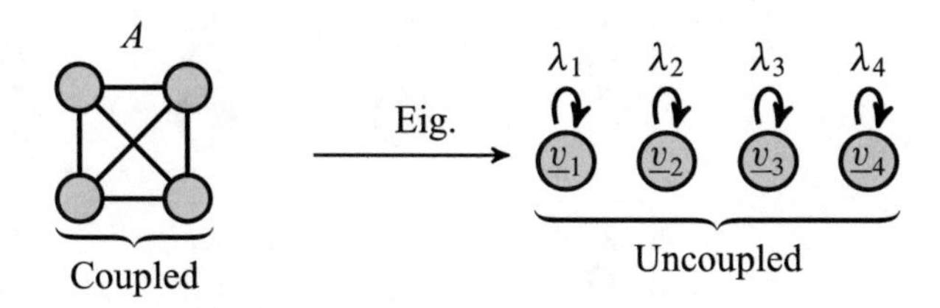

If A is not diagonalizable, this essentially implies that there is a coupling in the system which cannot be undone. One example would be a second-order system, such as the position of a car subject to a random acceleration

$$\ddot{x}(t) = w(t) \tag{5.23}$$

for a random (white noise) process w. Although this dynamic can be converted to first order via state augmentation,

$$\underline{z}(t) = \begin{bmatrix} \dot{x}(t) \\ x(t) \end{bmatrix} \qquad \dot{\underline{z}}(t) = \begin{bmatrix} 0 & 0 \\ 1 & 0 \end{bmatrix} \underline{z}(t) + \begin{bmatrix} 1 \\ 0 \end{bmatrix} w(t) \tag{5.24}$$

the resulting dynamics matrix cannot be diagonalized since the underlying dynamic really *is* second order, including $te^{\lambda t}$ terms, which cannot be represented as a weighted sum of first-order exponentials.

A dynamic A which is not diagonalizable can, at best, be reduced to a standardized coupling, known as the Jordan form, something like

$$A \xrightarrow{\text{Eig.}} \begin{bmatrix} \lambda_1 & & & \\ & \lambda_2 & 1 & \\ & & \lambda_2 & \\ & & & \lambda_3 \end{bmatrix} \tag{5.25}$$

graphically corresponding to a maximally reduced system of the form

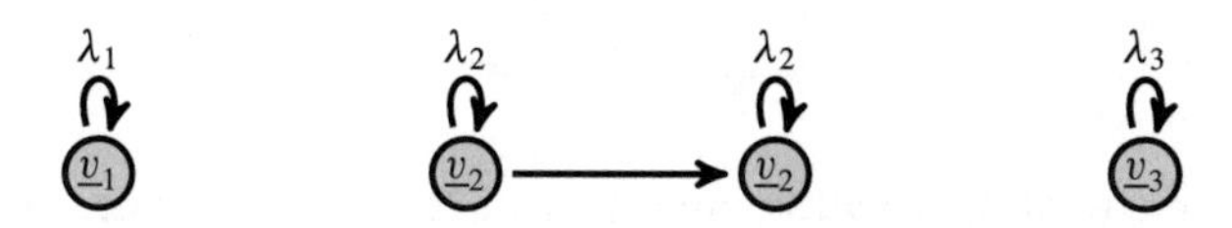

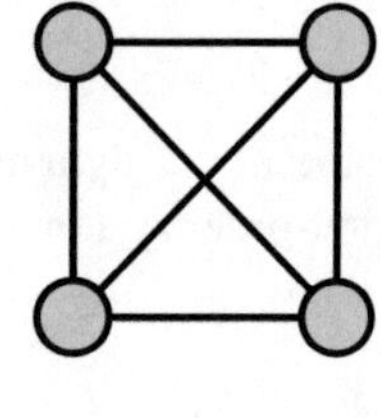

can transform to one of ...

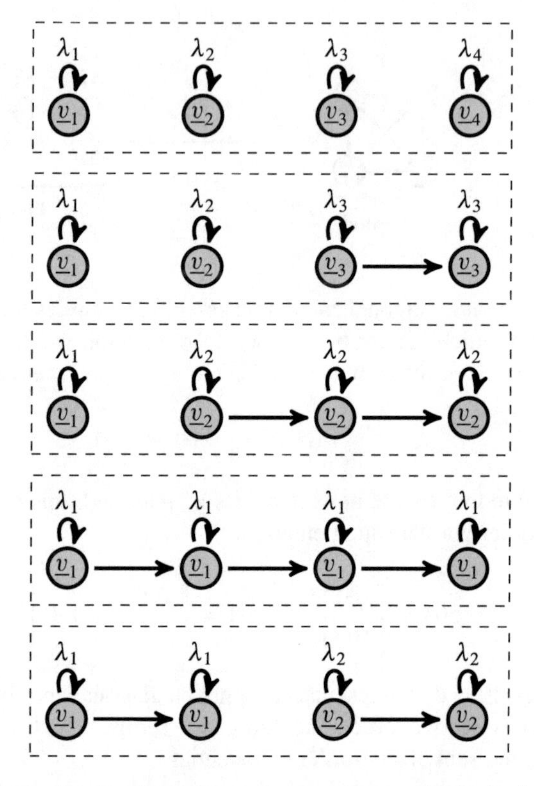

Fig. 5.2 System decoupling: given a fourth-order state, top, there are five possible forms that the decoupling can take, right.

In general a coupled problem has a variety of possible ways in which it could be decoupled, as illustrated in Figure 5.2. However the intuition remains unchanged from before:

An eigendecomposition decomposes a coupled system into uncoupled modes, however some of the modes may be higher order $q > 1$ and have a characteristic behaviour based on some linear combination of

$$e^{\lambda t} \quad te^{\lambda t} \quad \dots \quad t^{q-1}e^{\lambda t} \tag{5.26}$$

That is, the simplification/decomposition of (5.17) still applies, however the transformed states $\underline{y}_i$ are no longer all necessarily scalar, rather some of them will be vectors, based on the order of the coupling.

Example 5.2: Coupled Systems

Suppose we have two ponds, possibly polluted, with water volumes S_1 and S_2, with a water flow of F from one to the other:

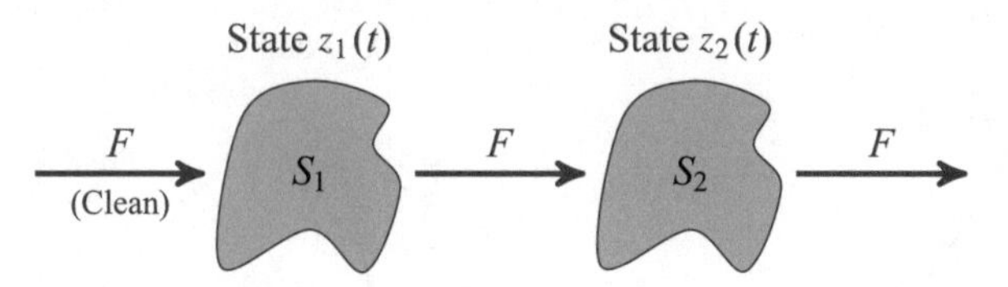

Clearly there is a coupling between the ponds, since any pollution from Pond 1 flows into Pond 2. Let $z_1(t)$ and $z_2(t)$ be the concentration, over time, of a pollutant in the ponds. The system dynamics are

$$\dot{z}_1(t) = -\frac{F}{S_1} z_1(t) \qquad \dot{z}_2(t) = -\frac{F}{S_2}\Big(z_2(t) - z_1(t)\Big) \qquad A = \begin{bmatrix} -F/S_1 & 0 \\ F/S_2 & -F/S_2 \end{bmatrix}$$

Taking the eigendecomposition of A, we find that

$$\begin{aligned} S_1 = S_2 &\longrightarrow \text{A single eigenvalue } \lambda = -F/S, \text{ a second-order mode} \\ S_1 \neq S_2 &\longrightarrow \text{Two eigenvalues, } \lambda_1 = -F/S_1, \lambda_2 = -F/S_2 \end{aligned}$$

Thus the second-order Jordan form appears only when the two ponds are *identical* in volume, which is very unlikely, thus we refer to the second-order case as *degenerate*. Nevertheless the coupled/Jordan concept has value. Suppose that

$$F = 1, \quad S_1 = 10, \quad S_2 = 11 \qquad z_1(0) = 5, \quad z_2(0) = 0$$

such that the ponds are similar in volume. The eigendecomposition of A gives two first-order modes y_1, y_2, and from (5.15) the transformation is

$$z_1(t) = 20y_1(t) \quad z_2(t) = 200y_2(t) - 199y_1(t)$$

Thus the state of the second pond is determined by the small difference between two large numbers y_1, y_2, which does not offer much insight. If we numerically simulate the dynamics, we obtain

Example continues . . .

5.4 Dynamics

Since the preceding discussion looked at the decoupling of coupled systems, we can now focus on decoupled systems, where the simplest possible case is the first-order, scalar, time-invariant one:

$$\dot{z} = \alpha z + \beta \tag{5.27}$$

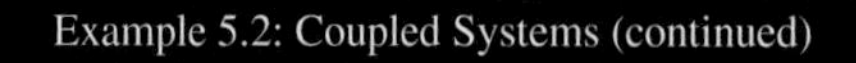
Example 5.2: Coupled Systems (continued)

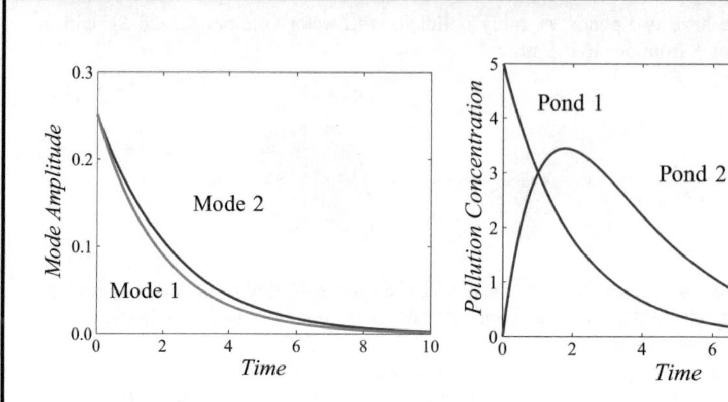

Note that the two modes (left) are, indeed, both simple first-order exponentials, but without intuitive meaning (in this case). On the other hand, the actual concentration states (right) look very nearly like

$$z_1 \simeq e^{-\lambda_1 t} \quad z_2 \simeq te^{-\lambda_2 t}$$

where from the Jordan form (5.26) we understand the te^t term of z_2 to be due to the fact that we essentially have a coupled second-order system.

Recall from Section 4.1 the concept of a fixed point, which is trivial to evaluate for the system of (5.27):

$$0 = \alpha\bar{z} + \beta \quad \longrightarrow \quad \bar{z} = -\frac{\beta}{\alpha} \tag{5.28}$$

The dynamic "matrix" $A = \alpha$ is just a scalar, with eigenvalue $\lambda = \alpha$, for which we very well know the behaviour of the system as

$$z(t) = \bar{z} + (z(0) - \bar{z}) \cdot e^{\alpha t}, \tag{5.29}$$

thus, consistent with (5.18), asymptotically the dynamics evolve as

$$\begin{array}{lll} \alpha < 0 & z \longrightarrow \bar{z} & \text{(stable)} \\ \alpha > 0 & |z| \longrightarrow \infty & \text{(unstable)} \end{array} \tag{5.30}$$

Indeed, Figure 5.3 shows the same behaviour graphically. The graphical interpretation is important, since for many nonlinear systems an exact analytical understanding will be exceptionally challenging, whereas a qualitative graphical interpretation will be relatively straightforward.

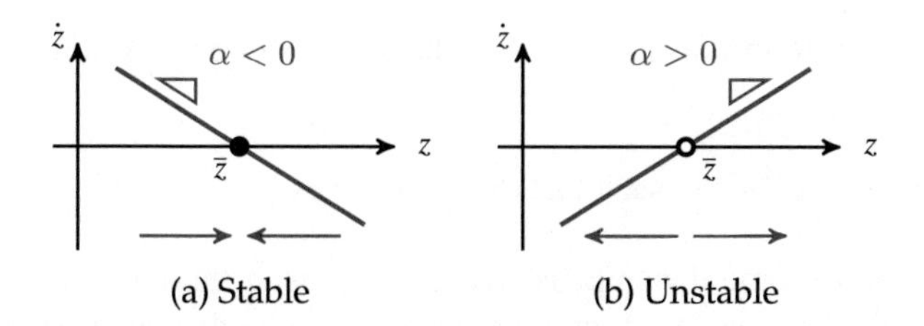

Fig. 5.3 Graphical interpretation of a simple linear dynamic system: wherever $\dot{z} > 0$ we know that z is increasing, and similarly that $\dot{z} < 0$ implies that z is decreasing. Therefore the system to the left, with negative slope α, will tend to return to $\bar{z}$ if disturbed, and is *stable*, whereas the system to the right, with positive slope α, will tend to diverge and is *unstable*.

With an understanding of one-dimensional dynamics, we wish to generalize this understanding to multiple dimensions. Given a dynamic system

$$\dot{\underline{x}} = A\underline{x} + \underline{b}, \tag{5.31}$$

as with (5.27) there is a *single* fixed point, easily evaluated as

$$\underline{0} = A\bar{\underline{x}} + \underline{b} \quad \longrightarrow \quad \bar{\underline{x}} = -A^{-1}\underline{b}. \tag{5.32}$$

Suppose the system experiences a disturbance at time $t = 0$, such that $\underline{x}(0) \neq \bar{\underline{x}}$. We assume the fixed point to be stable, so the system dynamics will return the disturbed state to the fixed point. We can consider two difference influences

$$\begin{aligned} \underline{x}(0) &= \bar{\underline{x}} + \delta & &\rightarrow & \underline{x}_\delta(t) & & \lim_{t \to \infty} \underline{x}_\delta(t) &= \bar{\underline{x}}, \\ \underline{x}(0) &= \bar{\underline{x}} + \epsilon & &\rightarrow & \underline{x}_\epsilon(t) & & \lim_{t \to \infty} \underline{x}_\epsilon(t) &= \bar{\underline{x}}. \end{aligned} \tag{5.33}$$

If the system is subject to *both* influences, the system clearly still converges to the fixed point:

$$\underline{x}(0) = \bar{\underline{x}} + \delta + \epsilon \quad \rightarrow \quad \underline{x}_{\delta\epsilon}(t) \qquad \lim_{t \to \infty} \underline{x}_{\delta\epsilon}(t) = \bar{\underline{x}} \tag{5.34}$$

meaning that

$$\underline{x}_{\delta\epsilon}(t) \neq \underline{x}_\delta(t) + \underline{x}_\epsilon(t), \tag{5.35}$$

appearing to contradict superposition. However it is the *response* to *disturbance* that superposes, not the constant fixed point. That is, superposition must be understood *relative* to the fixed point. As a result we will, in general, prefer to translate a linear dynamic problem to shift its fixed point to the origin, so we will study the simplified state

$$\underline{z} = \underline{x} + A^{-1}\underline{b} \qquad \text{such that (5.31) becomes} \qquad \dot{\underline{z}} = A\underline{z}. \tag{5.36}$$

That is, $\underline{z}$ has the same dynamic behaviour as $\underline{x}$, just a shifted version placing its fixed point at the origin. Assuming the dynamics to be diagonalizable, which we now know to mean that the individual modes are all first order, then given the decomposition $A = V\Lambda V^{-1}$, from (5.17) we know that the dynamics of $\underline{z}$ can be written as

$$\underline{z}(t) = \begin{bmatrix} | & & | \\ \underline{v}_1 & \cdots & \underline{v}_n \\ | & & | \end{bmatrix} \begin{bmatrix} e^{\lambda_1 t} & & \\ & \ddots & \\ & & e^{\lambda_n t} \end{bmatrix} \underbrace{V^{-1}\underline{z}(0)}_{= \text{vector } \underline{c}} \tag{5.37}$$

$$= \begin{bmatrix} | & & | \\ \underline{v}_1 & \cdots & \underline{v}_n \\ | & & | \end{bmatrix} \begin{bmatrix} e^{\lambda_1 t}c_1 & & \\ & \ddots & \\ & & e^{\lambda_n t}c_n \end{bmatrix} \tag{5.38}$$

$$\underline{z}(t) = \sum_{i=1}^{n} \underline{v}_i e^{\lambda_i t} c_i \tag{5.39}$$

That is, over time, $\underline{z}(t)$ is nothing more than a weighted sum of growing or shrinking eigenvectors, where the weights c_i are determined by the system initial condition $\underline{z}(0)$.

For example, suppose we are given the dynamic system

$$\dot{\underline{z}} = \begin{bmatrix} 0 & \text{-1} \\ \text{-1} & 0 \end{bmatrix} \underline{z} \tag{5.40}$$

for which the eigendecomposition is

$$\begin{bmatrix} 0 & \text{-1} \\ \text{-1} & 0 \end{bmatrix} \longrightarrow \lambda_1 = -1 \quad \underline{v}_1 = \begin{bmatrix} 1 \\ 1 \end{bmatrix} \quad \lambda_2 = +1 \quad \underline{v}_2 = \begin{bmatrix} 1 \\ \text{-1} \end{bmatrix} \tag{5.41}$$

From (5.39) we therefore know the general form of the solution to this system to be

$$\underline{z}(t) = c_1 e^{-t} \begin{bmatrix} 1 \\ 1 \end{bmatrix} + c_2 e^{t} \begin{bmatrix} 1 \\ \text{-1} \end{bmatrix}. \tag{5.42}$$

Thus the system tends to grow in one direction and shrink in another, known as a *saddle*, illustrated in the phase portrait of Figure 5.4:

> *A phase plot or phase portrait is a plot of state trajectories of a dynamic system.*

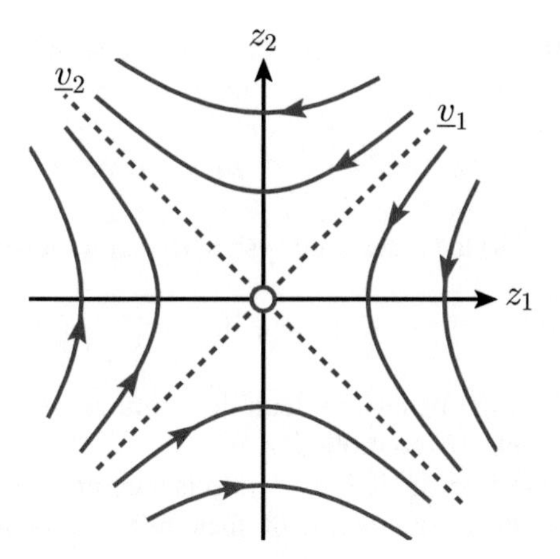

Fig. 5.4 Phase plot of a saddle: if a linear system has one stable mode ($\underline{v}_1$) and one unstable mode ($\underline{v}_2$) then the resulting mixture of convergent and divergent behaviour is known as a saddle.

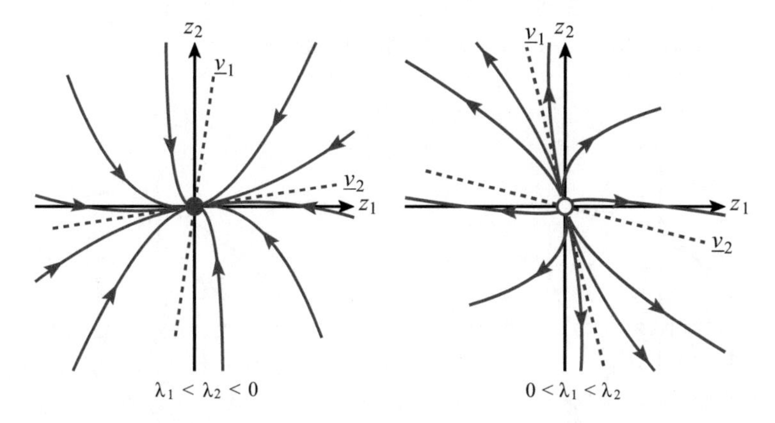

Fig. 5.5 Two illustrations of linear dynamic systems: stable (left) and unstable (right). In these examples the eigenvalues are not equal. For example, in the stable system $\lambda_1 < \lambda_2$, therefore the $\underline{v}_1$ component of the state converges more rapidly to the fixed point than the $\underline{v}_2$ component, causing curved trajectories.

Once we understand the saddle example well, the concept is relatively easy to generalize to other dynamic matrices, as illustrated in Figure 5.5: for a two dimensional problem we will always have two eigenvectors, where the associated eigenvalues may be positive (unstable), negative (stable), or zero (a degenerate[2] case). The eigendecomposition is so effective at characterizing a system, that we refer to the pattern or distribution of eigenvalues as the *spectrum*, however we will see (Section 5.6) that there are circumstances where the spectrum does not quite tell the whole story.

[2]Meaning highly sensitive to perturbation.

It is possible for the eigenvalues to be complex, but because the dynamics matrix is real the eigenvalues must be present in conjugate pairs

$$\lambda_1 = a + ib \qquad \lambda_2 = a - ib$$

Consequently from (5.39) the associated system dynamics must have the form

$$e^{\lambda_1 t} = \exp\bigl(a + ib\bigr) = e^{at}\bigl(\cos bt + i \sin bt\bigr), \tag{5.43}$$

thus we have spiral/circular behaviour, with the spiral growth/shrinkage controlled by a, and the angular velocity controlled by b.

Given a 2D problem $\dot{\underline{z}} = A\underline{z}$, if $\lambda = a + ib$ is a complex eigenvalue of A, with associated complex eigenvector $\underline{v} = \underline{\alpha} + i\underline{\beta}$, then there are two real-valued solutions

$$\underline{z}(t) = e^{at}\bigl(\cos(bt)\underline{\alpha} - \sin(bt)\underline{\beta}\bigr) \quad \text{and} \quad \underline{z}(t) = e^{at}\bigl(\sin(bt)\underline{\alpha} + \cos(bt)\underline{\beta}\bigr) \tag{5.44}$$

Two examples of this case are shown in Figure 5.6.

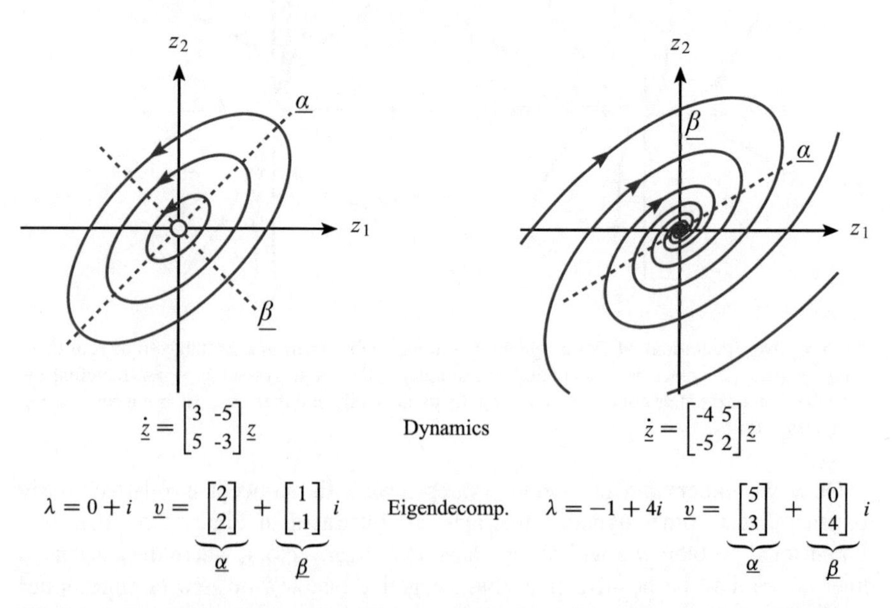

Fig. 5.6 Systems with complex eigenvalues: complex eigenvalues lead to oscillatory behaviour, where the real part of the eigenvalue determines whether the system is stable (right), a marginally stable centre (left), or unstable.

5.5 Control of Dynamic Systems

Although our focus is on the actual dynamics governing a system, most systems do also admit an input, some sort of deliberate influence. Obvious examples include the gas pedal and brake in your car, or the volume control on a radio.

Manipulating a system input in order to regulate the behaviour of a system is known as *control theory*, a vast topic of which we can only begin to discuss here. There may be two sorts of human-driven "input" to the system:

Deliberate feedback control in which the output of the system is observed, and that some function of that output is fed back into the input in order to produce the desired regulation. These inputs are explicitly designed to control the system, such as the re-introduction of native species or the strengthening of wetland protection.

Accidental inputs, usually as a side-effect of other human activity, such as the leaching of fertilizers into waterways or the emission of CO_2 into the atmosphere. These inputs clearly influence the system, however normally to perturb or imbalance the system rather than to regulate it.

We will not be designing control strategies in this book, however we will definitely want to understand the human influence on systems, the manner in which systems respond to inputs, whether deliberate or accidental, as illustrated in Figure 5.7.

Thus given a linear dynamic system,

$$\dot{\underline{z}}(t) = A(t)\underline{z}(t) \tag{5.45}$$

as studied throughout this chapter, really there are control inputs $\underline{u}(t)$ and other accidental/disturbance inputs $\underline{w}(t)$ present,

$$\dot{\underline{z}}(t) = A(t)\underline{z}(t) + B_c(t)\underline{u}(t) + B_o(t)\underline{w}(t), \tag{5.46}$$

where B_c, B_o describe the manner in which inputs $\underline{u}$ and $\underline{w}$, respectively, affect the system state.

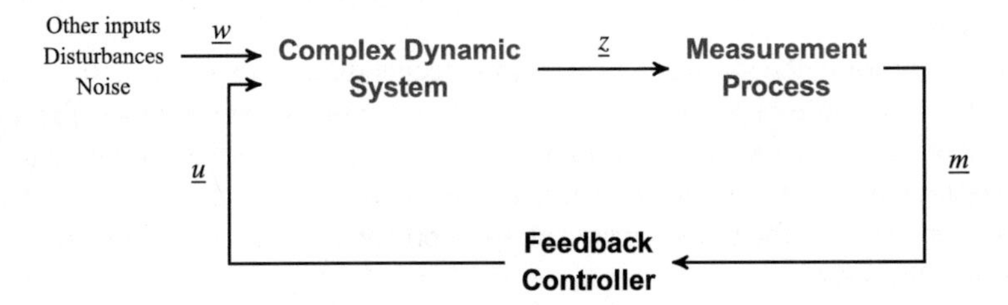

Fig. 5.7 Feedback control: given some dynamic system, we can attempt to control or regulate it by observing the unknown system state $\underline{z}$, indirectly, via measurements $\underline{m}$, and producing a control input $\underline{u}$. In most cases the system will be subject not only to our deliberate signal $\underline{u}$, but also to other disturbances and influences, whether human or otherwise.

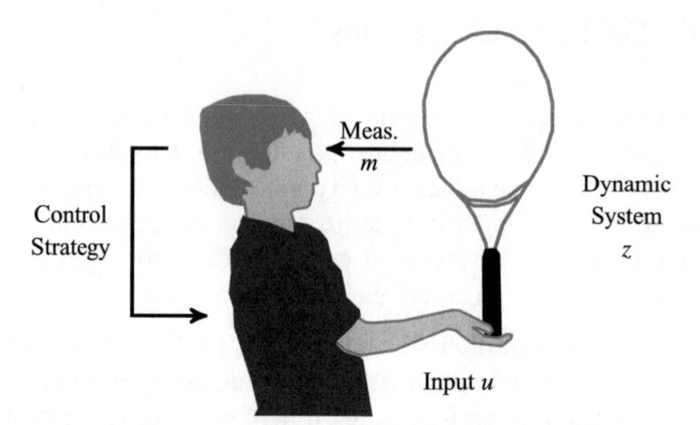

Fig. 5.8 Control of an unstable system: an inverted pendulum, or tennis racket, has an unstable fixed point when held upright, however the unstable system can be stabilized given an appropriate feedback control input u, based on measurements m of the system state z.

Some control problems are not difficult to solve, such as balancing a tennis racket upside-down on your hand, as in Figure 5.8, or regulating the temperature of a house[3] in the winter. In contrast, most socio-environmental systems are challenging to control for one of three reasons:

1. **Bounded Inputs:** For many large-scale systems the size of human input which can be asserted is tiny, such that the input signal is simply too small to control the system. In contrast to accidental inputs, such as deforestation, which can be very large, deliberate inputs need to be undertaken *on top of* or in opposition to other human activity, which therefore requires time, political will, money, and/or energy, and may be limited by any or all of these reasons.
2. **Long Time Constants:** In classical control, systems having long time constants (i.e., those with slow dynamics or small eigenvalues) are actually rather *easy* to control: the system is slow, making it quite simple to design a sufficiently fast controller. However most socio-environmental systems rely on political will or human urgency, which appears only once the system has actually begun to manifest noticeable changes (e.g., global warming, ozone hole), at which point the long time constants ensure that further change will continue to take place for decades.
3. **Observability:** The control of a system depends on the state of the system being observable or inferrable in some way, a topic to be discussed further in Chapter 11. In some cases we may have an exceptionally hard time acquiring measurements, such as in the deep ocean, or we may just not know what to measure, such as the unanticipated atmospheric chemistry surrounding the arctic ozone holes.

[3]Although, for practical reasons, temperature control turns out to have nonlinearities present, as will be discussed in Example 6.4.

These three issues, when combined, can lead to terribly challenging predicaments. In the context of global warming, for example, the time constant for atmospheric CO_2 is on the order of decades, so the current rate of temperature rise will continue for many years; at the same time, limiting the temperature rise would require very large inputs, beyond the scope of what humans are willing or politically able to do; finally it is difficult to measure global CO_2, and it isn't necessarily obvious what to even measure (methane?, temperature?, permafrost?, arctic ice?).

We will briefly return to the control problem in the context of nonlinear dynamic systems in Section 7.4.

5.6 Non-normal Systems

From the general solution to a linear system in (5.39), we know that for a two-dimensional system the dynamic behaviour behaves as

$$\underline{z}(t) = c_1 e^{\lambda_1 t}\underline{v}_1 + c_2 e^{\lambda_2 t}\underline{v}_2. \tag{5.47}$$

For a stable system, where both eigenvalues have negative real parts, it is therefore natural to assume that a given state converges smoothly to the fixed point at the origin.

It is, therefore, perhaps surprising that even if stable, (5.47) can *diverge* before converging to the fixed point. Such a situation can occur if the system is *non-normal*, meaning that the eigenvectors are not at right angles to one another. There is nothing invalid about non-orthogonal eigenvectors; indeed, we have seen such examples, as in Figure 5.5. The orthogonal case is *so* commonly studied that we subconsciously assume that all systems must be so; however, perversely, the orthogonal case is actually *degenerate*, and can therefore really only be expected in contrived circumstances!

The issue boils down to taking differences of large numbers. If

$$z = (1001) - (1000) \tag{5.48}$$

then z is small, of course, even though there are large numbers involved. However, if we now add *differing* dynamics,

$$z(t) = 1001 \cdot e^{-t} - 1000 \cdot e^{-2t} \tag{5.49}$$

then $z(t)$ starts small and grows to a large value before converging back to zero, as illustrated by the red curve in Figure 5.9.

The mechanics of the behaviour are examined in detail in Figure 5.10.

So the distribution of eigenvalues, the so-called spectrum of a system, clearly reveals whether a system is stable, but does *not* tell us whether there may be transients present. Testing whether a system is normal or not requires us to also know the eigenvectors; if the eigenvectors are stacked in V, as in (5.13), then system

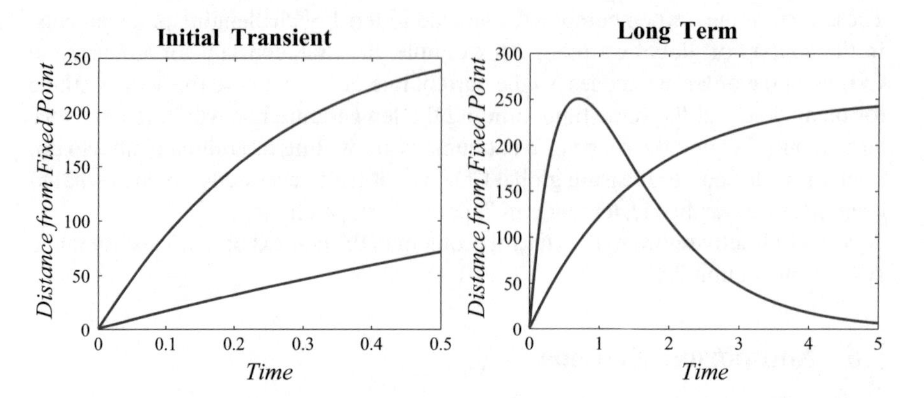

Fig. 5.9 Non-normal systems: examining the initial behaviours of two systems, left, which is stable and which is unstable? For a *normal* system, both red and blue systems would be considered unstable. After a longer period of time, right, we observe that the non-normal red system of (5.49) is, indeed, stable. It is not possible to say whether the blue system is stable or not, and would require an even longer period of observation, related to the *baseline* question of Chapter 4.

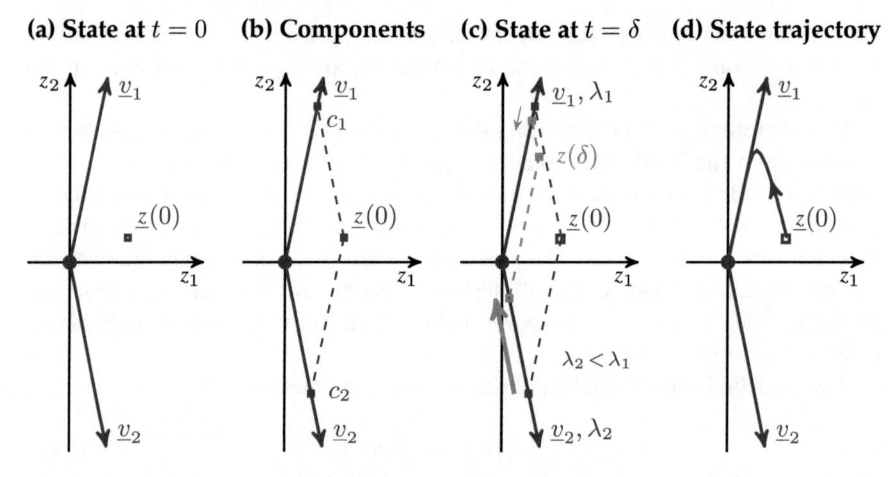

Fig. 5.10 Non-normal systems: suppose we are given *non*-orthogonal eigenvectors $\underline{v}_1, \underline{v}_2$ and a system state $\underline{z}(0)$, for which we can find the components c_1, c_2 by projecting the state onto the eigenvectors. Even if both eigenvalues are stable, if they are sufficiently different (**c**) the mode components will shrink at very different rates, causing the state trajectory initially to *grow* (**d**).

normality is easily checked by taking dot-products of the eigenvectors, as

$$\text{Let } Q = V^T V, \quad \text{then} \quad \begin{array}{ll} Q \text{ diagonal} & \longrightarrow \text{System normal} \\ Q \text{ not diagonal} & \longrightarrow \text{System non-normal} \end{array} \tag{5.50}$$

The lesson, discussed in Figure 5.9, is that we need to be quite cautious in assessing a system on the basis of observed behaviour. Similar to the baseline question of Chapter 4, asking when to *start* measuring, we now have a related question: how *long* to measure.

The issue of non-normality plays a significant role in the assessment of resilience (see Example 5.3). Resilience is very much an empirical concept, in that we wish to observe a given system, such as a forest, and then determine the impact of a disturbance, such as building a road through the forest or a dam nearby. Because the system needs to be assessed on the basis of observation, over a finite period of time, it is quite important to appreciate the possibility of non-normality. There are many examples involving the resilience of non-normal systems, including aquatic food chains, predator-prey models, and nutrient flows in tropical systems [6], where the last of these is explored in some detail in Problem 5.10.

Case Study 5: System Decoupling

Problem: Describe, in general, the analytical motion of the following spring–mass system:

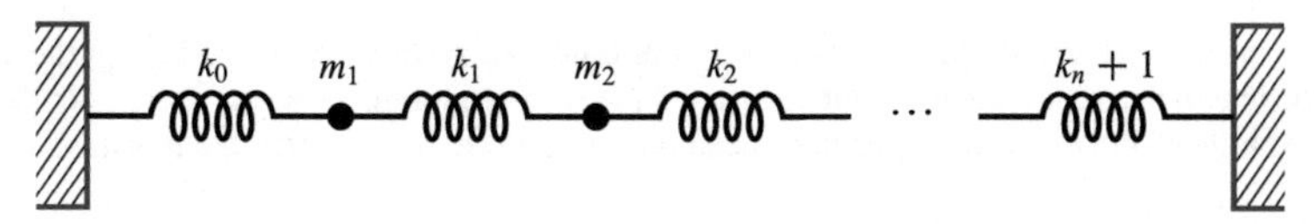

Not easy!!

It is easy enough, however, to write down the system dynamics. It is most convenient for the state z_i to represent the deviation of mass i from its rest position, thus in the preceding figure $z_i = 0$. Applying Newton's $F = ma$ to each mass:

$$m_i\ddot{z}_i = -k_i\big(z_i - z_{i-1}\big) + k_{i+1}\big(z_{i+1} - z_i\big) \tag{5.54}$$

At the end points the equations look a bit different, since the springs are attached to fixed walls, rather than to a movable mass:

$$\begin{aligned} m_1\ddot{z}_1 &= -k_1\big(z_1 - 0\big) + k_2\big(z_2 - z_1\big) \\ m_n\ddot{z}_n &= -k_n\big(z_n - z_{n-1}\big) + k_{n+1}\big(0 - z_n\big) \end{aligned} \tag{5.55}$$

The problem, of course, in trying to solve these equations is that n masses can oscillate in all kinds of complicated, interacting ways.

The key idea is that the system is exceptionally easy to solve when we have only one mass. Let us first collect Eqs. (5.54) and (5.55) in matrix–vector form:

$$\underbrace{\begin{bmatrix} m_1 & & \\ & \ddots & \\ & & m_n \end{bmatrix}}_{M} \underbrace{\begin{bmatrix} \ddot{z}_1 \\ \vdots \\ \ddot{z}_n \end{bmatrix}}_{\ddot{\underline{z}}} \; = \; \underbrace{\begin{bmatrix} (-k_1 - k_2) & k_2 & & \\ & \ddots & \ddots & \\ & & k_n & (-k_n - k_{n+1}) \end{bmatrix}}_{K} \underbrace{\begin{bmatrix} z_1 \\ \vdots \\ z_n \end{bmatrix}}_{\underline{z}} \tag{5.56}$$

Example 5.3: Resilience I — Linear Systems

Resilience measures how robust or tolerant a given social or ecological system is to some sort of disturbance, of which there are many:

Ecological Disturbances:	acid rain, logging, pollution, invasive species
Social Disturbances:	terrorism, electrical blackouts, recessions

Asymptotically, after a long time, it is the slowest mode (least negative eigenvalue) which limits how slowly a linear system returns to the fixed point after a disturbance. Therefore the classic measure of system resilience is

$$\mathit{Resilience}(A) = \min_i \Big\{ -\mathrm{Real}\big(\lambda_i(A)\big) \Big\} \tag{5.51}$$

The more rapid the return to the stable fixed point, the larger the resilience. It is only the real part of the eigenvalue which is needed here, since it controls the exponential decay; the complex part controls only the frequency of any oscillation.

However, as we know from Section 5.6, for non-normal systems there can be significant transient behaviour which is not measured in the asymptotic $t \longrightarrow \infty$ assessment of (5.51), therefore other measures have been proposed [6] to characterize the transient.

The *reactivity* measures the greatest rate of divergence at time $t = 0$ of the transient relative to the size of perturbation:

$$\mathit{Reactivity}(A) = \max_{\underline{z}_o} \left\{ \frac{1}{\|\underline{z}_o\|} \left. \frac{d\,\|\underline{z}\|}{dt} \right|_{t=0} \right\} = \max_i \left\{ \mathrm{Real}\left(\lambda_i\Big(\frac{A + A^T}{2}\Big)\right)\right\} \tag{5.52}$$

A reactivity greater than zero implies a non-normal further *growth* in the initial state beyond the initial perturbation. Whereas the reactivity is an instantaneous assessment *at* $t = 0$, to assess the system *over* time an alternative would be the maximum amplification,

$$\kappa_{\max} = \max_t \kappa(t) \quad \text{where} \quad \kappa(t) = \max_{\underline{z}_o} \left\{ \frac{\|\underline{z}(t)\|}{\|\underline{z}_o\|} \right\}. \tag{5.53}$$

Given a definition of resilience, there are at least three socio-environmental contexts of interest:

Ecological Resilience: Given a list of alternatives, we wish to select a human action which minimizes ecological disruption. That is, to which disturbance is the ecological system most robust?

Social Resilience: What strategies can a government put in place to make a society resilient to social disruptions and other challenges? That is, we wish to transform the societal dynamics to increase resilience.

Example continues ...

Example 5.3: Resilience I — Linear Systems (continued)

Social Policy: An effective policy needs to maximize its impact for a given amount perturbation (money or energy). That is, we seek the *opposite* of resilience: what timing or direction of perturbations *maximizes* the response?

Thus, although (5.51)–(5.53) may seem abstract, the study of resilience is of considerable importance, to be followed up in Example 7.3.

Further Reading: There are many good references on the topic of resilience: [3, 5, 6, 9, 12].

thus

$$\ddot{\underline{z}} = M^{-1}K\underline{z}. \tag{5.57}$$

The matrix $(M^{-1}K)$ describes the interactions between the masses; we can find the eigendecomposition of $(M^{-1}K)$

$$(M^{-1}K) = V\Lambda V^{-1} \tag{5.58}$$

to rewrite the system of equations:

$$\ddot{\underline{z}} = M^{-1}K\underline{z} = V\Lambda V^{-1}\underline{z}. \tag{5.59}$$

If we then let $\underline{y} = V^{-1}\underline{z}$, we have

$$\underline{y} = V^{-1}\underline{z} = V^{-1}V\Lambda V^{-1}\underline{z} = \Lambda V^{-1}\underline{z} = \Lambda\underline{y}. \tag{5.60}$$

However our new, transformed state variables $\underline{y}$ are non-interacting, since the spring–mass system is diagonalizable, meaning that Λ is a diagonal matrix, giving us an ensemble of simple, periodic *single* mass systems

$$\ddot{y}_i = \lambda_i y_i \tag{5.61}$$

Interestingly, our decoupled system has normalized unit mass with a spring constant controlled by the eigenvalue:

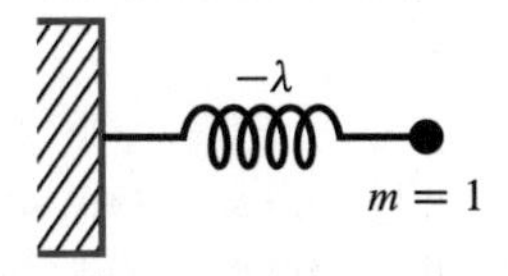

Let us now consider a simplified version of the problem, with equal masses and spring constants:

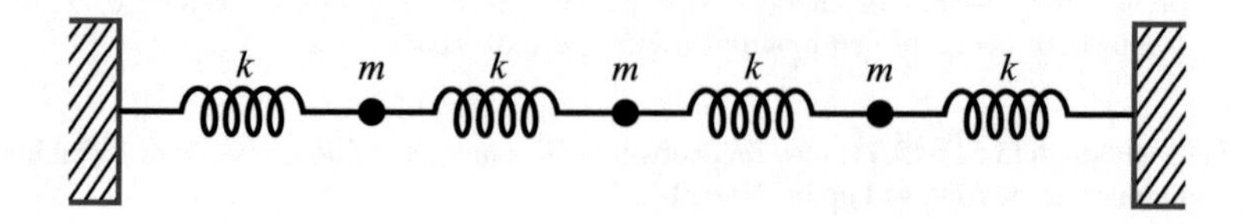

The dynamic system from (5.54) and (5.55) can neatly be written as

$$m\ddot{\underline{z}} = \begin{bmatrix} \text{-}2k & k & 0 \\ k & 2k & k \\ 0 & k & 2k \end{bmatrix} \underline{z} \tag{5.62}$$

so k factors out, allowing the interactions to be expressed as a simple matrix for which the eigendecomposition is

$$\underbrace{\begin{bmatrix} \text{-}2 & 1 & 0 \\ 1 & 2 & 1 \\ 0 & 1 & 2 \end{bmatrix}}_{\text{Matrix } K} = \underbrace{\begin{bmatrix} \frac{1}{2} & \frac{1}{\sqrt{2}} & \frac{1}{2} \\ \frac{1}{\sqrt{2}} & 0 & -\frac{1}{\sqrt{2}} \\ \frac{1}{2} & -\frac{1}{\sqrt{2}} & \frac{1}{2} \end{bmatrix}}_{\text{Eigenvect. in Columns}} \underbrace{\begin{bmatrix} -0.6 & & \\ & -2 & \\ & & -3.4 \end{bmatrix}}_{\text{Eigenvalues on Diagonal}} \begin{bmatrix} \frac{1}{2} & \frac{1}{\sqrt{2}} & \frac{1}{2} \\ \frac{1}{\sqrt{2}} & 0 & -\frac{1}{\sqrt{2}} \\ \frac{1}{2} & -\frac{1}{\sqrt{2}} & \frac{1}{2} \end{bmatrix}^{-1} \tag{5.63}$$

Recall that the eigendecomposition tells us the *modes* of the system, those vibrations which will persist over time. For example, the vibration of m_1 on its own is not a mode, since vibrating m_1 will quickly cause m_2 and m_3 to vibrate. However all three weights sliding back and forth — that's a mode. The eigenvectors in (5.63) actually identify the three modes for us:

Mode	Eigenvector	Motion Pattern	Spring Constant
I.	$\left(\frac{1}{2} \quad \frac{1}{\sqrt{2}} \quad \frac{1}{2}\right)$		Slow $k = 0.6$
II.	$\left(\frac{1}{\sqrt{2}} \quad 0 \quad -\frac{1}{\sqrt{2}}\right)$		Medium $k = 2$
III.	$\left(\frac{1}{2} \quad -\frac{1}{\sqrt{2}} \quad \frac{1}{2}\right)$		Fast $k = 3.4$

However our goal was to identify a closed-form solution to the coupled system. How can we use the above decomposition to do this? Suppose we are given an initial condition; let's deliberately select something that doesn't look like a mode, such as a disturbance of m_1 on its own:

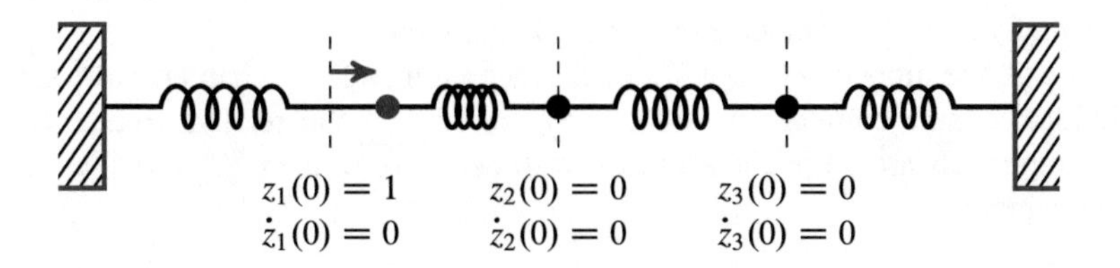

Then, as in (5.17), we have

$$\begin{array}{ccc} \underline{z}(0), \dot{\underline{z}}(0) & \xrightarrow[\text{Eig.}]{V^{-1}} & \begin{array}{l} \underline{y}(0) = V^{-1}\underline{z}(0) \\ \dot{\underline{y}}(0) = V^{-1}\dot{\underline{z}}(0) \end{array} \\ & & \Big\downarrow \text{Easy} \\ \underline{z}(t) = V\underline{y}(t) & \xleftarrow[\text{Eig.}]{V} & \begin{array}{l} y_i(t) = y_i(0)\cos\left(\sqrt{\lambda_i} t\right) + \\ \quad\quad \dot{y}_i(0)\sin\left(\sqrt{\lambda_i} t\right) \end{array} \end{array} \tag{5.64}$$

The initial conditions determine the constant scalar weights $y_i, \dot{y}_i$; in our particular case

$$\underline{y}(0) = V^{-1}\begin{bmatrix} 1 \\ 0 \\ 0 \end{bmatrix} = \begin{bmatrix} 1/2 \\ 1/\sqrt{2} \\ 1/2 \end{bmatrix} \qquad \dot{\underline{y}}(0) = V^{-1}\begin{bmatrix} 0 \\ 0 \\ 0 \end{bmatrix} = \begin{bmatrix} 0 \\ 0 \\ 0 \end{bmatrix}. \tag{5.65}$$

Our general solution for $z(t)$ can be written analytically, consisting of three sinusoids at different frequencies controlled by λ_i, each multiplying a fixed mode pattern $\underline{v}_i$:

$$\begin{aligned} \underline{z}(t) = \underbrace{y_1}_{\text{weight}} \cdot \underbrace{\cos\left(\sqrt{0.6}t\right)}_{\text{dynamic}} \cdot \underbrace{\begin{bmatrix} 1/2 \\ 1/\sqrt{2} \\ 1/2 \end{bmatrix}}_{\text{shape}} + \underbrace{y_2}_{\text{weight}} \cdot \underbrace{\cos\left(\sqrt{2}t\right)}_{\text{dynamic}} \cdot \underbrace{\begin{bmatrix} 1/\sqrt{2} \\ 0 \\ 1/\sqrt{2} \end{bmatrix}}_{\text{shape}} \\ + \underbrace{y_3}_{\text{weight}} \cdot \underbrace{\cos\left(\sqrt{3.4}t\right)}_{\text{dynamic}} \cdot \underbrace{\begin{bmatrix} 1/2 \\ -1/\sqrt{2} \\ 1/2 \end{bmatrix}}_{\text{shape}} \end{aligned} \tag{5.66}$$

The eigendecomposition only needs to be taken once; changing the initial conditions only changes the relative contributions (weights y_i) of the different modes, but does not actually in any way change the mode shapes themselves.

As was mentioned on page 67 at the beginning of the chapter, although the individual cosine components are clearly periodic, if the relative frequencies of the components are an irrational ratio, such as $\sqrt{2/0.6}$, then $\underline{z}(t)$ itself will never actually repeat.

Further Reading

The references may be found at the end of each chapter. Also note that the textbook further reading page maintains updated references and links.

Wikipedia Links: Linear system, Linear Time Invariant Systems Theory, Eigendecomposition, Jordan normal form, Control Theory

There are a great many excellent texts on linear dynamic systems, particularly those of Gajic [2] and Scheinerman [10].

The field of control theory is vast, and there are many undergraduate texts to consider as next steps, two of which are the books by Friedland [1] and Nise [7].

The topic of non-normal systems is relatively specialized. The book by Trefethen [11] is really the definitive treatment of the subject, but at an advanced level. If the reader is interested, Problems 5.9 and 5.10 develop concrete numerical examples to try.

Sample Problems

Problem 5.1: Short Questions

(a) For a given $n \times n$ matrix A, how do we define the eigenvalues and eigenvectors of A?
(b) What do we mean by the *mode* of a linear dynamic system?
(c) What does it mean for a given $n \times n$ dynamics matrix A to *not* be diagonalizable?

Problem 5.2: Analytical — Phase Plot

For each of the following systems the associated eigendecomposition is given:

$$\begin{bmatrix}\dot{x}\\ \dot{y}\end{bmatrix} = \begin{bmatrix}0 & 2\\ 2 & 0\end{bmatrix}\begin{bmatrix}x\\ y\end{bmatrix} + \begin{bmatrix}3\\ \text{-}1\end{bmatrix} \qquad \underline{v}_1 = \begin{bmatrix}\text{-}1\\ 1\end{bmatrix}\ \underline{v}_2 = \begin{bmatrix}1\\ 1\end{bmatrix}\ \begin{matrix}\lambda_1 = -2\\ \lambda_2 = 2\end{matrix}$$

$$\begin{bmatrix}\dot{x}\\ \dot{y}\end{bmatrix} = \begin{bmatrix}2 & 1\\ 0 & 3\end{bmatrix}\begin{bmatrix}x\\ y\end{bmatrix} + \begin{bmatrix}2\\ 3\end{bmatrix} \qquad \underline{v}_1 = \begin{bmatrix}1\\ 0\end{bmatrix}\ \underline{v}_2 = \begin{bmatrix}1\\ 1\end{bmatrix}\ \begin{matrix}\lambda_1 = 2\\ \lambda_2 = 3\end{matrix}$$

$$\begin{bmatrix}\dot{x}\\ \dot{y}\end{bmatrix} = \begin{bmatrix}\text{-}3 & 4\\ \text{-}8 & 5\end{bmatrix}\begin{bmatrix}x\\ y\end{bmatrix} + \begin{bmatrix}3\\ 8\end{bmatrix} \qquad \underline{v} = \begin{bmatrix}1\\ 2\end{bmatrix} \pm \begin{bmatrix}\text{-}1\\ 0\end{bmatrix} i \quad \lambda = 1 \pm 4i$$

For each system,

(a) Draw the phase plot.
(b) What sort of dynamic behaviour does this system possess?

Problem 5.3: Analytical — System Analysis

We are given a system and its corresponding eigendecomposition:

$$\begin{bmatrix}\dot{x}\\ \dot{y}\end{bmatrix} = \begin{bmatrix}\text{-}3 & 1\\ 1 & \text{-}3\end{bmatrix}\begin{bmatrix}x\\ y\end{bmatrix} \qquad \underline{v}_1 = \begin{bmatrix}1\\ \text{-}1\end{bmatrix}\ \underline{v}_2 = \begin{bmatrix}1\\ 1\end{bmatrix}\ \begin{matrix}\lambda_1 = -2\\ \lambda_2 = -2\end{matrix} \tag{5.67}$$

which we now wish to analyze in three different ways:

(a) *Graphically*: Sketch the phase plot.
(b) *Analytically*: Starting from state (x_o, y_o), use the strategy of (5.17) to analytically find the solution $\big(x(t), y(t)\big)$.
(c) *Numerically*: Apply a simple Forward–Euler discretization [see (8.22)] to (5.67) and simulate the system, starting from several initial points, such as

$$\big(x_o, y_o\big) = \Big(5 \cdot \cos(k\pi/4), 5 \cdot \sin(k\pi/4)\Big) \qquad 0 \le k < 8.$$

Problem 5.4: Analytical — Diagonalization

Suppose that A, B are both $n \times n$ dynamics matrices,

$$\dot{\underline{z}} = A\underline{z}(t) \qquad \dot{\underline{y}} = B\underline{y}(t)$$

where B is diagonalizable but A is not. Briefly describe the qualitative (conceptual, high level) differences between the modes of $\underline{z}$ and those of $\underline{y}$.

Problem 5.5: Analytical — Discrete Time

Let A be the dynamic matrix for a discrete-time linear system

$$\underline{z}(t+1) = A\underline{z}(t).$$

Then we know that $\underline{z}(t) = A^t \underline{z}(0)$. Similar to the continuous-time case discussed in the text, how does having the eigendecomposition of A simplify the computation of A^t?

Problem 5.6: Conceptual — Jordan Form

The eigendecomposition tries to decouple the dynamics of a system. Suppose that the eigendecomposition is *not* able to diagonalize a 6×6 matrix A, such that we are left with the partially-decoupled system

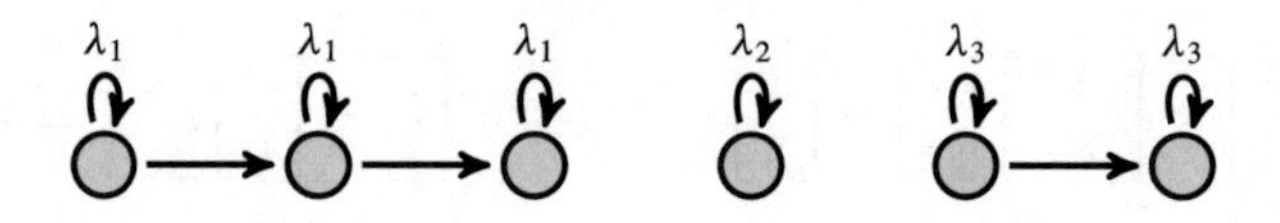

(a) Give one possible Jordan-form matrix corresponding to this system.
(b) Sketch typical impulse responses for each of the three sub-systems.

Problem 5.7: Numerical/Computational — Eigendecompositions

Let's investigate eigendecompositions numerically. In some numerical mathematics program (Matlab, Octave) answer the following:

(a) Suppose we create random matrices

$$A = \texttt{randn}(20); \qquad B = \frac{A + A^T}{2}$$

Both A and B are random, however B is symmetric whereas A is not. Compute the eigenvalues of A and B; what do you observe?

(b) Are most matrices diagonalizable? Take a random matrix

$$A = \texttt{randn}(20);$$

find its eigendecomposition, apply the eigenvectors to compute D, the diagonalization of A, and let d_{max} be the largest off-diagonal element in D. Perform this test 10,000 times and plot the distribution of d_{max}.

(c) It turns out that non-diagonalizability is a *degenerate* property, meaning that if A is not diagonalizable, $A + \epsilon_A$ almost certainly is, for nearly any infinitesimal perturbation ϵ_A.

Look carefully at the Jordan form in (5.25) and also at the eigenvalues you generated in part (a). Develop a strong argument why nearly all matrices must be diagonalizable.

Problem 5.8: Numerical/Computational — Non-normal Systems

Non-normal systems are not particularly mysterious. Any system that has momentum or inertia, for example, will continue to diverge for a period of time, given an initial kick.

Specifically, suppose we are given a point mass, having position $x(t)$ and velocity $v(t)$ in the state

$$\underline{z}(t) = \begin{bmatrix} x(t) \\ v(t) \end{bmatrix}.$$

The mass is subject to a weak return force, pulling its position to zero, and is also subject to friction, which pulls its velocity to zero. The dynamics could look something like

$$\dot{\underline{z}}(t) = \begin{bmatrix} \dot{x}(t) \\ \dot{v}(t) \end{bmatrix} = \begin{bmatrix} 0 & \overset{\dot{x} \equiv v}{1} \\ \underset{\text{Return}}{-0.01} & \underset{\text{Friction}}{-0.02} \end{bmatrix} \begin{bmatrix} x(t) \\ v(t) \end{bmatrix}$$

For the given dynamics,

(a) Compute the eigendecomposition and demonstrate that the system is

  (i) Stable,
  (ii) Oscillatory, and
  (iii) Non-Normal.

(b) Determine that the system can exhibit a transient divergence by computing *Reactivity*(A) in (5.52).

Problem 5.9: Numerical/Computational — Non-normal Systems

It is relatively easy to construct stable non-normal systems if we directly specify the eigendecomposition. Let

$$\underline{v}_1 = \begin{bmatrix} \cos(\theta) \\ \sin(\theta) \end{bmatrix} \qquad \underline{v}_2 = \begin{bmatrix} \cos(-\theta) \\ \sin(-\theta) \end{bmatrix} \qquad 0 < \theta < \pi/2$$

then for $\theta = \pi/4$ the two eigenvectors are normal (orthogonal), and for other values of θ the eigenvectors are non-normal.

Aside from θ, the other aspect of our system which affects the dynamics are the system eigenvalues. The main question concerns the *relative* values of the eigenvalues, so without any loss of generality we will set $\lambda_1 = -1$, and then test values $\lambda_2 < 0$. Since both λ_1, λ_2 are negative, the system is clearly stable.

Suppose we always initialize our system as $\underline{z}(0) = \begin{bmatrix} 1 \\ 0 \end{bmatrix}$

Table 5.1 The dynamics matrix A for Problem 5.10, describing the dynamic ecology of nutrients in a tropical rainforest, modified from [6].

...on the rate of growth of	Influence of ...								
	Leaves	Stems	Litter	Soil	Roots	Fruits	Detr.	Herb.	Carn.
Leaves	-1.56	0.669	0	0	0	0	0	0	0
Stems	0	-0.712	0	0	2.56	0	0	0	0
Litter	1.46	0.036	-6.41	0	0	1.14	0	55.8	17.3
Soil	0	0	0	-0.022	0	0	316	0	0
Roots	0	0	0	0.020	-2.56	0	0	0	0
Fruits	0	0.007	0	0	0	-2.03	0	0	0
Detritivores	0	0	6.41	0	0	0	-316	0	0
Herbivores	0.10	0	0	0	0	0.89	0	-62.6	0
Carnivores	0	0	0	0	0	0	0	6.8	-17.3

As in (5.53), let us measure the degree of amplification $\kappa_{\max}$ in the trajectory $\underline{z}(t)$ as the furthest extent from the origin:

$$\kappa_{\max} = \max_t \sqrt{z_1(t)^2 + z_2(t)^2}$$

where $z_1(t)$ and $z_2(t)$ are evaluated at discrete points in time t using (5.39). Numerically evaluate $\kappa_{\max}(\theta, \lambda_2)$ for a range of values for θ and λ_2:

(a) Produce a contour plot for $\kappa_{\max}$, plotting the contour which separates $\kappa_{\max} = 1$ (no transient growth) from $\kappa_{\max} > 1$ (non-normal growth in the transient).
(b) Discuss your observations. What are the circumstances which lead to significant transient behaviour?

Problem 5.10: Numerical/Computational — Non-normal Systems

The paper [6] by Neubert and Caswell has a nice example of a non-normal ecological system, for which a simplified version of the dynamics A appears in Table 5.1.

We begin with a look at the values in A:

(a) Why are all of the diagonal elements negative? Why should the presence of leaves imply a negative growth rate for leaves, for example?
(b) Note that most of the columns of A sum to exactly zero. What is the ecological explanation of this observation?
(c) What is the ecological significance of the choice of one column which does *not* sum to zero?
(d) Following on the two preceding questions (b,c), prove that if all of the columns of A sum to zero the system cannot be stable.

Next, we want to understand the resilience of this system, per the discussion in Example 5.3, in terms of eigendecompositions:

(e) Find the eigenvalues of A and determine that the system is stable. From these eigenvalues, compute *Resilience*(A) as in (5.51).
(f) Find the eigenvectors of A and determine that the system is non-normal.
(g) Compute *Reactivity*(A) in (5.52) from the eigenvalues of the symmetric sum $(A+A^T)/2$. If you examine all of the eigenvalues you will observe that there are *two* positive ones, meaning that the non-normal transient may have a combined behaviour of two time constants.

Problem 5.11: Reading

Given that linear systems are characterized by exponentials, it is worth better understanding the properties of the exponential function and the connections to human thinking and behaviour.

Read or watch[4] one of

- *Chapter 5: Dangerous Exponentials* in [4]
- *Crash Course Chapter 3: Exponential Growth* by Chris Martenson
- *Arithmetic, Population, and Energy* by Albert Bartlett

and, building on Example 5.1, prepare a half-page discussion summarizing the key attributes of exponential functions and why they pose such a difficulty for people to understand.

Problem 5.12: Policy

As discussed in Example 5.1, we hear regularly about references to *modest* economic growth or monetary inflation of 1 % or 2 %, however any fixed percentage growth rate, regardless how modest, is still exponential over time.

Following up on Problem 5.11, discuss the public policy implications of a political system which appears to be premised, or at least relying upon, indefinite growth.

References

1. B. Friedland. *Control System Design: An Introduction to State-Space Methods*. Dover, 2005.
2. Z. Gajic. *Linear Dynamic Systems and Signals*. Prentice Hall, 2003.
3. L. Gunderson and C. Holling. *Panarchy: Understanding Transformations in Human and Natural Systems*. Island Press, 2012.
4. C. Martenson. *The Crash Course: The Unsustainable Future of our Economy, Energy, and Environment*. Wiley, 2011.
5. D. Meadows, J. Randers, and D. Meadows. *Limits to Growth: The 30-Year Update*. Chelsea Green, 2004.

[4]The book and videos may be found online; links are available from the textbook reading questions page.

6. M. Neubert and H. Caswell. Alternatives to resilience for measuring the responses of ecological systems to perturbations. *Ecology*, 78(3), 1997.
7. N. Nise. *Control Systems Engineering*. Wiley, 2015.
8. A. Oppenheim, A. Willsky, and H. Nawab. *Signals & Systems*. Prentice Hall, 1997.
9. M. Scheffer. *Critical transitions in nature and society*. Princeton University Press, 2009.
10. E. Scheinerman. *Invitation to Dynamical Systems*. Prentice Hall, 1995.
11. L. Trefethen and M. Embree. *Spectra and Pseudospectra: The Behaviour of Nonnormal Matrices and Operators*. Princeton, 2005.
12. B. Walker and D. Salt. *Resilience Thinking: Sustaining Ecosystems and People in a Changing World*. Island Press, 2006.

Chapter 6
Nonlinear Dynamic Systems: Uncoupled

In July, 1992 the Canadian government imposed a moratorium on the Grand Banks cod fishery, abruptly closing an international industry of nearly 500 years. The annual cod harvests in the 1980s had been 250,000 tons, a level consistent with most of the previous hundred years, yet suddenly in 1992 there was a precipitous decline in cod biomass. Furthermore, despite the fact that cod mature over a relatively short time frame of 2–4 years, today, over 20 years later, the cod fishery has nowhere near recovered.

The cod fishery, along with nearly every physical, ecological, or social system, is *nonlinear*, therefore to not understand nonlinear systems is to not understand the world. The absolutely essential points are that superposition no longer applies, and that the system can display a persistence, known as hysteresis:

You apply a force to a stick, it bends.
You double the force, you double the bend. So far, this is still linear.

You increase the force further, the stick breaks. That's nonlinear.
Unbending the stick does not unbreak it. That's hysteresis.

A response out of proportion with the input is the hallmark of a system pushed to its limit or breaking point:

- Fish stocks collapsing fairly suddenly, after many years of fishing;
- A spontaneous bank run or financial collapse after many years of normal operations;
- Rapid changes in polar ice coverage.

Even without a limit or breaking point, how do we understand phenomena like ice ages, which appear to quite suddenly switch on and off, despite the fact that the heating from the sun is very nearly constant?

P. Fieguth, *An Introduction to Complex Systems*,
DOI 10.1007/978-3-319-44606-6_6

6.1 Simple Dynamics

We wish to begin by studying $\dot{z} = f(z)$, the simplest possible nonlinear system:

- One dimensional (scalar state),
- First order (only the first derivative of z is present),
- Time stationary ($f()$ does not vary with time),
- Deterministic (no noise term and no external inputs).

We'll begin with a nice example from Strogatz [12]:

$$\dot{z} = \sin(z) \qquad \therefore \frac{dz}{dt} = \sin z \qquad \therefore dt = \frac{dz}{\sin z} \tag{6.1}$$

We would *like* to solve for $z(t)$, but in general for nonlinear $f()$ we won't know how. In principle the latter equation can be integrated to give us the implicit solution

$$t = \int \csc z \cdot dz = -\ln\left|\csc z + \cot z\right| + \text{constant} \tag{6.2}$$

however this really gives us no insight into the behaviour of $z(t)$.

Instead, if we begin with a system plot of $\dot{z}$ versus z, from Section 5.4 we know that

- The fixed points are located at the roots of the dynamic equation,
- The stability at each fixed point is determined by the slope at that point.

giving us the system and phase plots of Figure 6.1. Given a phase plot, we can certainly qualitatively draw a sketch of the evolution of the system over time, as shown in Figure 6.2, much simpler and easier to understand than (6.2).

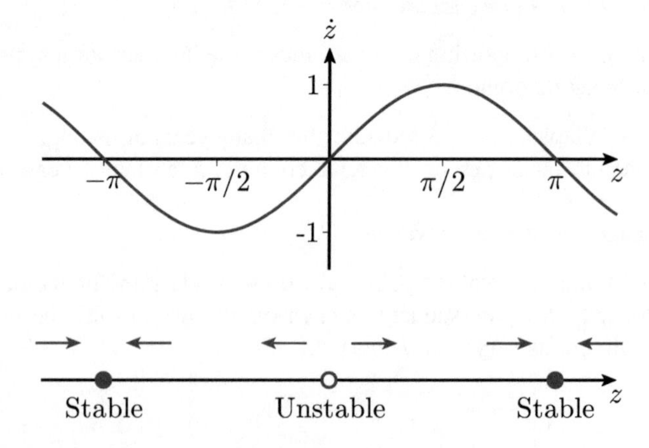

Fig. 6.1 The system and phase plots corresponding to (6.1).

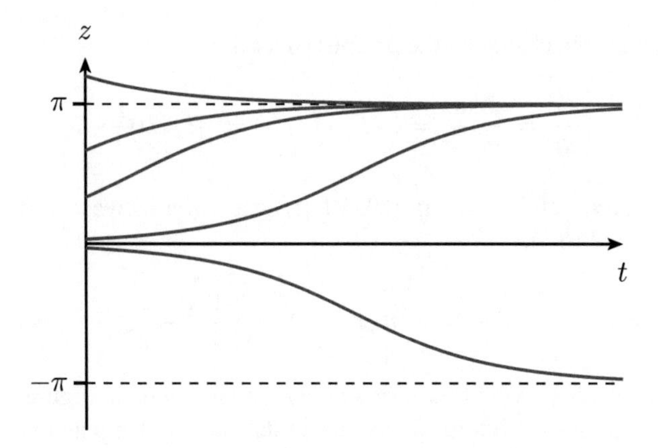

Fig. 6.2 The time evolution deduced, by inspection, from Figure 6.1.

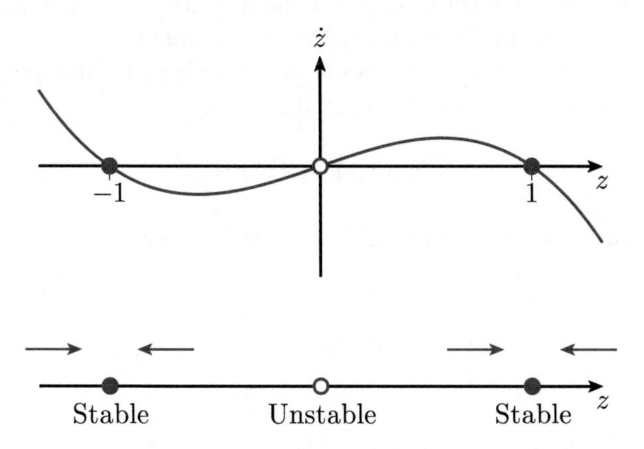

Fig. 6.3 The system and phase plots corresponding to the cubic equation of (6.3).

In principle, we could apply this approach to *any* nonlinear dynamic $f()$, as long as we can sketch $f()$, find its roots, and determine the slope. For example, given

$$\dot{z} = f(z) = -z^3 + z, \tag{6.3}$$

this cubic polynomial is easy to sketch, as shown in Figure 6.3, and for which the corresponding phase plot follows readily.

An alternative representation of system behaviour is that of a potential function $V(z)$,

$$f(z) = -\frac{dV}{dz} \tag{6.4}$$

We can evaluate the change of the potential over time:

$$\frac{dV}{dt} = \frac{dV}{dz}\frac{dz}{dt} = \big(-f(z)\big) \cdot f(z) = -\big(f(z)\big)^2 \leq 0 \tag{6.5}$$

That is, along *any* valid trajectory $z(t)$, $V\big(z(t)\big)$ can never increase. Returning to the cubic example of (6.3),

$$\text{Given } \dot{z} = f(z) = -z^3 + z \qquad \text{then } V(z) = +\frac{1}{4}z^4 - \frac{1}{2}z^2 + \text{constant} \tag{6.6}$$

the potential function is the fourth-order polynomial plotted in Figure 6.4. Observe how the fixed points are intuitively located at flat spots of the potential, with stable points in the valleys and unstable points at the tops of peaks. In interpreting a potential plot, it is important to keep in mind that the state (the moving particle) does not have momentum: it isn't really "rolling" downhill.

Finally there are a few detailed cases to clear up; Figure 6.5 illustrates four cases with ambiguous notions of stability. In all four cases

$$\dot{z} = f(z) \quad f(\bar{z}) = 0 \tag{6.7}$$

thus we do, indeed, have a fixed point $\bar{z}$, however the slope

$$\left.\frac{df}{dz}\right|_{z=\bar{z}} = 0, \tag{6.8}$$

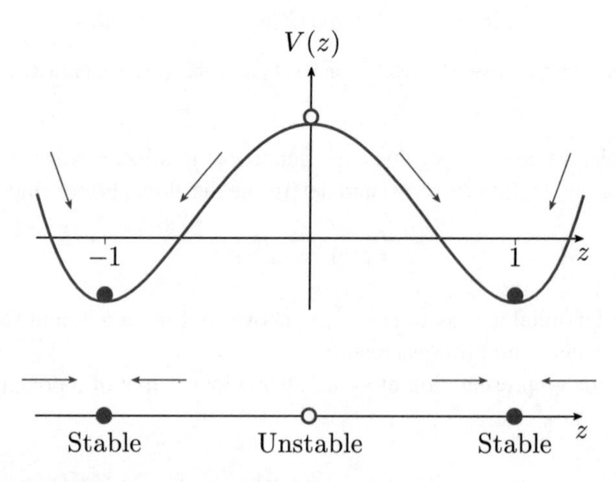

Fig. 6.4 Potential function: a plot of the potential function corresponding to Figure 6.3, as derived in (6.6). Over time, the system can only move to a lower potential, never higher. The stable points correspond to the bottoms of the potential basins.

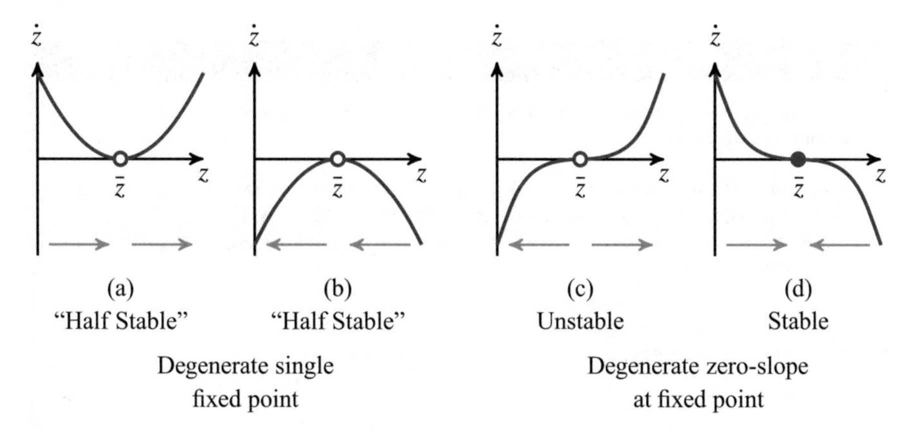

Fig. 6.5 Degeneracy: four degenerate cases and their respective stabilities are shown. The two parabolic cases (**a**), (**b**) have a degenerate fixed point, which would either disappear or split into two if the parabola were shifted up or down. In contrast, the cubic dynamics (**c**), (**d**) retain their single fixed point and stability behaviour under perturbation, it is only the zero–slope at the fixed point which is degenerate.

suggesting that the system stability is on the boundary between stable and unstable.

> *A system attribute is said to be degenerate at parameter $\theta = \bar{\theta}$, if for even the smallest perturbation $\bar{\theta} + \epsilon$ the system attribute is no longer present.*

We have encountered degeneracy before, in the coupled systems of Example 5.2 and studied in Problem 5.7. Similarly the razor-thin margin between stable and unstable, where a slope or an eigenvalue is *exactly equal* to zero[1], as we have in (6.8), is degenerate since this is a circumstance that will disappear with the slightest perturbation.

In any real system, there is *zero* probability of the slope in (6.8) being *exactly* zero, therefore in practice we will not concern ourselves in detailing whether a degenerate system is half-stable or unstable, however in the interest of clarity the possible circumstances and associated conclusions are summarized in Figure 6.5 and Table 6.1.

[1]Although, as we will see in Chapter 10, there are many *complex systems* which *are* balanced robustly exactly at the boundary between stable and unstable, but not based on the perfect tuning of a parameter.

Example 6.1: Population Growth

In Example 3.4 we saw a few qualitative models of the interplay of societal and complexity dynamics.

The limits which complexity sets on a society very much parallel the limits that an ecosystem places on the population of animals that live within it. At this point, following our discussion in Section 6.1, we can examine a few quantitative models of population dynamics. In all cases,

n is the number of organisms,
r is the reproduction rate (per individual per unit time).

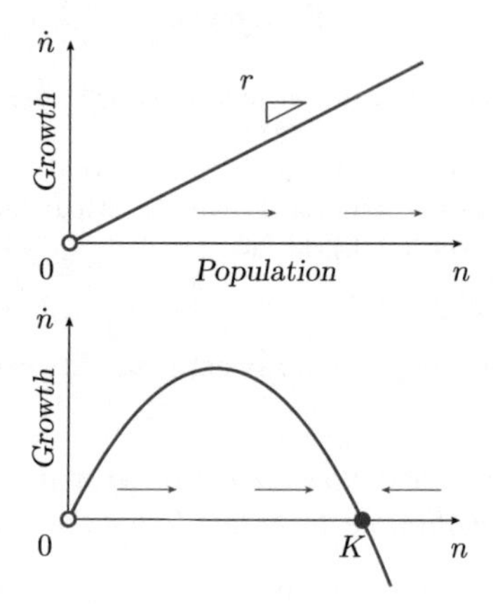

Case 1: r is fixed, so $\dot{n} = rn$. This is a linear system, with a single unstable fixed point, and the population grows exponentially without bound.

Case 2: In practice there is a limited carrying capacity K. Widely used is a nonlinear logistic model

$$\dot{n} = rn\left(1 - \frac{n}{K}\right)$$

which introduces a stable fixed point at K.

Case 3: There is a so-called Allee Effect, whereby animals may need a certain critical density of population to protect against predators, leading the unstable fixed point to move away from the origin, such that for populations below the fixed point the species heads towards extinction.

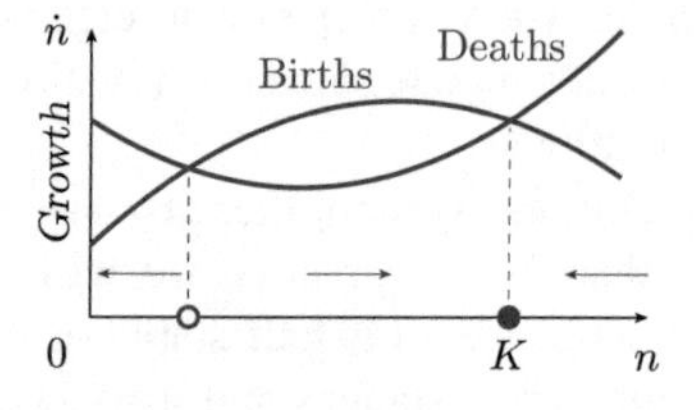

Case 4: Of course, the carrying capacity K itself could be eroded by an excess population: too many animals overgraze, kill the meadow, and now the number of animals which can be fed has been reduced. This is actually a two-dimensional nonlinear system, since we are suggesting that *both* $n(t)$ and $K(t)$ are evolving with time, the classic predator-prey model. Particularly interesting is when the resulting system has no stable fixed point, leading to indefinite oscillations. Two-dimensional nonlinear systems will be the subject of Chapter 7, with oscillating systems in particular in Section 7.3.

Further Reading: *Limits to Growth* [7], *Complex Population Dynamics* [13]
Population Dynamics, Allee Effect in *Wikipedia*.

Table 6.1 The interpretation of dynamics at fixed points, paralleling the illustration in Figure 6.5.

$f(z) \neq 0$ Not a fixed point
$f(\bar{z}) = 0$ Fixed point at $z = \bar{z} \ldots$

- $f'(\bar{z}) > 0$ Unstable fixed point — Figure 5.3b
- $f'(\bar{z}) < 0$ Stable fixed point — Figure 5.3a
- $f'(\bar{z}) = 0$ Degenerate situation — Figure 6.5 …
  - $f''(\bar{z}) \neq 0$ Bifurcation, Half Stable, Degenerate fixed point
  - $f''(\bar{z}) = 0$ No Bifurcation, Regular fixed point …
    - $f'''(\bar{z}) > 0$ Unstable fixed point — Figure 6.5c
    - $f'''(\bar{z}) < 0$ Stable fixed point — Figure 6.5d

6.2 Bifurcations

To this point we have considered a given system dynamics model $f(\underline{z})$ and studied the corresponding fixed points. Most models $f(\underline{z}, \theta)$ will be subject to a parameter θ, such as the CO_2 level in the atmosphere or an animal reproduction rate, and we are interested how the fixed points and dynamics are affected by θ:

> *A bifurcation corresponds to an abrupt or discontinuous change in system behaviour as a system parameter is changed.*

Keep in mind that a linear system is linear in the *state*, but not necessarily in the parameter, so that

$$\dot{z} = \cos(\theta)\, z - \theta^3, \tag{6.9}$$

although unusual, is perfectly linear in z. A linear system will, except at moments of degeneracy, always have exactly *one* fixed point $\bar{z}$, so the only dependence of $\bar{z}$ on θ is the value of $\bar{z}$ and whether that fixed point is stable or unstable.

In contrast, for nonlinear f both the locations and the *number* of fixed points can change as a function of θ.

For example, Figure 6.6 shows an ensemble of dynamic systems corresponding to

$$\dot{z} = f(z, \theta) = z^2 + \theta \tag{6.10}$$

for which the roots (the fixed points) are easily calculated:

$$f(z, \theta) = 0 \quad \longrightarrow \quad z = \begin{cases} \pm\sqrt{-\theta} & \text{for } \theta < 0 \\ \text{No solutions} & \text{for } \theta > 0 \end{cases} \tag{6.11}$$

So for $\theta > 0$ the function f has no roots, and therefore no fixed points, whereas for all $\theta < 0$ there is a pair of roots, as plotted in the *fixed point* or *bifurcation plot* of Figure 6.6. At $\theta = 0$ we observe the spontaneous appearance/disappearance of

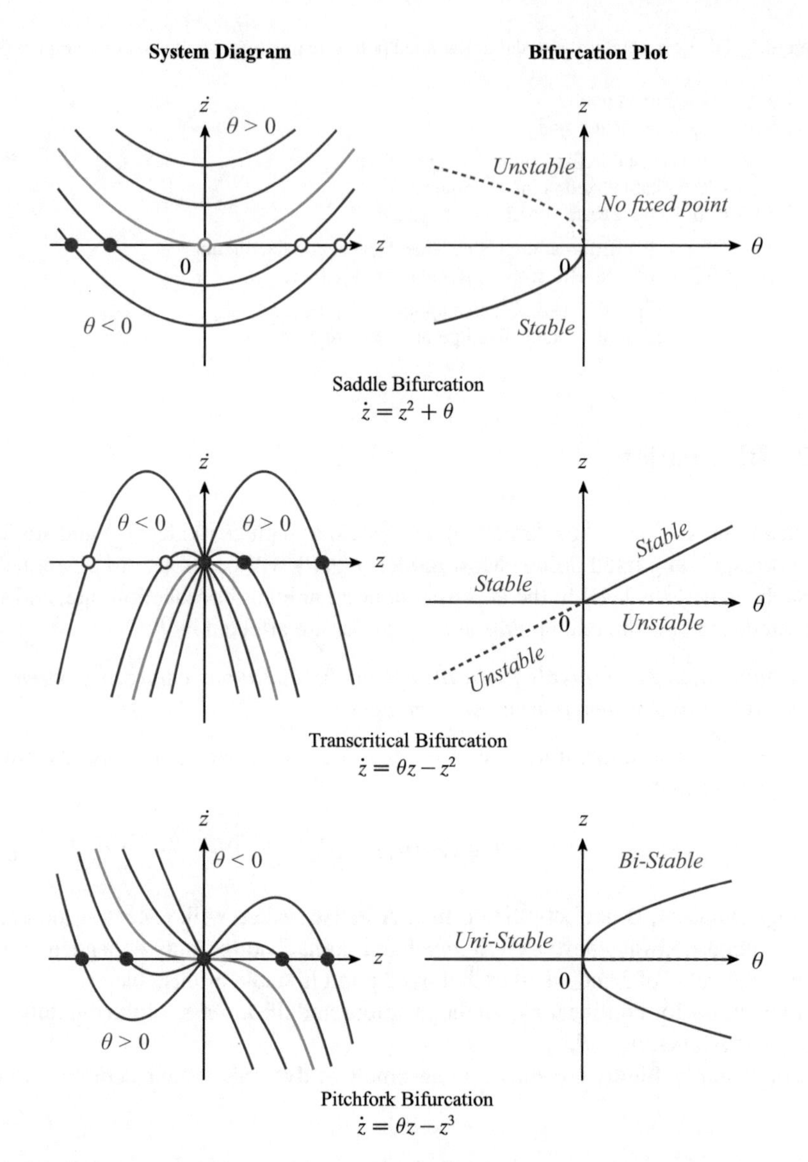

Fig. 6.6 Bifurcations: a bifurcation is a sudden change in system behaviour as a function of a system parameter. Each of these three nonlinear systems has a sudden change in the number and/or stability of fixed points at $\theta = 0$. In all cases, blue corresponds to $\theta < 0$, green to the degenerate case right at $\theta = 0$, and red to $\theta > 0$.

a pair of fixed points, known as a *saddle bifurcation*. Two further bifurcations are shown in Figure 6.6: a transcritical bifurcation and a pitchfork bifurcation.

Figure 6.6 examined only individual bifurcations; in practice a nonlinear system may have multiple bifurcations of different types, with a double saddle bifurcation

Example 6.2: Bead on a Hoop

A frictionless sliding bead on a rotating hoop (or, perhaps more familiar, twirling a string between your finger and thumb) is a classic example of a pitchfork bifurcation. We have a small bead which is free to slide on a hoop, which is rotating about the vertical axis with angular velocity ω. The bead is characterized by its state $\underline{z} = (\dot{\phi}, \phi)$.

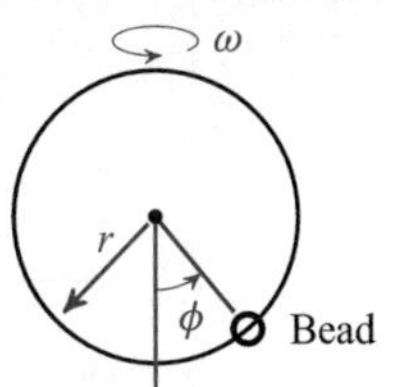

The forces acting upon the bead are those of gravity and centripetal motion. We can find the components of these forces in the tangential direction, since the radial forces just press against the hoop and do not affect the motion of the bead:

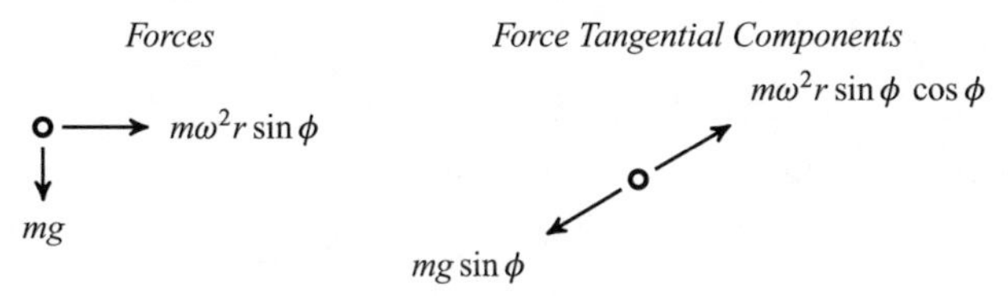

From Newton, $F = ma$, thus

$$mr\ddot{\phi} = -mg\sin\phi + m\omega^2 r \sin\phi\cos\phi \tag{6.12}$$

A fixed point occurs when $\dot{\underline{z}} = 0$; that is, when $\ddot{\phi} = \dot{\phi} = 0$. The bead is frictionless, so there is no $\dot{\phi}$ term, leaving us with

$$0 = mg\sin\phi\Big(-1 + (\omega^2 r/g)\cos\phi\Big) \tag{6.13}$$

The $\sin\phi$ term gives us fixed points at $\phi = -\pi, 0, \pi$. If we let $\gamma = (\omega^2 r/g)$, then the latter term in (6.13) has fixed points at $\cos\phi = \gamma^{-1}$:

If $\gamma^{-1} > 1$, then there are no further fixed points, since $|\cos\phi| < 1$
If $\gamma^{-1} < 1$, then there is a pair of fixed points at $\pm\cos^{-1}\gamma^{-1}$

with the stability of the fixed points determined by the sign of $\partial\ddot{\phi}/\partial\phi$. The result is a pitchfork bifurcation: at a slow spin ω the bead rests at the bottom, but above a critical velocity ($\gamma = 1$) the bead jumps to the side, going further up the hoop as the velocity is increased.

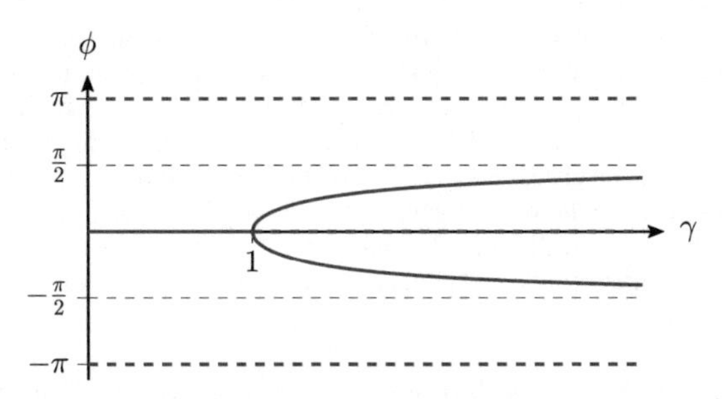

Far from being an abstract or artificial system, a similar bifurcation (although with more complex equations due to aerodynamics) occurs in the familiar dynamics of a maple key, in the transition from falling to rapidly spinning.

Further Reading: Two journal papers, a bit advanced, but showing interesting connections:
Varshney et al., "The kinematics of falling maple seeds ...," *Nonlinearity* (25), 2012
Andersen et al., "Analysis of transitions between fluttering ...," *Fluid Mech.* (541), 2005

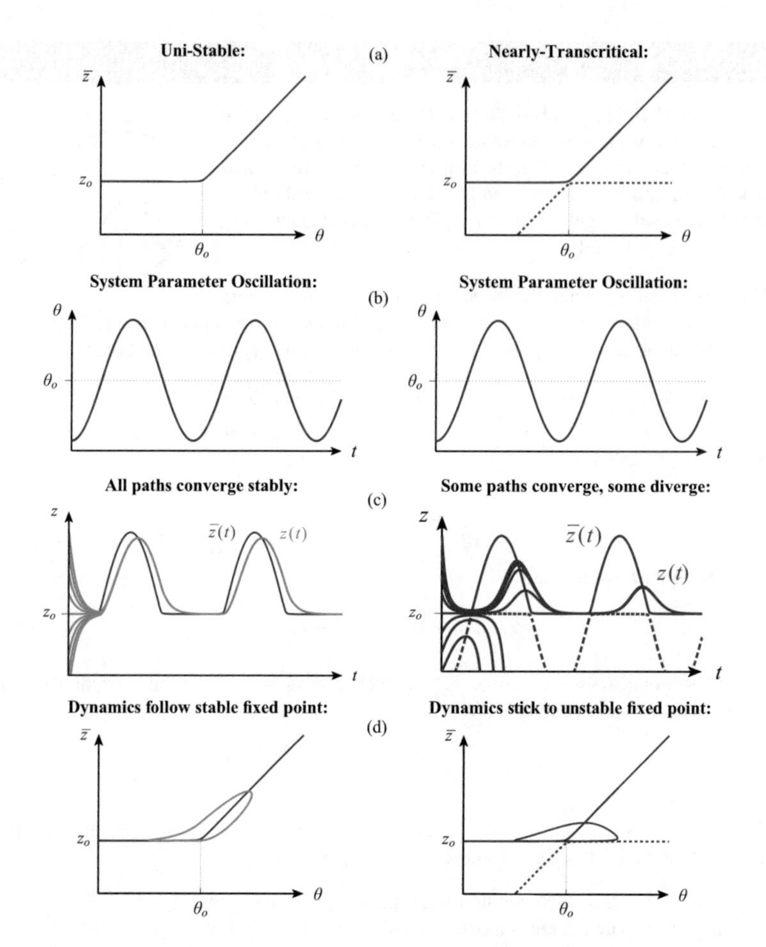

Fig. 6.7 Transient behaviour: the *uni-stable* (left) and *nearly-transcritical* (right) systems have the same stable fixed point (**a**). Suppose (**b**) that system parameter θ oscillates about θ_o: we would expect the system state $z(t)$ to follow the stable fixed point. In (**c**), for the unistable system $z(t)$ does indeed follow $\bar{z}$, with a slight lag. In contrast, in the transcritical system, having the *same* stable fixed point, the state exhibits a significant delay in moving towards the stable fixed point. Why? Recall that an unstable fixed point is still a *fixed* point; thus, quite unlike a magnet, *at* the unstable point there is *no* force pushing the state away (**d**), where the state (red) remains near the unstable fixed point (dashed blue). Therefore in bifurcations where a stable fixed point becomes unstable, as in the transcritical and pitchfork cases, there can be significant unexpected transient behaviour.

appearing quite commonly in the systems we will study (for example in Figure 6.10).

Building on the discussion of non-normal systems in Section 5.6, which can exhibit large transients by combining fast–slow dynamics, Figure 6.7 offers an example of unusual system transient behaviour in the presence of bifurcations. At any fixed point it must be true that $\dot{z}(t) = 0$, therefore there is actually *no* repulsive force at an unstable fixed point. Under normal circumstances the system state would not normally come close to an unstable fixed point, since there are repulsive forces all around it, however an important exception is a transition across a bifurcation, in which a stable point suddenly becomes *un*stable, and from which it can take significant time for the system to recover.

By definition, the number and/or stability of the fixed points is degenerate at a bifurcation, since the fixed point behaviour is discontinuous at the point of bifurcation. However the *type* of bifurcation, itself, may also be degenerate, if an infinitesimal perturbation in the system causes the overall bifurcation categorization to change.

In particular, it is easy to show that the transcritical and pitchfork bifurcations are both degenerate. Let us consider perturbed variations on the systems of Figure 6.6:

$$\begin{array}{cc} \text{Perturbed Transcritical:} & \text{Perturbed Pitchfork:} \\ \dot{z} = \theta z - z^2 + \epsilon & \dot{z} = \theta z - z^3 + \epsilon \end{array} \tag{6.14}$$

The perturbations are sketched in Figure 6.8, from which we can identify the fixed points, which are then plotted in bifurcation plots in Figure 6.9. It is clear that the transcritical and pitchfork bifurcations are degenerate, intuitively because they both require four fixed-point trajectories to coincidentally meet at a point, a coincidence which is not maintained under perturbation.

Degeneracy does not render a bifurcation irrelevant, however. For example the *nearly*–transcritical system of Figure 6.7 actually has *no* bifurcation, like the top–right perturbation of Figure 6.9, yet the system transient behaviour is characteristic of a transcritical bifurcation. Therefore transcritical and pitchfork bifurcations, although degenerate, may be useful in characterizing and understanding closely related systems.

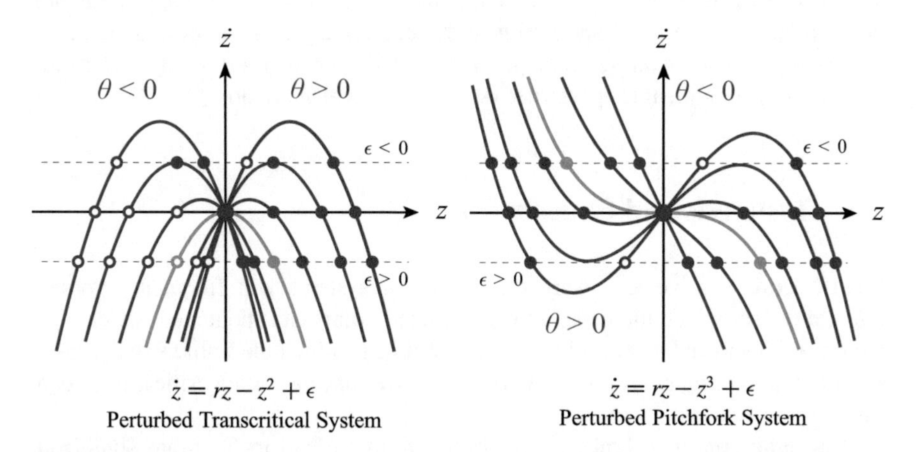

Fig. 6.8 System perturbation: here we examine how a bifurcation is affected by a small perturbation ϵ. The number and location of the fixed points will be a function of ϵ (dashed line), such that fixed points are the intersections of the solid curves with the dashed line as described by ϵ. There will be a maximum of two fixed points in the transcritical system (left) or three fixed points in the pitchfork system (right). The resulting bifurcation plots are shown in Figure 6.9.

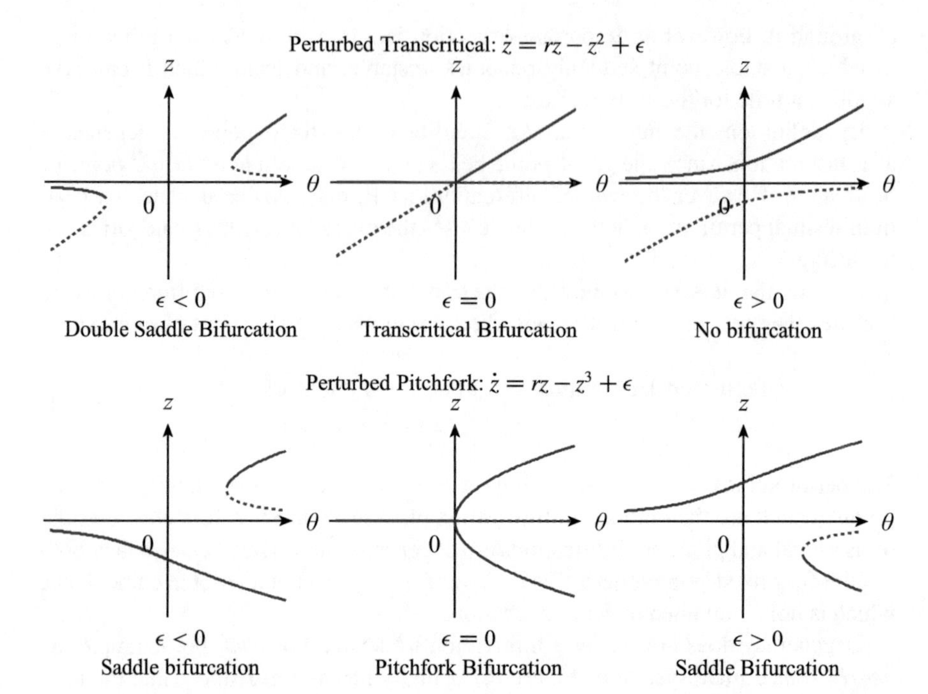

Fig. 6.9 Degeneracy: a small perturbation, shown in Figure 6.8, can cause a transcritical (top) or pitchfork bifurcation (bottom) to change.

Linearization, to be discussed in Section 7.1, is a common strategy for studying nonlinear systems. Since linearization does introduce some degree of approximation or perturbation to the original nonlinear system, an appreciation of degeneracy is important in understanding what properties of the linear (perturbed) system are present in the nonlinear (unperturbed) system, and which are not.

6.3 Hysteresis and Catastrophes

From Section 6.2 we are aware that nonlinear systems will frequently possess bifurcations, at which the behaviour of the fixed points can spontaneously change. Example 6.2 showed a very mild example of this, in which the stable fixed point of a bead on a rotating hoop suddenly began to move once a certain critical rotational velocity was reached.

It is quite common, however, for bifurcations to lead to far more significant impacts on the system dynamics. Consider the system bifurcation plots shown in Figure 6.10. As the slope of z versus θ increases, the location of the fixed point becomes increasingly sensitive to parameter θ. At some point the dynamics fold upon themselves, and we have a *fold* or double-saddle bifurcation. Such a system is said to be *bi-stable*, since for certain values of θ there are two possible stable fixed points for z. This bi-stability leads to profound system properties:

1. There will be large (so-called "catastrophic") state transitions, and
2. The system has memory, which leads to irreversible behaviour.

The three dynamics shown in Figure 6.10 should not necessarily be viewed as three different systems. As sketched in Figure 6.11, we may very well have a *single* system, where the presence or absence of a fold is controlled by an additional parameter ζ.

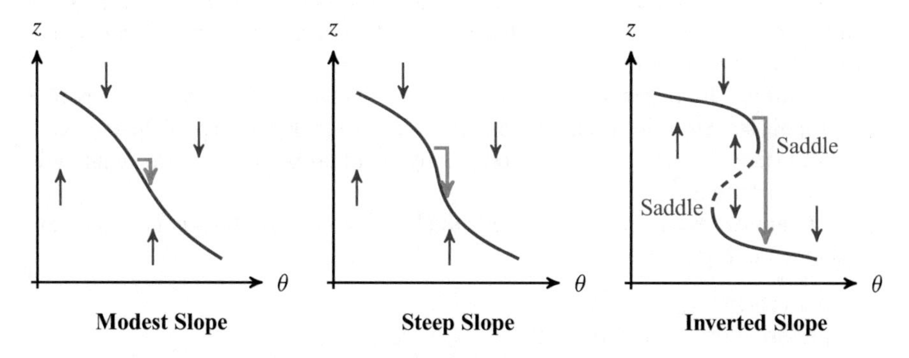

Fig. 6.10 Double-saddle bifurcation: as a mild slope (left) steepens (middle), a fixed perturbation in parameter θ leads to larger jumps (green arrow) in state z. If the slope inverts (right), a double-saddle bifurcation appears, in which case a tiny perturbation in θ can lead to a large jump in z, referred to as a catastrophic state transition.

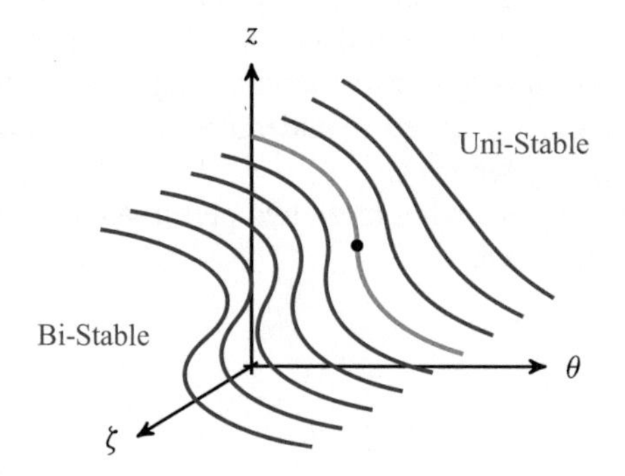

Fig. 6.11 The three panels in Figure 6.10 do not necessarily refer to three different systems, rather all to the same system parameterized by some *additional* parameter ζ, such that the presence (blue) or absence (red) of a fold (and associated catastrophic transitions) is controlled by ζ.

6.3.1 Catastrophic State Transition

In most cases, we expect a small perturbation to lead to a small influence on the system state. In particular, *all* linear systems exhibit such behaviour, because of two key properties:

- Except in degenerate cases, there is precisely one, unique fixed point;
- The location of the fixed point is a continuous function of the elements of the system matrix A.

Therefore there will never be a second, distant fixed point to transition to, and any system matrix perturbation[2] will only perturb the current fixed point and cannot cause a distant relocation.

In contrast, in response to a perturbation many nonlinear systems exhibit catastrophic state transitions, large state transitions from one fixed point to another, a change in state which appears to be completely out of proportion with the magnitude of the perturbation.

There are two forms of perturbation which can lead to a catastrophic transition, as sketched in Figure 6.12:

1. A perturbation of the *state* z,
2. A perturbation of the *system* $f()$.

A perturbation of the state does not alter any of the system fixed points, the system dynamics are unchanged. However if the state is perturbed past a nearby unstable fixed point, then the state has crossed a potential function boundary and is pushed away from the unstable point to a different, distant fixed point.

In contrast, a system perturbation does, indeed, change the system dynamics. Near a bifurcation a perturbation of the system may cause a stable fixed point to disappear entirely, causing the system to undergo a large transition to a very different stable fixed point, possibly far away.

A very simple example showing these two perturbations is illustrated for a toy wooden block in Example 6.5.

[2]To be precise, although the fixed point is a continuous function of the elements of A, from (6.9) we know that a system parameter could, in principle, appear *non*linearly. Furthermore, if the dependence on the system parameter were discontinuous, such as

$$\dot{z} = \tan(\theta)z \tag{6.15}$$

then a perturbation in θ could indeed cause a discontinuous jump in the fixed point.

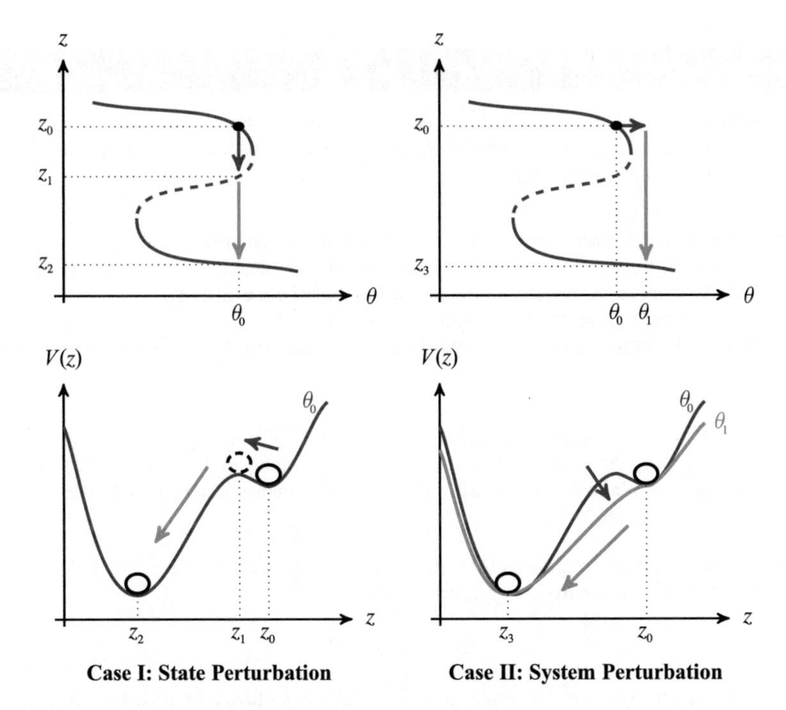

Fig. 6.12 Catastrophic transition: a catastrophic state transition may occur for two reasons: a state perturbation (left) or a system perturbation (right). In both cases, a perturbation (blue arrow) causes a change which leads to the catastrophic transition (green arrow).

6.3.2 System Irreversibility and Hysteresis

All stable linear systems are *reversible*: if the system or state is changed in some way, and then changed back, we are back where we started, as in Figure 6.13.

In many ways this is obvious: a linear system has a single, unique fixed point $\bar{z}$. Therefore regardless what change is made to the system dynamics, after the original system is restored we will converge to the original stable fixed point:

$$A_0, \bar{z}_0 \xrightarrow[\text{Change}]{\text{System}} A_1, \bar{z}_0 \xrightarrow{\text{Time}} A_1, \bar{z}_1 \xrightarrow[\text{Restore}]{\text{System}} A_0, \bar{z}_1 \xrightarrow{\text{Time}} A_0, \bar{z}_0 \tag{6.21}$$

In contrast, many nonlinear systems are irreversible, as illustrated in Figure 6.13. If the system is bi-stable, then a perturbation in the state or in the system (Figure 6.12) may induce a catastrophic transition from one stable state to another, a transition

Example 6.3: System Perturbations

The discussion in the text repeatedly makes reference to system *perturbation*, such as in (6.21), (6.22) or Figure 6.8. Where does this system perturbation come from? Indeed, a variety of scenarios are possible:

- A spontaneous unanticipated change, such as a system failure or break.
- Driven by inputs to the system, such as part of a deliberate control strategy (Section 5.5), or non-deliberate as a by-product of other human activity.
- Affected by random noise, as discussed in Section 4.2.
- Subject to additional dynamics, beyond those explicitly modelled for the system state z.

Clearly these explanations overlap. For example, a broken bearing in a motor (a system failure) may have been caused by earlier damage (an input) to the motor or due to the development of a crack in the bearing (based on the unknown dynamics to which the ball-bearing is subject).

To develop the above list in a bit more detail, a simple dynamic system consists of a state z with dynamics $f()$ controlled by a parameter θ,

$$\dot{z}(t) = f\big(z(t), \theta\big), \tag{6.16}$$

such that a change in θ, by whatever cause, leads to a change or perturbation in the dynamics of z. However "*by whatever cause*" seems arbitrary or vague, and in most cases θ will actually be subject to its *own* dynamics $g()$,

$$\dot{z}(t) = f\big(z(t), \theta(t)\big) \qquad \dot{\theta}(t) = g\big(z(t), \theta(t)\big), \tag{6.17}$$

a coupled nonlinear dynamic, to be explored further in Chapter 7. So although z remains the state of interest to us, we no longer pretend θ to be a fixed constant. For example, (6.12) specified the dynamics for a bead on a hoop, with ϕ capturing the state of the bead:

$$\ddot{\phi} = -{}^{g}/_{r} \sin\phi + \omega^2 \sin\phi \cos\phi \tag{6.18}$$

where ω was a system parameter, the angular velocity of the hoop. However if we were spinning the hoop with our fingers, then ω would be a *response* to the applied input torque τ, therefore ω would have its own dynamics,

$$\dot{\omega} = \tau / I(\phi), \tag{6.19}$$

where the moment of inertia $I(\phi)$ would be a function of the hoop and bead masses and radii. That is, the combined system (6.18), (6.19) is subject to an external input τ and a variety of parameters (masses and radii).

In general, a system will also be subject to some degree of noise (Section 4.2), leading to

$$\dot{z}(t) = f\big(z(t), \theta(t), u(t), \nu\big) + w_z(t) \qquad \dot{\theta}(t) = g\big(z(t), \theta(t), u(t), \nu\big) + w_\theta(t), \tag{6.20}$$

that is, we have a state of interest z, subject to dynamics $f()$, and attributes of the system θ, subject to dynamics $g()$. Both dynamics may or may not be subject to parameters ν, driven by an external input u, and subject to some sort of stochastic or random influence w_z, w_g. Changes in any of θ, u, ν, w_z, w_θ may lead to a system perturbation.

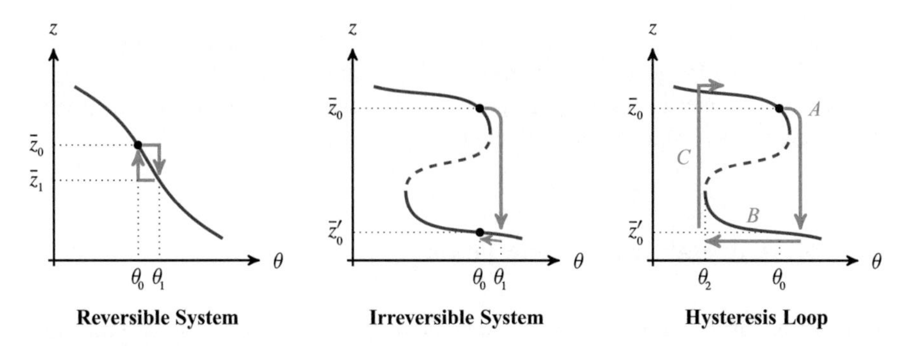

Fig. 6.13 Hysteresis: we normally expect systems to be reversible (left): perturbing a system parameter from θ_0 to θ_1 and back to θ_0 should leave us where we started. A system which fails to do so is *irreversible* (middle), in this case because the system is *bi-stable*, and the parameter perturbation leads to a catastrophic state transition which does not reverse. To recover the original state (right) would require a traversal around the hysteresis loop, involving two catastrophic state transitions A, C and a *far* greater parameter change B back to θ_2.

which is *not* un-done when the perturbation is removed:

$$A_0, \bar{z}_0 \xrightarrow[\text{Change}]{\text{System}} A_1, \bar{z}_0 \xrightarrow{\text{Time}} A_1, \bar{z}_1 \xrightarrow[\text{Restore}]{\text{System}} A_0, \bar{z}_1 \xrightarrow{\text{Time}} A_0, \bar{z}'_0 \tag{6.22}$$

The irreversibility is a form of system memory — the present state of the system is a function of the past. Clearly the state of a linear system is also a function of the past, however this is a transient dependence, since in the limit as $t \longrightarrow \infty$ the state converges to the single fixed point, and any memory of the input is lost:

$$z(t) = \bar{z} + e^{-\lambda t}\big(c - \bar{z}\big) \xrightarrow[t \longrightarrow \infty]{} \bar{z} \tag{6.23}$$

In contrast, a bi-stable system has a *persistent* memory, a memory which lasts indefinitely:

$$\underline{z}(t) \longrightarrow \begin{cases} \underline{\bar{z}} & \text{Given input history A} \\ \underline{\bar{z}}' & \text{Given input history B} \end{cases} \tag{6.24}$$

The challenge with hysteresis, as illustrated in Figure 6.13, is that a return to the original state can be very difficult. The wider the hysteresis loop, the further the system parameter needs to be pushed before the state is able to transition back to the original state.

There are many examples of hysteresis in natural systems, a few of which are shown in Example 6.6 and Case Study 6. Certainly there are very significant policy implications in dealing with systems having hysteresis; in particular, the precautionary principle (essentially, better safe than sorry) very much applies, since the "cost" of holding a system back from a catastrophic transition is much cheaper than the far greater cost of traversing a hysteresis loop. The topic is further nuanced by the question of *expected* cost, based on the likelihood of the catastrophic transition, further examined in Problem 6.12.

6.4 System Behaviour near Folds

In general, for a complex social or ecological system, we will not have an exact system model or a bifurcation diagram to refer to, so we may not actually know how close we are to a catastrophic state transition. It is therefore relevant to ask how the behaviour of a system may change as a fold (saddle bifurcation) is approached.

Consider first the bifurcation plot of Figure 6.14. The fixed point $\bar{z}$ changes as a function of θ; the slope β describes the sensitivity of the fixed point to parameter changes:

$$\beta = \frac{d\bar{z}}{d\theta} \qquad \text{where} \quad |\beta| \longrightarrow \infty \text{ at a fold.} \tag{6.25}$$

That is,

As a fold is approached, we expect amplified sensitivity to system parameter changes.

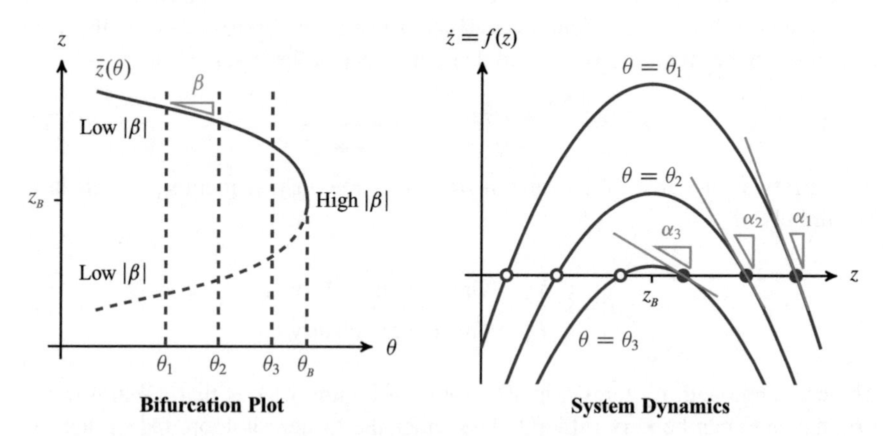

Fig. 6.14 Behaviour near folds: as a fold at (θ_B, z_B) is approached, the sensitivity β, the slope of the fixed point to the system parameter θ, is increased, left. Similarly the slope $\alpha = \dot{f}(z)$ of the dynamic at the fixed point is reduced, right.

Example 6.4: Deliberate Hystereses

Hysteresis definitely implies system properties of irreversibility, bi-stability, memory, and nonlinearity. However far from being a problem, there are many contexts in which hysteresis is *designed* into a system, deliberately.

Suppose you wish to press a switch to send an electrical signal. At the time instant t_c where the metal contacts first touch, the contacts are travelling towards each other at some velocity. Since metals are relatively elastic, the switch contacts may very well bounce a few times before making consistent contact.

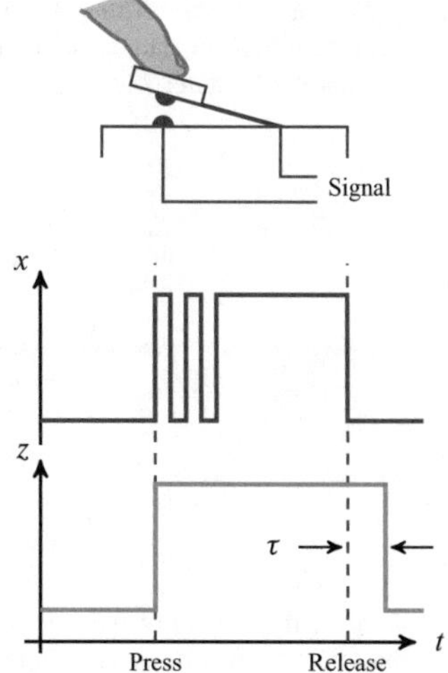

If the resulting switch signal goes to electronic circuitry which is edge–sensitive or edge–triggered, as most digital logic is, then the multiple bounces will look like repeated presses of the switch (red), rather than the intended single press.

The issue can be solved by *debounce* circuitry, which is essentially a hysteresis in time (green), here shown as a delay in accepting a transition from high to low. Given the switch input signal x, the debounced output z could be computed as

$$h(t+1) = \begin{cases} h(t) + 1 & x(t) = 0 \\ 0 & x(t) = 1 \end{cases} \qquad z(t+1) = \begin{cases} 1 & h(t) < \tau \\ 0 & h(t) \geq \tau \end{cases}$$

A second, exceptionally common example is that of any heating or cooling system. If a furnace were controlled by a single temperature setpoint, T_{set}, minor temperature fluctuations and measurement noise in the temperature sensor could lead to rapid *on–off* transitions.

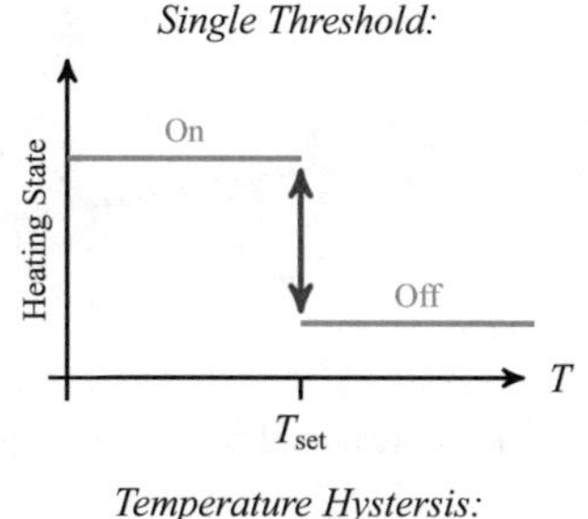

However with the exception of electric baseboard heaters, most heating and cooling systems, particularly compressors, do not tolerate being rapidly turned on and off. As a result, all thermostats have a built–in temperature hysteresis, such that the house will be warmed to slightly *above* the desired setpoint, and then allowed to cool to slightly *below* the setpoint. We then clearly see a tradeoff between comfort (a consistent temperature, implying small δ) and infrequent cycling time (large δ).

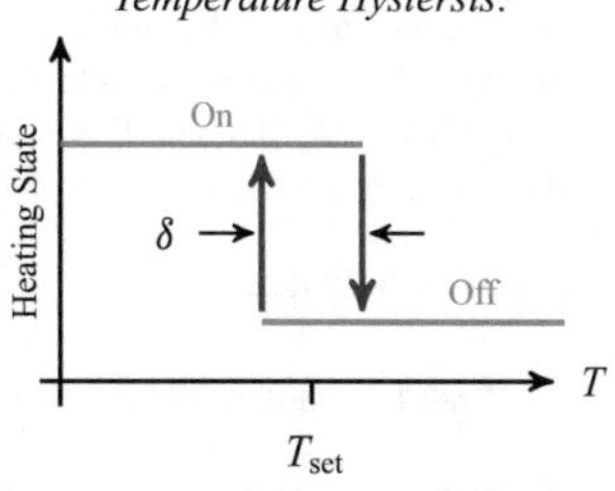

This sort of hysteresis in signal value is very commonly used in digital circuits to limit noise effects, and is known as a Schmitt Trigger.

Further Reading: H. Petroski, "Engineering: Designed to Fail," *American Scientist* (85) #5, 1997
Schmitt Trigger in *Wikipedia*.

Example 6.5: Playing with Blocks

Anyone who has ever played with wooden toy blocks, especially with a younger sibling, definitely understands catastrophic state transitions!

A tall block, placed on end, can be tricky to balance, because the stable fixed point angle z is so close to unstable points. A small touch with a hand, corresponding to a *state* perturbation, can cause the block to fall, which is the catastrophic state transition to another stable state:

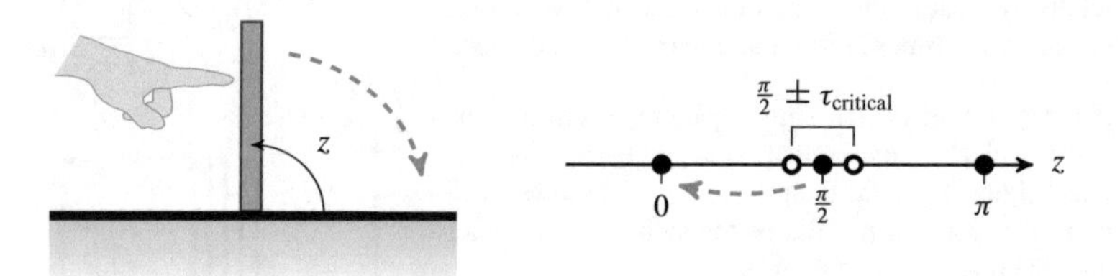

where

$$\tau_{\text{critical}} = \tan^{-1}(w/l)$$

is the critical angle of balance for a block of width w and length l, standing on end. On the other hand, tilting the whole table on which the block is standing corresponds to a *system* perturbation, since the nature of the overall system is actually being changed. For example, for tilt angle $\theta > \tau_{\text{critical}}$,

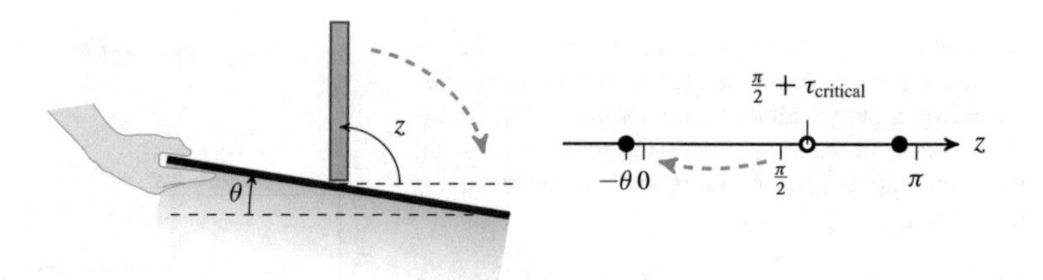

The bifurcation diagram for a balanced rectangular block is plotted below.

For $|\theta| > \theta_{\text{friction}}$ the angle of the table is too steep to allow friction to hold the block in place, even when lying flat, and there are no fixed points at all, even unstable ones.

Certainly both cases

$$\tau_{\text{critical}} > \theta_{\text{friction}} \quad \text{and} \quad \tau_{\text{critical}} < \theta_{\text{friction}}$$

are possible, since τ_{critical} is a function only of block geometry, unrelated to θ_{friction}, which is only a function of the material properties of the block and the table.

See Problems 6.6 and 6.7 to further build upon this example.

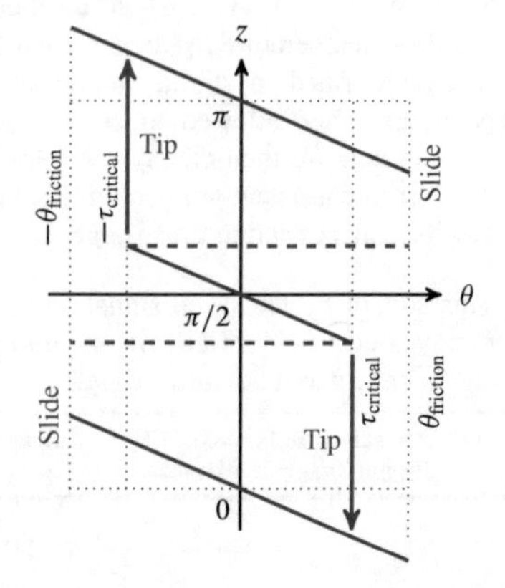

Example 6.6: Hysteresis in Freshwater Lakes

The book by Scheffer [9] presents a nice example of hysteresis. Freshwater lakes subject to phosphorus or nitrogen runoff (see Case Study 3) can undergo quite sudden transitions from clear to turbid (muddy). However, curiously, once stricter environmental controls are introduced, such that the phosphorus and nitrogen are reduced to historic levels, the turbid state remains.

At a risk of significantly oversimplifying lake ecology, vegetation and turbidity tend to oppose each other, since more plants hold down silt, whereas silty water blocks sunlight and prevents plants from growing, as shown.

As a consequence, each of vegetation and turbidity, on their own, are subject to positive feedback, meaning that the dynamics are unstable, with the instability blocked by a nonlinear limit (e.g., even with no silt, there is only room for so many plants). Such a limited unstable dynamic tends to lead to *bi-stable* systems, systems which have stable fixed points at two extremes, such as very clear and very turbid:

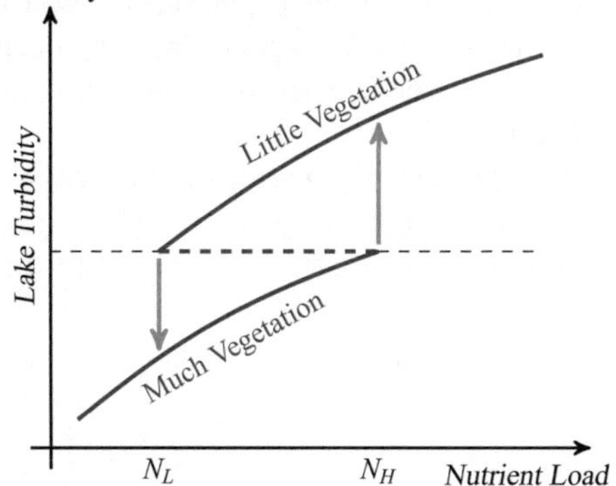

With an understanding of Section 6.3, we are now in a position to shed significant light on lake dynamics. The relatively sudden transition from clear to turbid is a catastrophic state transition, when the nutrient load exceeds a certain threshold N_H.

The inability to return the lake to a clear condition, despite reducing the nutrient load below N_H, is due to the bi-stability, such that the system is stuck in the upper turbid state, and much more effort is required to bring the nutrient load below N_L before the original state can be recovered.

In Case Study 6 we will encounter a similar example involving positive feedback and system bi-stability in the context of global climate.

Further Reading: Chapters 2 and 7 in *Critical Transitions in Nature and Society* by Scheffer [9].

Next, Figure 6.14 also plots the system dynamics as a fold is approached. Suppose we linearize the dynamics about the stable fixed point, as sketched in the figure:

$$\dot{z} = \alpha(\theta) \cdot (z - \bar{z}) \tag{6.26}$$

From Chapter 5 we know the time–behaviour of such a linear system:

$$\underbrace{z(t)}_{\text{State}} = \underbrace{\bar{z}}_{\text{Fixed Pt.}} + \underbrace{e^{\alpha(\theta)\cdot t}}_{\text{Rate of return}} \cdot \underbrace{c}_{\text{Disturbance}} \tag{6.27}$$

In general, for differentiable nonlinear dynamics, as θ approaches a fold $\alpha(\theta)$ goes to zero, meaning that the exponential decay becomes more gradual, thus

As a fold is approached, we expect slower returns to the stable state.

That the rate of return slows near a fold is intuitive. Recall from Example 5.3 that the system *resilience* was defined in (5.51) to be equal to the slowest eigenvalue. As a fold is approached, this slowest eigenvalue gets slower (closer to zero) and resilience is decreased, which is what we expect, since approaching the fold implies an increased likelihood of catastrophic transition (see Problem 6.9). A plot of this behaviour is shown in Figure 6.15.

For a given system of interest the rate of return is not necessarily easily measured, since the system will normally not be subject to single perturbations from which the return time can be observed. However a slower return time implies a system

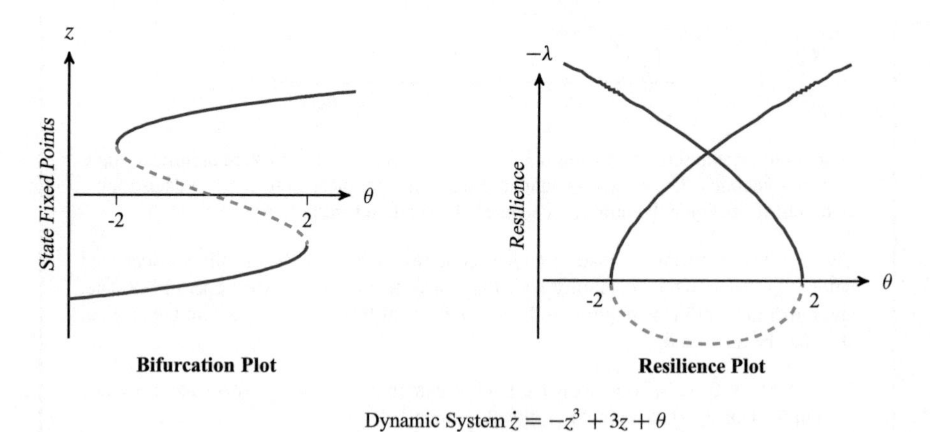

Fig. 6.15 Dynamics near a fold: the cubic nonlinear dynamic has the familiar bifurcation plot as shown, left. The negative eigenvalue for the fixed points is plotted at right. Observe how the eigenvalue approaches zero at the fold, thus the dynamics are slowest near the fold and faster elsewhere. The *resilience* of the system, measured as the negative eigenvalue, decreases as the folds are approached.

correlated over longer periods of time, an effect which *can* be observed directly from the autocorrelation of (4.19).

Finally, in practice the system state is subject to random fluctuations and disturbances. From (6.27) we know that

$$z(t) = \bar{z} + e^{-\alpha(\theta)\cdot t} \cdot c \tag{6.28}$$

Since the analysis is much simpler in discrete time, we observe that

$$z(t + \delta_t) = \bar{z} + e^{-\alpha(\theta)\cdot(t+\delta_t)} \cdot c = \bar{z} + \gamma e^{-\alpha(\theta)\cdot t} \cdot c \tag{6.29}$$

where $\gamma = e^{-\alpha(\theta)\delta_t}$, from which the discrete-time dynamic follows as

$$z(t + \delta_t) - \bar{z} = \gamma\Big(z(t) - \bar{z}\Big). \tag{6.30}$$

Let us now suppose that a unit-variance random disturbance ω takes place every δ_t seconds; then (6.30) becomes

$$z(t + \delta_t) - \bar{z} = \gamma\Big(z(t) - \bar{z}\Big) + \omega(t + \delta_t) \tag{6.31}$$

Finally, if we index our time series by n, such that $z_n \equiv z(t + n\delta_t)$, then

$$z_{n+1} - \bar{z} = \gamma(z_n - \bar{z}) + \omega_n \tag{6.32}$$

and therefore, taking the statistical variance of both sides,

$$\text{var}(z_{n+1}) = \gamma^2 \text{var}(z_n) + 1. \tag{6.33}$$

Assuming that the statistics of the process have settled down, meaning that the time series is in steady-state, the variance cannot be changing with time:

$$\text{var}(z_{n+1}) = \text{var}(z_n) \equiv \sigma^2 \tag{6.34}$$

Combining with (6.33) we find

$$\sigma^2 = \frac{1}{1 - \gamma^2} \tag{6.35}$$

As θ approaches a fold $\alpha(\theta) \longrightarrow 0$, thus $\gamma \longrightarrow 1$, thus $\sigma^2 \longrightarrow \infty$.

As a fold is approached, we expect greater amplification of random fluctuations.

6.5 Overview

We have seen a variety of different diagrams and it is important that we keep these clear in our head. Figure 6.16 summarizes the four main types of diagrams. In general, solid dots are stable points and open circles are unstable, and in bifurcation plots solid lines are stable and dashed lines are unstable.

Next, we have encountered a wide variety of system properties in our discussion. An overview, comparing and contrasting linear and nonlinear systems is shown in Table 6.2. In particular, all natural and social systems are bounded or finite, meaning that such systems are necessarily nonlinear, precluding any sort of indefinite exponential behaviour, further contrasted by the discussions in Example 5.1 and Example 6.1.

Near to a catastrophic transition it is certainly true that an infinitesimal perturbation could push a nonlinear system very far one way or the other, the so-called *Butterfly Effect*. However there is a significant misunderstanding, at least in popular culture, regarding this effect. Nonlinear systems are most definitely not *necessarily* sensitive; far from a catastrophic transition a nonlinear system can be highly robust and unwilling to change.

Furthermore many nonlinear systems will have no catastrophic transition at all, a common example being any linear system subject to limits or saturation. The resulting system would be unistable and *not* subject to hysteresis or catastrophic transitions, but nevertheless nonlinear and not obeying superposition such that, following up upon (5.5),

$$\text{Average}\Big(\text{System Response}\big(\text{Input}(t)\big)\Big) \neq \text{System Response}\Big(\text{Average}\big(\text{Input}(t)\big)\Big). \tag{6.36}$$

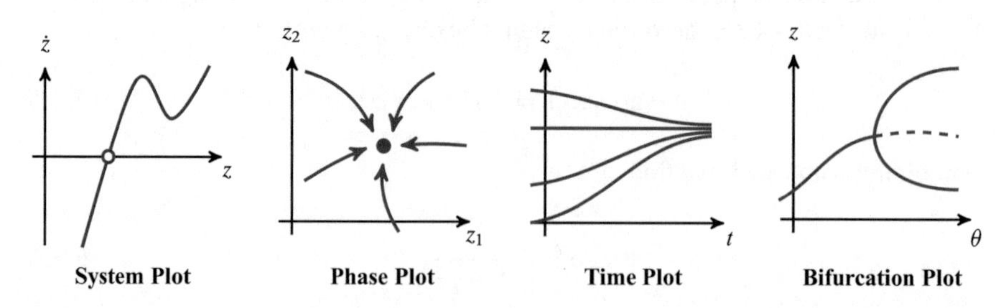

Fig. 6.16 Diagrams: an overview of the four main diagrams which we use in describing and studying nonlinear systems. We would normally start with a system plot, from which we derive phase and time plots. The bifurcation plot really summarizes many behaviours, since each value of parameter θ describes a different dynamic.

Table 6.2 Linear and nonlinear systems: an overview of the basic properties contrasting linear and nonlinear systems.

Context	Linear system	Nonlinear system
System of equations	Number of solutions is 0 Overconstrained 1 Invertible ∞ Underconstrained	Any number of solutions
Fixed points	Single fixed point	Any number of fixed points
Potential function	Two possibilities: Convex up (stable) Convex down (unstable)	Non-convex multiple minima possible
Dynamics characterized by	Unbounded complex exponentials	Bounded or unbounded bi-stable, chaotic
Given input $\cos(\omega t)$	A linear system must produce $a\cos(\omega t + \phi)$	A nonlinear system may produce Other frequencies Periodic non-sinusoids Non-periodic outputs
Superposition	Yes	Not in general
Reversible	Yes	Not in general
Hysteresis	No	Possible
Bi-stable	No	Possible

Finally, there are a few properties which are *not* unique to nonlinear systems:

- **High system sensitivity to inputs:** Nonlinear systems certainly have exceptional sensitivity at catastrophic transitions, however a linear system can have arbitrarily high amplifications.
- **Nonlinear dependence on system parameters:** A linear system is only linear in the *state*, however, as was already demonstrated in (6.9), a linear system can indeed have a nonlinear dependence on parameters.
- **Systems having memory:** As was described in (6.23), (6.24) a linear dynamic system *can* (and usually does) have memory, implying that the system output is a function of past inputs. The key difference is that nonlinear systems subject to hysteresis can have an indefinitely persistent memory, which a stable[3] linear system cannot.

[3]Technically an *unstable* linear system can have persistent memory, in the sense that the sign of the growing exponentials retain some memory of the initial state of or input to the system when it was initialized.

- **Positive feedback systems:** Positive feedback refers to a system whereby an increase in the state leads to further increases; that is, the system is unstable. Figure 5.3 in Chapter 5 certainly showed that there is no contradiction in having an unstable, linear system. What is true, however, is that real-world positive-feedback systems must be nonlinear, since unstable linear systems are unbounded.
- **Resonant or High-Q systems:** Anyone who has sung in the shower or heard the squealing feedback of a microphone too close to a speaker understands resonance. Resonance is caused by frequency selectivity, an amplification of certain frequencies and the attenuation of others, however linear systems can be constructed with such filtering behaviour.

Case Study 6: Climate and Hysteresis

Issues surrounding the earth's climate have become a controversial topic, in particular regarding the nature of humanity's current or projected future influence on climate.

Climate prediction may be an exceptionally challenging task, however it is quite easy to demonstrate that bifurcations, hystereses, and catastrophic state transitions appear repeatedly in the earth's climatic history. That is, far from being just a theoretical mathematical concept, bifurcations are present in the earth's climate in very measurable ways. In this case study we will survey four such examples:

Case 1: Historical Oxidation of the Atmosphere

The earth's atmosphere has not always had the composition of gases that it does today. The early earth had an atmosphere dominated by methane, and any oxygen which was present would have reacted with methane in the presence of UV light, or have reacted with other "reducing" minerals (those that react with oxygen), such as iron, which binds with oxygen in rusting. As photosynthetic life developed, oxygen was produced in greater quantities, eventually exhausting the supply of reducing minerals, leading oxygen to build up in the atmosphere. However the atmospheric oxygen increase was relatively sudden, suggesting a state transition.

For a given supply of reducing minerals there is a fixed rate of oxygen removal, however the atmospheric oxygen-methane system, sketched in Figure 6.17, is believed to possess bi-stability and hysteresis because of the system memory associated with the formation of an ozone layer:

- Beginning with an atmosphere high in methane, there is no oxygen to form an ozone layer, leading to the lower persistent stable state:

No ozone layer $\longrightarrow$ High UV penetration
$\longrightarrow$ High rates of methane–oxygen reaction
$\longrightarrow$ Suppressed atmospheric oxygen

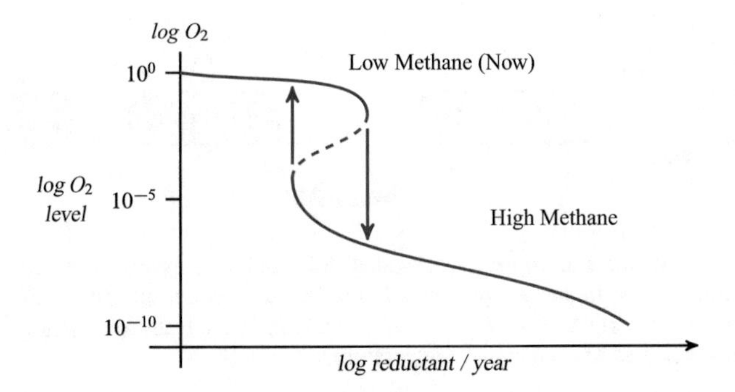

Fig. 6.17 Oxidation of the atmosphere: the oxygen levels in the atmosphere did not build up gradually, rather at some point *(the "Great Oxidation")* there was a catastrophic state transition to an atmosphere with much higher oxygen content. The horizontal axis measures the capacity for minerals to absorb (reduce) oxygen; what is of greater interest is the presence of a bi-stability, due to the presence/absence of an ozone layer.

Adapted from C. Goldblatt et al., "Bistability of atmospheric oxygen . . . ," *Nature* (443), 2006.

- Beginning with an atmosphere high in oxygen, a protective ozone layer forms, leading to the upper persistent stable state:

  Ozone layer → Low UV penetration
  → Low rates of methane–oxygen reaction
  → Preserved atmospheric oxygen

The historic "Great Oxidation" may thus have been a catastrophic state transition from low to high oxygen concentrations. Nevertheless, as implied by Figure 6.20, the complete process of adding oxygen to the atmosphere did take place over an extended period of time.

Case 2: Planet–Wide Climate Shifts

At a very basic level, global ice coverage and temperature are unstable, as sketched in Figure 6.18:

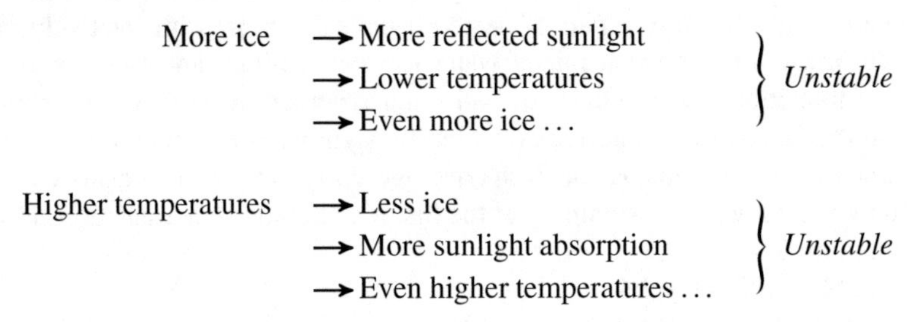

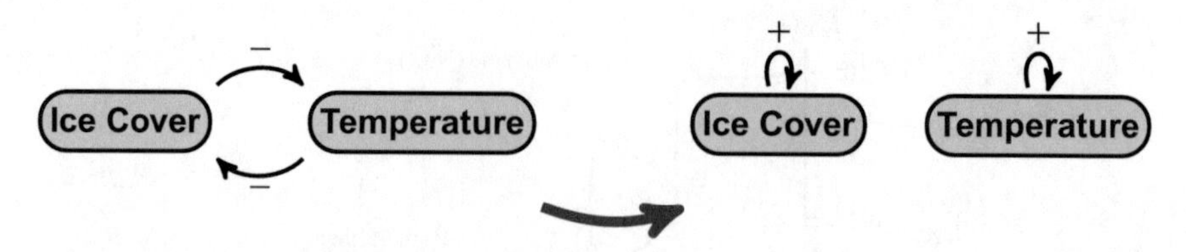

Fig. 6.18 Ice and temperature oppose each other, left: more ice means more reflected sunlight, thus lower temperatures; higher temperatures imply less ice. As a result, the single-state systems for ice and temperature, right, involve positive feedback: more ice leads to even more ice, similarly higher temperatures lead to even higher temperatures.

similar to the freshwater lake dynamics which we saw in Example 6.6. We therefore appear to have two stable states at the climate extremes:

Temperature increase until all ice has melted → Ice-free earth
Ice coverage increase until the planet is ice–covered → Snowball earth

Under this perspective our current climate, consisting of a mix of ice-free and glaciated regions, is therefore something of a more delicate balance, all the more reason to be somewhat prudent or cautious regarding human influence on climate. The stability of our current climate is caused by more subtle temperature–ice interactions which lead to negative feedback behaviour, such as

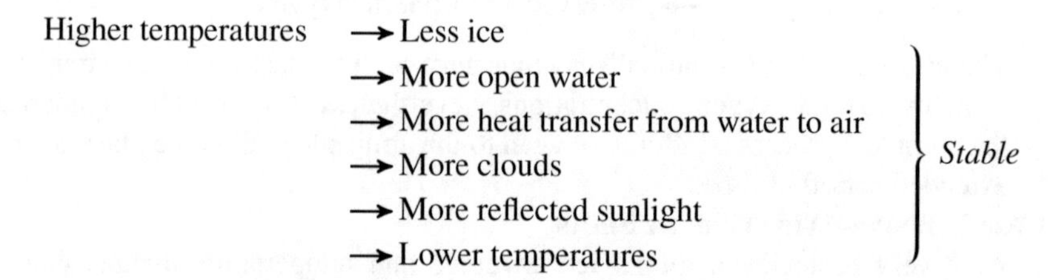

The three long-term stable global climate states are shown in Figure 6.19. Interestingly we have a pair of nested hysteresis loops, with relatively modest state transitions between mixed water-ice and ice-free, and truly catastrophic state transitions to/from the snowball state. Whether the earth was ever fully ice covered is controversial, however there is significant evidence that there were three periods of global ice or slush coverage, early in the earth's history when the sun was less warm. A summary of the historic climatic state transitions is shown in Figure 6.20.

Case 3: Glaciation Cycles

Figure 6.20 certainly makes clear the issue of baseline, which we discussed in Chapter 4, especially in Case Study 4. In particular, most people would consider it "normal" to have the arctic and antarctic ice-covered, an assumption which has been valid throughout the entirety of human history, but which has *not* been the case for most of the earth's history.

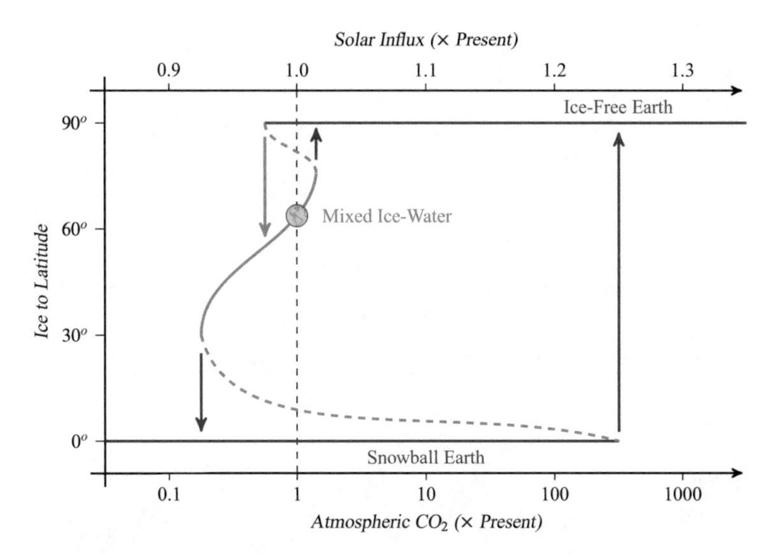

Fig. 6.19 Global climate states: the earth may have no ice (top), polar ice to some latitude (middle), or be completely ice covered (bottom), thus we have two nested hysteresis loops. The states are driven by heat, which could be caused by changes in the brightness of the sun (top axis) or the carbon-dioxide level in the atmosphere (bottom axis).

See P. Hoffman, D. Schrag, "The snowball Earth hypothesis," *Terra Nova* (14), 2002.

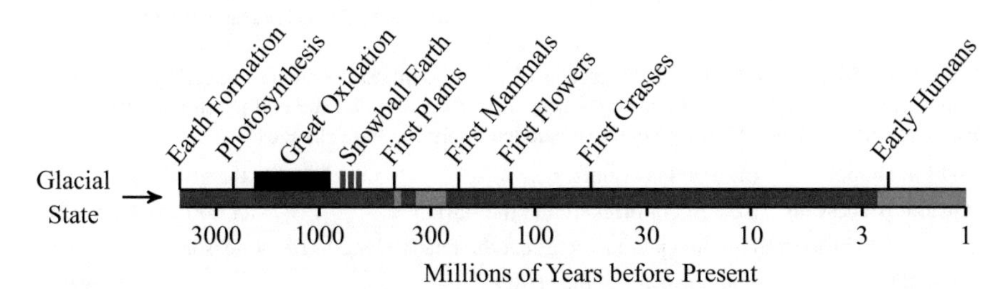

Fig. 6.20 Geologic timeline: the earth has repeatedly cycled between the stable states of Figure 6.19, although certainly the snowball cycling hypothesis remains uncertain. Since humans first appeared the earth has been in the mixed water-ice state, however really this is historically unusual. Since mammals first appeared, 225 million years ago, for only 1 % of that time did the earth have polar ice! The colour of the *Glacial State* corresponds to the states of Figure 6.19.

Figure 6.21 takes a more detailed look at recent climate history, where the baseline issue is emphasized further — temperature and atmospheric CO_2 appear to have been subject to constant change. The temperature (and corresponding glaciation) patterns seem quite periodic, and indeed the history is understood to be forced by periodic variations in the earth's orbit, known as the Milankovich Cycles, summarized in Table 6.3.

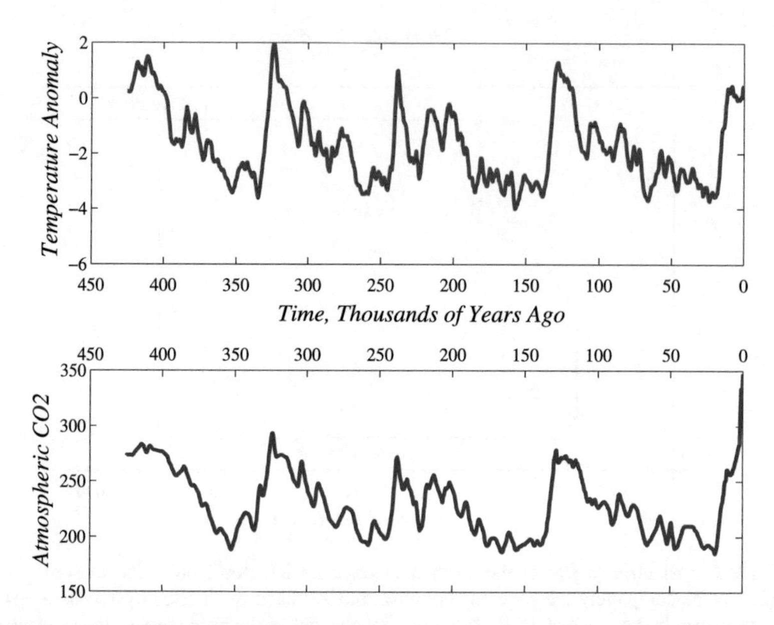

Fig. 6.21 Glaciation cycles: in the last half million years, the earth has repeatedly gone through periods of ice ages and glacial retreat. The time series repeatedly show relatively sudden warming jumps, the hallmark of a catastrophic state transition.

Data from NASA Goddard Institute for Space Studies.

Table 6.3 Milankovich cycles: many aspects of earth's climate are slowly nudged by a variety of changes in the earth's orbit. The changes here are very slow; that the earth's climate sometimes responds very suddenly (ice-ages) is due to sudden nonlinear state changes.

Orbital shape	(more/less eccentric)	100,000 and 400,000 years
Apsidal precession	(rotation of orbit around the sun)	21,000–26,000 years
	Both eccentricity and precession affect the relative lengths of the seasons.	
Axial tilt	(more/less tilt of the earth's axis)	41,000 years
	A greater tilt leads to more pronounced seasons, a lower tilt to milder differences between winter and summer	
Axial precession	(the direction in which north points)	26,000 years
	Determines whether the northern or southern winter is closer to/further from the sun.	
Orbital inclination	(orbital tilt above/below the solar plane)	100,000 years
	There may be more dust and debris in the plane of the solar system, so orbital inclination may affect how much of this dust arrives on earth.	

Case 4: Thermo-Haline Circulation

The Thermo-Haline Circulation refers to the global-scale water circulation pattern in the oceans, driven by differing water densities, which are primarily determined by temperature and salinity. A simple view of the circulation, ignoring salinity, is shown in Figure 6.22.

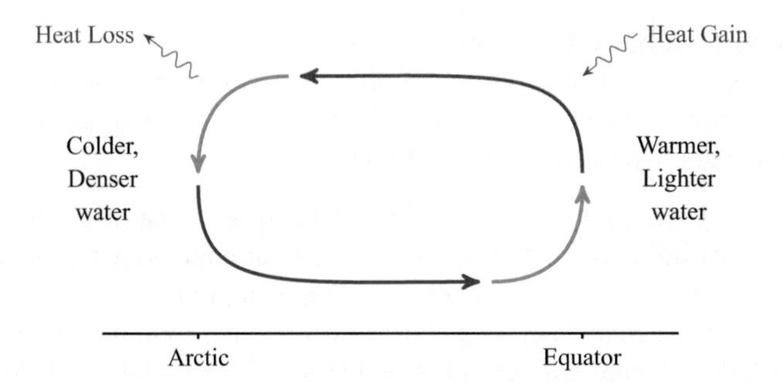

Fig. 6.22 Thermo-haline circulation: warm equatorial and cold high-latitude (northern and southern) temperatures create differences in sea-water density which lead to ocean circulations.

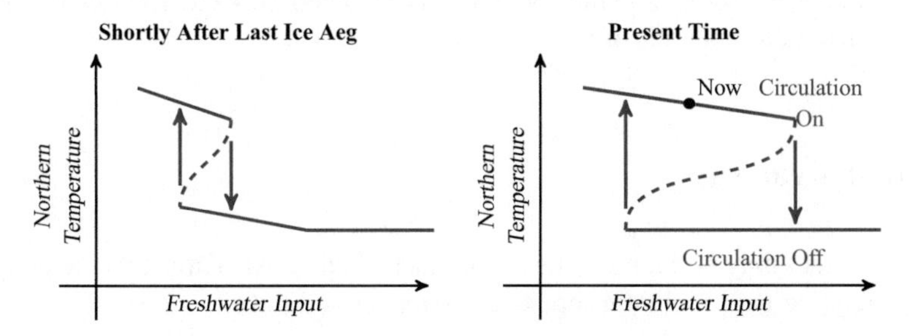

Fig. 6.23 Thermo-haline hysteresis: the ocean circulation of Figure 6.22 is driven by water density. An excessive input of freshwater, which is light, can turn off the circulation. The circulation is currently quite stable, right, but is likely to have repeatedly cycled after the last ice age because of the narrow hysteresis, left, at that time.

See D. Paillard, "Glacial Hiccups," *Nature*, 2001.

Because fresh water is much less dense that salty sea water, if enough fresh water is introduced in the North Atlantic, for example due to glacial melting, then the ocean circulation is shut down. Indeed, the ocean circulation is bi-stable with hysteresis, as shown in Figure 6.23, an effect which will be examined later in Example 8.3. There is evidence of wide climate swings after the last ice age, consistent with the circulation repeatedly turning on and off, however as shown in the figure, the nature of the climate after the last ice age led to a rather narrow hysteresis loop, where repeated state catastrophes were somewhat more likely, as compared to the present time, where the hysteresis loop is quite broad.
It is worth clarifying the role of the thermo-haline circulation to the Gulf Stream, a massive transfer of heat from the tropics to the north Atlantic, without which Europe would be significantly colder. The Gulf Stream is actually the combination of two effects:

1. In part the *temperature-salt driven* thermo-haline circulation of Figure 6.23 and Example 8.3;
2. But primarily a *wind-driven* ocean gyre, to be discussed in Section 8.2.

Case 5: Recent Climatic Transitions

Most people consider climate to be relatively constant (whatever we believe "normal" weather to be). Although climate typically changes slowly, sudden changes are not just historic, but continue to the present:

- *1000's of years ago*: The relatively sudden appearance of the Sahara Desert 5500 years ago, which had previously had abundant vegetation, a nonlinear climate response to slowly changing radiative input [9].
- *100's of years ago*: The so-called Little Ice Age, roughly between 1550CE and 1850CE, brought much colder winters to Europe and North America.
- *10's of years ago*: The North-American Dust Bowl, an extended drought for much of the 1930s in many parts of North America.
- *1's of years ago*: El–Niño/La–Niña, as sketched on page 135, which appear on irregular intervals of a few years.

Further Reading

The references may be found at the end of each chapter. Also note that the textbook further reading page maintains updated references and links.

Wikipedia Links — Nonlinear Systems: Nonlinear System, Bifurcation theory, Hysteresis, Catastrophe Theory

Wikipedia Links — Climate Systems: Great Oxidation, Snowball Earth, Greenhouse and Icehouse Earth, Glacial Period, Milankovitch Cycles, Thermohaline Circulation, Little ice age, Medieval warm period

The book by Scheffer [9] is very engaging and follows a philosophy similar to this present text, creating connections between environmental and nonlinear systems. The book is highly recommended, particularly to follow up on the climate and hysteresis examples.

Probably the most widely cited and read book on nonlinear systems is that by Strogatz [12], which is highly recommended. There are really a very large number of books published on the subject, and a few others the reader may wish to consider include [1, 2, 5].

Bifurcations and transitions occur in nearly all complex nonlinear systems, as we will see in Chapter 10. Thus in addition to contexts of climate (Case Study 6) [8] and ecology (Example 6.6), similar transitions are encountered in human physiology (epilepsy, asthma) and finance (market crashes, asset bubbles [11]).

The issue of resilience is one of the key questions in the intersection of society and nonlinear dynamics. Issues of resilience are pertinent on all scales, from a community, to a city, to globally. Many books and articles have been written on the subject; those that treat resilience from more of a quantitative/systems approach include the works by Deffuant and Gilbert [3], Meadows et al. [7], Gunderson and

Holling [4], Loring [6], Scheffer [9], and Walker and Salt [14]. The specific question of assessing resilience by *anticipating* an approaching bifurcation is examined in detail in [10].

Sample Problems

Problem 6.1: Short Questions

(a) Identify and briefly discuss a population dynamic model.
(b) What is the significance of *hysteresis* in the study of global climate?
(c) Under what circumstances is a *catastrophe* present in a nonlinear system?
(d) There are many different bifurcations present in earth climate systems. Give one example, draw a labelled diagram, and clearly explain the climatological significance.
(e) Draw a sketch of a bifurcation plot of a nonlinear system *not* subject to hysteresis.
(f) Draw a sketch of a bifurcation plot of a nonlinear system having *three* fold bifurcations.

Problem 6.2: Analytical — Nonlinear Dynamics

Suppose we have the continuous-time system

$$\dot{z} = f(z) = z^4 - z^2$$

(a) Find the fixed points and whether they are stable via *graphical* means, i.e., by sketching.
(b) Find the fixed points and whether they are stable via *analytical* means, i.e., by calculating by hand.
(c) Identify and sketch a potential function corresponding to f.
(d) Now suppose the system is parameterized,

$$\dot{z} = f(z) = z^4 - \theta z^2$$

Draw the bifurcation plot for this system, showing the fixed point(s) as a function of θ. Identify the type of bifurcation which is present.

Problem 6.3: Analytical — Bifurcations

We are given two different nonlinear dynamic systems:

$$\dot{x} = 6x - x^3 + \theta$$

and

$$\dot{y} = y^4 - 4y + \zeta$$

for states x, y respectively controlled by parameters θ, ζ. For each system:

(a) Sketch the system diagram.
(b) Sketch the bifurcation plot.
(c) Determine the (parameter,state) pair corresponding to each bifurcation.
(d) Over what range of parameter θ, ζ is the nonlinear system bi-stable?

Problem 6.4: Analytical — Stability

In the bead-on-hoop model of Example 6.2 we had the dynamics

$$mr\ddot{\phi} = -mg\sin\phi + m\omega^2 r\sin\phi\cos\phi$$

where the fixed points were found to be at

$$\bar{\phi} = \sin^{-1}(0) \quad \text{and} \quad \bar{\phi} = \cos^{-1}(\gamma^{-1}), \gamma = (\omega^2 r/g)$$

Reconstruct the bifurcation plot of page 105 by determining the stability for each fixed point as a function of γ, which can be inferred from the sign of the derivative

$$\frac{\partial\ddot{\phi}}{\partial\phi}.$$

Problem 6.5: Analytical — Nonlinear Dynamics (Challenging)

Suppose we have the nonlinear dynamic system

$$\dot{z} = \sin(z) + \alpha z$$

We can approach this problem in two ways:

(a) Solve the problem graphically, by sketching:

  (i) Draw the system plot. In your sketch, superimpose three curves, corresponding to $\alpha = 0$, α slightly positive, and α slightly negative. Identify stable and unstable fixed points.
  (ii) Draw the fixed point diagram, showing the fixed points of z as a function of α. Label any bifurcations.

(b) Solve the problem analytically:

  (i) Identify the relationship between α and fixed points in z.
  (ii) Analytically determine the domains of stability and instability.

Problem 6.6: Analytical — Bifurcations

The bifurcation plot corresponding to a simple wooden–block system was developed in Example 6.5. In that example, the diagram was drawn for the case where

$$\tau_{\text{critical}} < \theta_{\text{friction}}$$

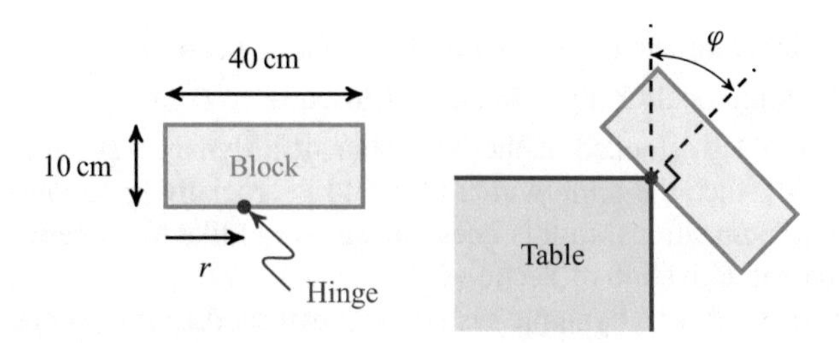

Fig. 6.24 Configuration of Problem 6.7: a wooden block is hinged on the corner of a table, such that the state of the block is the angle φ, with the overall system controlled by a single parameter r, the location of the hinge.

Now redraw the bifurcation plot corresponding to

$$\tau_{\text{critical}} > \theta_{\text{friction}}$$

Are there actually any catastrophic state transitions (block tipping) present in this case?

Problem 6.7: Analytical — Bifurcations

This problem will consider a variation of the simple wooden–block system of Example 6.5. Figure 6.24 shows a wooden block at the edge of a table. The block is subject to gravity (it gets pulled down) and it is hinged at the corner of the table, so that it can only rotate; it can't slide sideways or fall off the table.

Let r represent the location of the hinge along the bottom of the block, as shown, and φ the rotation angle of the block, where the table and hinge limit the angle as $0 \le \varphi \le \pi/2$.

Produce a clear, well labelled bifurcation diagram for this system.

Problem 6.8: Numerical/Computational — Bifurcations

We would like to study bifurcations; for this problem we will focus on the pitchfork bifurcation (see Figure 6.6):

$$\frac{dz}{dt} = \zeta + \theta z - z^3$$

Develop code to generate two plots:

(a) Let $\zeta = 0$. Generate a 2D array, storing the values of dz/dt for a range of values for θ and z. Produce a plot of the critical points, clearly showing the bifurcation, probably as a contour plot.
(b) We would like to see the more complex dependence on θ *and* ζ. Generate a 3D array, storing the values of dz/dt for a range of values for θ, ζ and z. We're looking for a surface in 3D, the complex folded shape at which dz/dt is zero. Produce an isosurface plot of this surface.

Ideally, rotate the plot to make the fold more easily visualized.

Offer a brief commentary on your observations of the plots.

Problem 6.9: Numerical/Computational — Closeness to Fold

In Section 6.4 we looked at the behaviour of a dynamic system as a fold is approached, since the *anticipation* of a fold is crucially important to avoiding catastrophic transitions. In this question we would like to numerically validate the analytical derivation of Section 6.4.

Suppose we have a dynamic system with two saddle bifurcations:

$$\dot{z} = -z^3 + 3z + \theta \tag{6.37}$$

for system parameter θ. First we need to understand the system a little; Figure 6.15 would probably be a helpful place to start. Either numerically or analytically,

(a) For $\theta = 0$, identify the fixed points for z.

(b) Identify the values $\theta_{\text{min}}, \theta_{\text{max}}$ where the bifurcations take place; thus the dynamic in (6.37) is bi-stable for

$$\theta_{\text{min}} \leq \theta \leq \theta_{\text{max}}$$

To simulate (6.37) we need to discretize the system,

$$z_{n+1} = z_n + \delta_t \dot{z} \tag{6.38}$$

$$= z_n + \delta_t\big[-z_n^3 + 3z_n + \theta\big] + \omega_n \tag{6.39}$$

for time step δ_t and for random fluctuations ω_n.

(c) Let us experimentally validate fluctuation amplification near a fold. We will use a tiny noise variance to prevent any catastrophic state transition.

Let $\delta_t = 0.01$ and noise variance $\text{var}(\omega_n) = \sigma_\omega^2 = 10^{-6}$

For one-hundred values of θ between θ_{min} and θ_{max}:

- Initialize $z = \bar{z}_1(\theta)$, the smaller of the two fixed points
- Simulate 1000 time steps of (6.39)
- We want to give the process time to settle down, so compute the variance only from the second half of the data points:

$$\sigma_z^2(\theta) = \text{var}\big(z_{500}, \ldots, z_{1000}\big)$$

If we let the degree of amplification be given by

$$\kappa(\theta) = \frac{\sigma_z^2 \cdot \delta_t}{\sigma_\omega^2}, \tag{6.40}$$

then produce a plot of $\kappa(\theta)$ as a function of θ.

(d) Let us experimentally validate the likelihood of catastrophic state transition near a fold.

Let $\delta_t = 0.01$ and noise variance $\text{var}(\omega_n) = \sigma_\omega^2 = 10^{-2}$
For one-hundred values of θ between $\theta_{\min}$ and $\theta_{\max}$:

- Do 1000 simulations, for each one ...
  - Initialize $z = \bar{z}_1(\theta)$, the smaller of the two fixed points
  - Simulate 500 time steps of (6.39)
  - After 500 time steps, see whether z_{500} is closer to $\bar{z}_1(\theta)$ or $\bar{z}_2(\theta)$
- Let $j(\theta)$ count the number of times z_{500} was closer to $\bar{z}_2$

Produce a plot of $j(\theta)$ as a function of θ.

(e) Interpret your results.

Problem 6.10: Reading — Transitions

Part III of Scheffer's book [9] is titled *Dealing with Critical Transitions*, five chapters examining the existence and nearness of catastrophic transitions in a system.

Select one of the chapters[4] in Part III. What are the challenges that we face in dealing with such systems, and what suggestions does Scheffer have to offer?

Problem 6.11: Policy — Extractive Industries

In Example 6.1 we saw a simple population model given a finite carrying capacity:

$$\dot{n} = rn\left(1 - \frac{n}{K}\right)$$

Since the population, whether of fish or trees, is able to reproduce over time, it should be possible to harvest or extract some amount:

$$\dot{n} = rn\left(1 - \frac{n}{K}\right) - E \qquad \text{or} \qquad \dot{n} = rn\left(1 - \frac{n}{K}\right) - \bar{E}n \tag{6.41}$$

where E represents the rate of extraction, or $\bar{E}$ the extraction rate as a fraction of the population.

(a) Compare the fixed points of the two systems in (6.41).
(b) Plot the yield curve, the stable fixed point(s) of population as a function of the extraction rate E.
(c) What is the maximum sustainable (i.e., dynamically stable) extraction rate E?
(d) In general there are economic motivations to extract as much as possible. At the same time, for most natural systems there are very likely uncertainties in

[4] Reading links are available online at from the textbook reading questions page.

the values of r and K. What are the policy risks here? Keeping Section 6.4 in mind, what would be some prudent policy strategies for managing this risk?

A great deal has been written on this subject. You might wish to consider looking up Maximum sustainable yield or Ecological yield.

Problem 6.12: Policy — Catastrophic Transitions

Any catastrophic state transition is, by definition, *not* business as usual, and is therefore of concern to policy makers, or at least *should* be.

The statistically expected cost of any transition or event is given by

$$E\big[\text{Cost}\big] = \big(\text{Cost of transition}\big) \cdot \big(\text{Probability of transition}\big)$$

What are the challenges for policy makers in estimating these two quantities?

You could answer this question in general, or in the context of some specific issue (soil salination, global warming, ...).

References

1. K. Alligood, T. Sauer, and J. Yorke. *Chaos: An Introduction to Dynamical Systems*. Springer, 2000.
2. V. Arnol'd. *Catastrophe Theory*. Springer, 1992.
3. G. Deffuant and N. Gilbert (ed.s). *Viability and Resilience of Complex Systems: Concepts, Methods and Case Studies from Ecology and Society*. Springer, 2011.
4. L. Gunderson and C. Holling. *Panarchy: Understanding Transformations in Human and Natural Systems*. Island Press, 2012.
5. R. Hilborn. *Chaos and Nonlinear Dynamics: An Introduction for Scientists and Engineers*. Oxford, 2000.
6. P. Loring. The most resilient show on earth. *Ecology and Society*, 12, 2007.
7. D. Meadows, J. Randers, and D. Meadows. *Limits to Growth: The 30-Year Update*. Chelsea Green, 2004.
8. J. Rockström, W. Steffen, et al. Planetary boundaries: exploring the safe operating space for humanity. *Ecology and Society*, 14, 2009.
9. M. Scheffer. *Critical transitions in nature and society*. Princeton University Press, 2009.
10. M. Scheffer, S. Carpenter, V. Dakos, and E. van Nes. *Generic Indicators of Ecological Resilience: Inferring the Chance of a Critical Transition*. Annual Reviews, 2015.
11. D. Sornette. *Why Stock Markets Crash*. Princeton, 2004.
12. S. Strogatz. *Nonlinear Dynamics and Chaos*. Westview Press, 2014.
13. P. Turchin. *Complex Population Dynamics*. Princeton, 2003.
14. B. Walker and D. Salt. *Resilience Thinking: Sustaining Ecosystems and People in a Changing World*. Island Press, 2006.

Chapter 7
Nonlinear Dynamic Systems: Coupled

The 1998 North American Ice Storm was a succession of ice storms in Ontario, Quebec, the Canadian Maritimes, and northeastern United States. The amount of ice deposited was completely unprecedented, with widespread tree damage and an astonishing collapse of over 1000 electrical towers, leading to impassable roads and power outages to over four million people for weeks. The cities of Ottawa and Montreal were effectively shut down, and 16,000 Canadian Forces personnel were deployed, the most in over 40 years.

It is believed that many extreme weather events are associated with the Southern Oscillation, the warming or cooling of parts of the tropical eastern Pacific Ocean which affects the location and strength of the jet stream, leading to so-called El Niño/La Niña events …

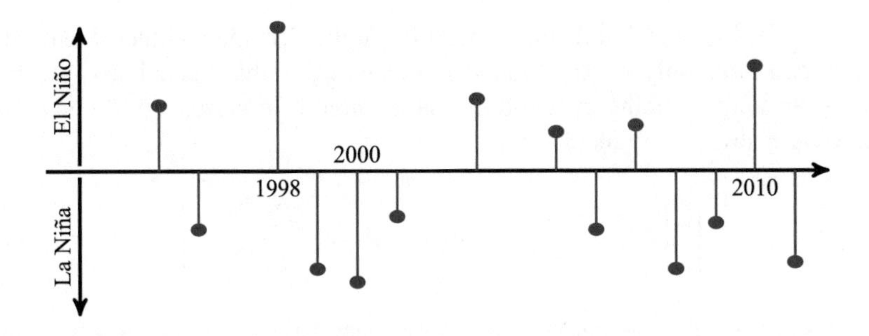

We certainly see an unusually large El Niño event in the winter of 1998, and similarly in 2010, where the Vancouver Winter Olympic Games suffered from unseasonably warm weather. However what is also striking is the irregular, but persistent, cycling back and forth. What causes such a variability, such that 1 year can be so different from the next?

P. Fieguth, *An Introduction to Complex Systems*,
DOI 10.1007/978-3-319-44606-6_7

It is actually quite common for oscillations to appear in natural and social systems — economic boom–bust cycles, ice ages, and the predator-prey dynamics of rabbits and foxes.

How do we explain persistent, stable oscillations? Spirals in linear systems (Figure 5.6) have an oscillation tendency, but are either stable, converging to a point, or unstable and diverging, neither of which represent a persistent oscillation. A *centre* is a linear system which oscillates, however it is a degenerate system, requiring precisely tuned eigenvalues, therefore essentially impossible to create in a physical system.

As we shall see, coupled *nonlinear* systems can very readily give rise to stable oscillations.

7.1 Linearization

Two-dimensional nonlinear systems are very attractive to study, since they are the highest dimensional system that can easily be sketched on a piece of paper, yet they do have multiple state elements, so it is possible to examine interactions *between* state variables.

The two-dimensional problem is characterized by a state $\underline{z} = \left[x\ y\right]^T$, such that we now have two first-order dynamically evolving state elements

$$\begin{aligned} \dot{x}(t) &= f(x, y) \\ \dot{y}(t) &= g(x, y) \end{aligned} \tag{7.1}$$

As we saw in Chapter 5 and summarized in Figure 7.1, multi-dimensional linear systems can have only a very limited number of possible behaviours, therefore most of our understanding of two-dimensional nonlinear systems will come via a *linearization* about every fixed point:

$$\begin{bmatrix} \dot{x} \\ \dot{y} \end{bmatrix} \simeq A \begin{bmatrix} x - \bar{x} \\ y - \bar{y} \end{bmatrix} \quad \text{where} \quad A = \begin{bmatrix} \partial f/\partial x & \partial f/\partial y \\ \partial g/\partial x & \partial g/\partial y \end{bmatrix} \tag{7.2}$$

A is known as a *Jacobian* matrix, and locally characterizes the system behaviour in the vicinity of $(\bar{x}, \bar{y})$.

The challenge, then, is how to associate a given matrix A with one of the ten types of dynamics shown in Figure 7.1. In principle we already know the answer, in that Figure 7.1 shows us how each behaviour relates to the system eigenvalues, and we can express those behaviours graphically, as shown in Figure 7.2 a. The problem is that, in practice, the eigenvalues are relatively complicated and inconvenient to compute, at least by hand.

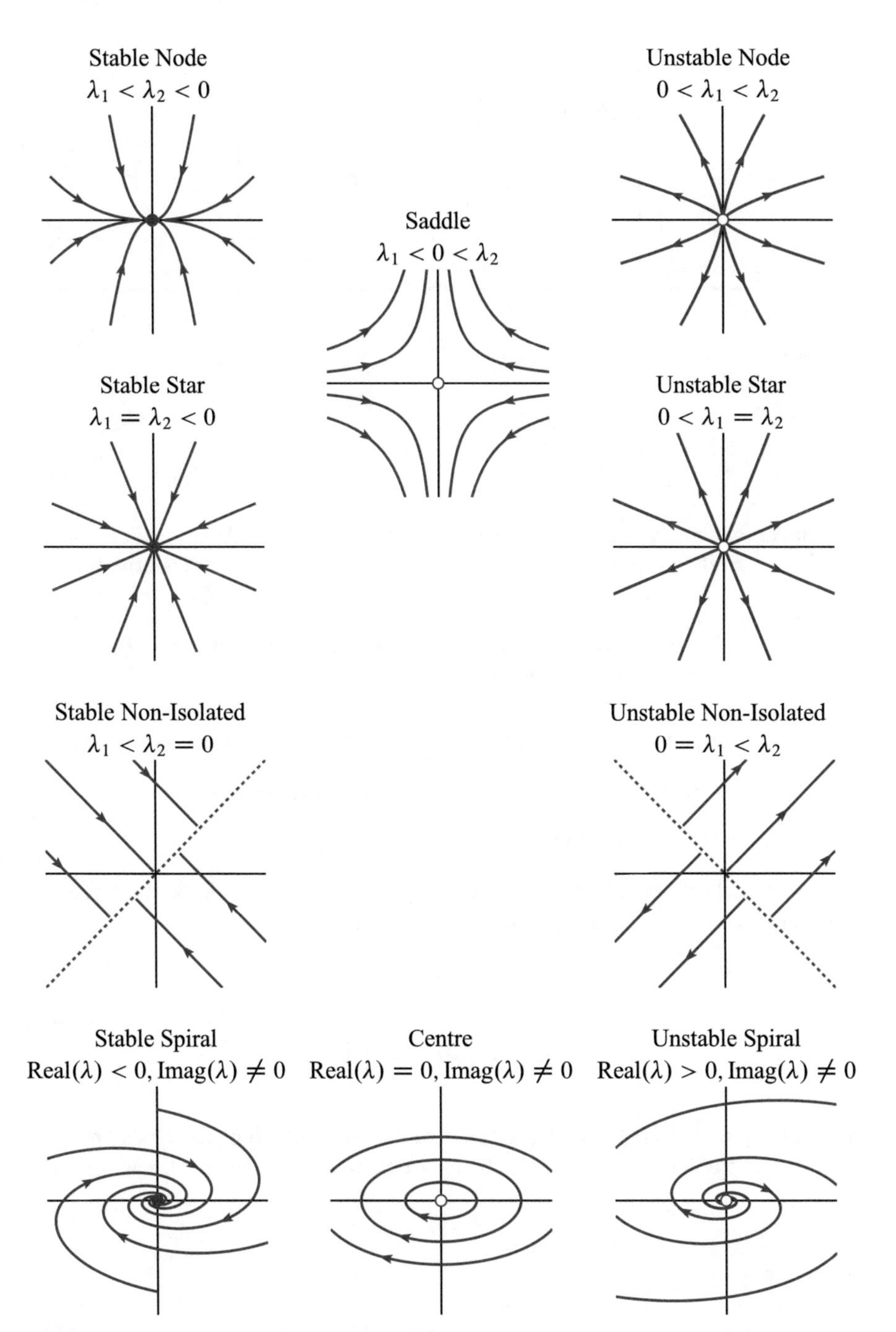

Fig. 7.1 The ten basic behaviours of linear systems: the behaviours of two-dimensional linear systems, shown here as phase diagrams, are characterized by their eigenvalues and form the basis of understanding *non*linear systems.

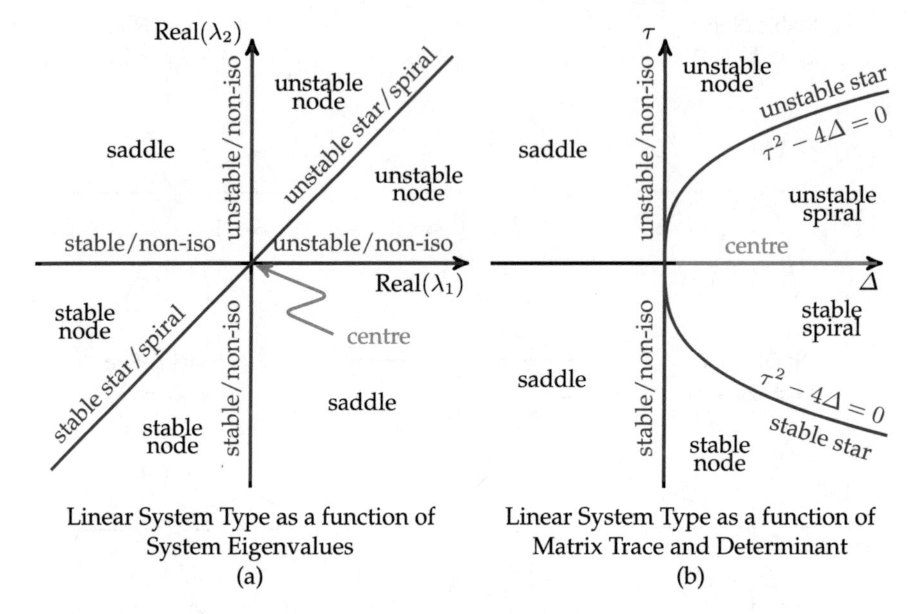

Fig. 7.2 Overview of behaviour classification: we can identify a linear system based on its eigenvalues (**a**), or based on the trace and determinant (**b**).

A much simpler alternative is to consider the trace and determinant,

$$\begin{aligned}\text{Trace}(A) &= \text{tr}(A) = \text{sum of diagonal elements of } A \\ \text{Determinant}(A) &= \det(A)\end{aligned} \tag{7.3}$$

both of which are simple to compute. Although it is not obvious, the trace and determinant are closely related to the matrix eigenvalues,

$$\begin{aligned}\tau(A) &= \text{tr}(A) = \sum_i \lambda_i \\ \Delta(A) &= \det(A) = \prod_i \lambda_i\end{aligned} \tag{7.4}$$

The corresponding graphical dependence of the ten linear dynamic types on τ and Δ is shown graphically in Figure 7.2b. In particular, for the degenerate case of a star

$$\lambda_1 = \lambda_2 = \lambda, \tag{7.5}$$

such that

$$\tau = 2\lambda \quad \Delta = \lambda^2 \quad \longrightarrow \quad \tau^2 - 4\Delta = 0 \tag{7.6}$$

which explains the presence of the parabola in Figure 7.2b. If the eigenvalues are complex, then they must appear as a conjugate pair

$$\lambda_1 = a + ib \quad \lambda_2 = a - ib, \tag{7.7}$$

in which case

$$\tau = 2a \quad \Delta = a^2 + b^2 \quad \longrightarrow \quad \tau^2 - 4\Delta < 0 \tag{7.8}$$

meaning that the parabola $\tau^2 - 4\Delta = 0$ separates the spirals (having complex eigenvalues) from the nodes (having real eigenvalues), leaving us with a simple test to distinguish the five fundamental non-degenerate behaviours:

$$\begin{array}{llll}
\Delta > 0 & \tau > 0 & \tau^2 - 4\Delta > 0 & \longrightarrow \text{ Unstable Node} \\
\Delta > 0 & \tau > 0 & \tau^2 - 4\Delta < 0 & \longrightarrow \text{ Unstable Spiral} \\
\Delta > 0 & \tau < 0 & \tau^2 - 4\Delta < 0 & \longrightarrow \text{ Stable Spiral} \\
\Delta > 0 & \tau < 0 & \tau^2 - 4\Delta > 0 & \longrightarrow \text{ Stable Node} \\
\Delta < 0 & & & \longrightarrow \text{ Saddle}
\end{array}$$

The reader is encouraged to select a few eigenvalue pairs, based on Figure 7.1, and to validate the partitioning in Figure 7.2b.

To be sure, it is not only the basic behaviours of Figure 7.1 which can be inferred from the linearized system, rather it is *any* attribute of a linear system, such as its resilience (5.51), its non-normality (Section 5.6), or its reactivity to perturbations (5.52).

7.2 2D Nonlinear Systems

For a given two-dimensional nonlinear system, we begin to characterize it by understanding its fixed points, linearized dynamics, and basins of attraction:

The Fixed Points of a joint system are those points where *all* state elements are fixed, unchanging over time:

$$\underline{\bar{z}} = \begin{bmatrix} \bar{x} \\ \bar{y} \end{bmatrix} \quad \text{is a fixed point if} \quad \begin{aligned} \dot{x} &= f(\bar{x}, \bar{y}) = 0 \\ \dot{y} &= g(\bar{x}, \bar{y}) = 0 \end{aligned} \tag{7.9}$$

The Linearized Dynamics are determined by computing the Jacobian, (7.2), from which the linearized behaviour can be categorized on the basis of Figures 7.1 and 7.2. As was discussed in Section 6.2 the linearization does, of course, introduce some degree of approximation to the original nonlinear system. Therefore if the linearized system type is degenerate (star, non-isolated, centre), there is more than one possible behaviour of the corresponding nonlinear system.

A Basin of Attraction describes that subset of the state space, such that initializing the dynamics anywhere within the basin leads to convergence to a single stable fixed point. Thus all points $\underline{z}$ in basin $\mathcal{B}_i$ converge to stable fixed point $\underline{\bar{z}}_i$:

$$\mathcal{B}_i = \left\{ \underline{z}(0) \quad \text{such that} \quad \lim_{t \longrightarrow \infty} \underline{z}(t) = \underline{\bar{z}}_i \right\} \tag{7.10}$$

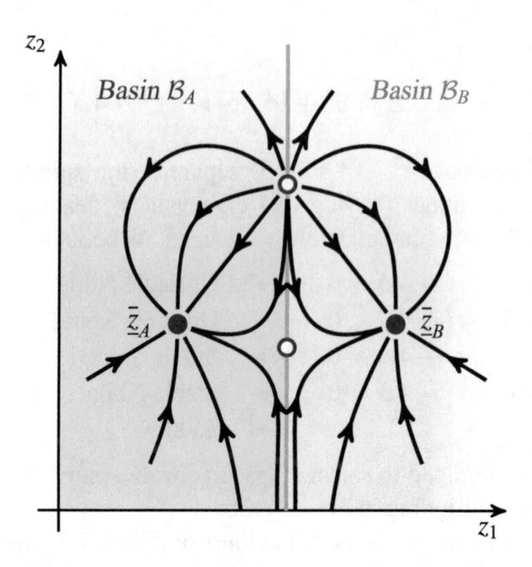

Fig. 7.3 Basins of attraction: in Figure 4.1 we saw a two-dimensional nonlinear dynamic, reproduced here. Each of the stable points has associated with it a *basin*, a subset of the domain that ends up at that stable point. The basin boundary is the vertical grey line.

Examples of basins are shown in Figure 7.3 and in Examples 7.1 and 7.2.

A Limit Cycle is a new phenomenon, one which is not seen in linear systems or in one-dimensional nonlinear systems. A limit cycle is a periodic orbit $\mathcal{C}$ in state space, such that if the state is on the cycle it remains there indefinitely:

$$\underline{z}(0) \in \mathcal{C} \quad \longrightarrow \quad \underline{z}(t) \in \mathcal{C} \text{ for all } t,$$

very much analogous to a fixed point. As shown in Figure 7.4, it is possible for a limit cycle to be stable or unstable, depending on whether a perturbation of a state on the cycle leads to convergence or divergence.

It is important to recognize the differences between limit cycles and the centres/spirals of linear systems in Figures 5.6 and 7.1. A spiral always converges to or from a *point*, whereas a limit cycle converges to or from an *orbit*. A centre gives the appearance of a limit cycle, however a centre is a degenerate dynamic and consists of an infinity of concentric elliptical cycles, whereas a limit cycle is a *single* cycle, non-degenerate, and not necessarily elliptic. Indeed, although every limit cycle is also associated with a fixed point inside of it,[1] the linearized dynamics associated with that fixed point are normally *not* a centre.

A significant challenge in studying limit cycles is detecting them. Whereas a fixed point can be determined based on a mathematical test at a *point*, the behaviour of a limit cycle is necessarily distributed over an orbit:

Explicitly: we need to somehow hypothesize an orbit $\mathcal{C}$, and then show that the system dynamics stay on the orbit,

[1] With the exception of discontinuous nonlinear systems, as shown in Example 7.5.

Example 7.1: Nonlinear Competing Dynamics

Suppose we have a meadow with some number of rabbits x and goats y. Both species reproduce and die over time, and so obey dynamics:

$$\begin{aligned} \text{Rabbits} \quad \dot{x}(t) &= f(x,y) = \big(3000 - x - 4y\big)\,x \\ \text{Goats} \quad \dot{y}(t) &= g(x,y) = \big(1000 - x/2 - y\big)\,y \end{aligned} \tag{7.11}$$

The rabbits multiply more quickly, however goats are larger and eat more food. The meadow is limited in size, so both animal species are limited in number. We can solve for the fixed points by inspection:

$$(0,0) \quad (3000,0) \quad (0,1000) \quad (1000,500)$$

The dynamics of the linearized system are computed by finding the Jacobian matrix:

$$A = \begin{bmatrix} \frac{\partial f}{\partial x} & \frac{\partial f}{\partial y} \\ \frac{\partial g}{\partial x} & \frac{\partial g}{\partial y} \end{bmatrix} = \begin{bmatrix} 3000 - 2x - 4y & -4x \\ -y/2 & 1000 - x/2 - 2y \end{bmatrix} \tag{7.12}$$

We now need to evaluate A at each of the fixed points, find its trace and determinant, and then to look up the associated behaviour in Figure 7.2:

Fixed point	A	$\tau(A)$	$\Delta(A)$	Linearized behaviour
$(0,0)$	$\begin{bmatrix} 3000 & 0 \\ 0 & 1000 \end{bmatrix}$	4000	$3.0 \cdot 10^6$	Unstable node
$(3000,0)$	$\begin{bmatrix} -3000 & -12000 \\ 0 & -500 \end{bmatrix}$	-3500	$1.5 \cdot 10^6$	Stable node
$(0,1000)$	$\begin{bmatrix} -1000 & 0 \\ -500 & -1000 \end{bmatrix}$	-2000	$1.0 \cdot 10^6$	Stable node
$(1000,500)$	$\begin{bmatrix} -1000 & -4000 \\ -250 & -500 \end{bmatrix}$	-1500	$-0.5 \cdot 10^6$	Saddle

All that is left is to sketch a phase plot, essentially by connecting the dots:

Example continues ...

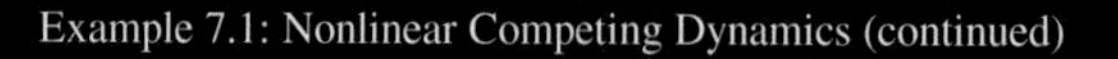

Example 7.1: Nonlinear Competing Dynamics (continued)

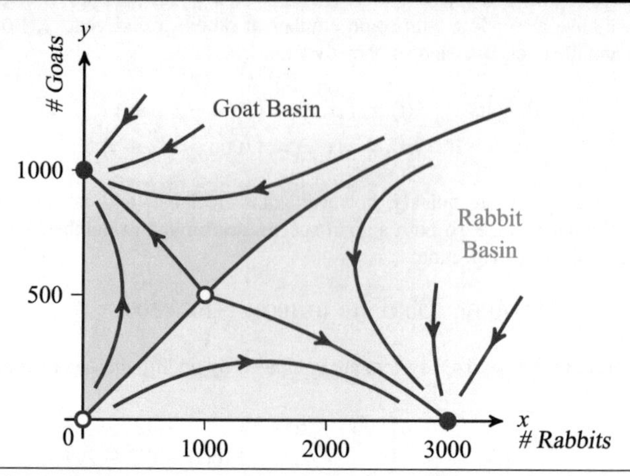

Further Reading: Related Predator-Prey models are discussed in nearly every nonlinear dynamics text; see [12].
See also Problem 7.10, Problem 7.15, and Lotka-Volterra Equation in *Wikipedia*.

Example 7.2: Basins for Root Finding

In continuous time, the basin associated with each stable fixed point is a single, connected region surrounding the fixed point. Not too surprising.

In contrast, in discrete time

$$\underline{z}(t+1) = f\big(\underline{z}(t)\big)$$

the state experiences a jump, possibly a big jump, from one time step to the next, so a given basin may very well consist of unconnected regions.

Discrete-time nonlinear systems are not just abstract and theoretical; there are well-known discrete time nonlinear dynamics, one of the most familiar being Newton's method,

$$x(t+1) = f\big(x(t)\big) = x(t) - \frac{q\big(x(t)\big)}{q'\big(x(t)\big)} \quad \text{where} \quad q' = \frac{dq}{dx}$$

an iterative approach to finding the roots of $q()$. Suppose we consider a simple polynomial $q(x) = x^6 + 1$, meaning that we wish to find the six roots of $x^6 + 1 = 0$. The roots are complex, so we have a two-dimensional state

$$z(t) = \begin{bmatrix} \text{real}\big(x(t)\big) \\ \text{imag}\big(x(t)\big) \end{bmatrix}$$

Example continues ...

Example 7.2: Basins for Root Finding (continued)

There are six fixed points, the six roots of $q()$, thus there are six basins:

Amazingly, the basins of attraction are *fractal*, for as basic an idea as finding the roots of a simple polynomial. *Now* you know why initializing Newton's method to find a specific root is not so easy!

Further Reading: Newton's method is an example of an iterated nonlinear function; these often produce beautiful, quite fascinating images. See Problem 7.13, [12] and Newton's Method, Attractor, and Fractal in *Wikipedia*.

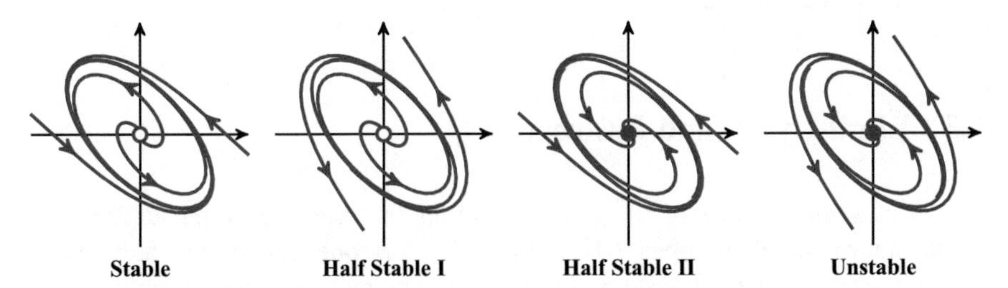

Fig. 7.4 Stability of limit cycles: there are four types of limit cycle behaviour, depending on whether the cycle attracts or repels, on the inside and on the outside. These phase plots are based on numerical simulations of actual limit cycles, using the technique of Example 7.4.

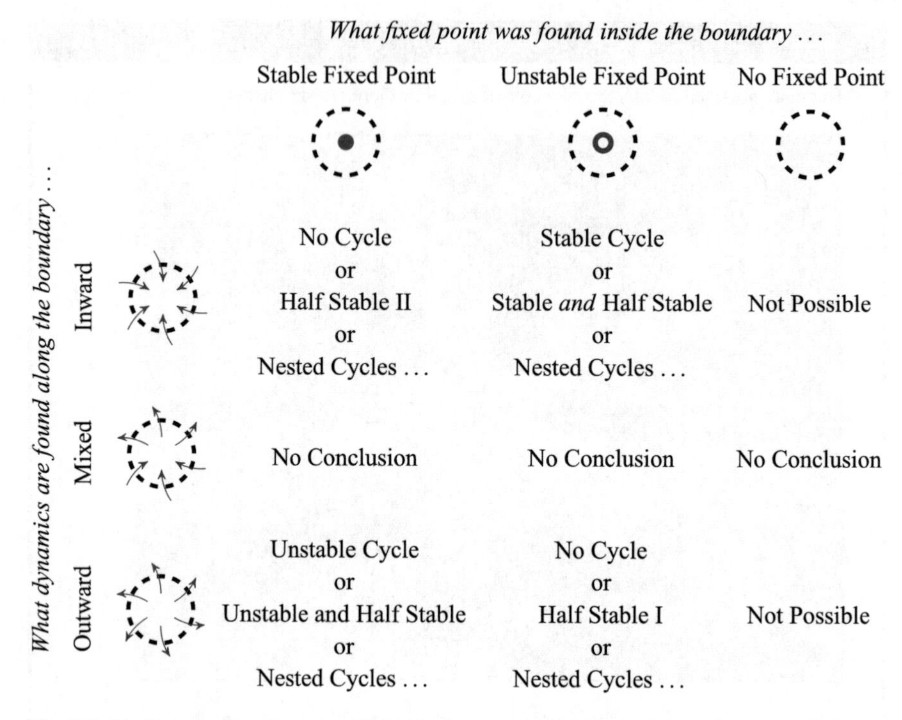

Fig. 7.5 Limit cycle detection: given the behaviour *around* a boundary, left, and the type of fixed point *within* the boundary, top, what can we say about the presence or absence of a limit cycle within the domain? In those cases where a limit cycle is possible there could be a single limit cycle or several nested concentrically.

Implicitly: we select some path enclosing an hypothesized limit cycle, and study the dynamics along the path. The basic possibilities and corresponding inferences are illustrated in Figure 7.5.

7.3 Limit Cycles and Bifurcations

So we now know some of the basic properties of limit cycles, but under what circumstances do they appear? Example 7.4 offers one illustration, however in ecological/environmental systems limit cycles are more commonly induced via systems with bifurcations and differing dynamic rates.

Figure 7.6a shows the dynamics of food consumption F as a function of population P, a common double–saddle system which we saw in Chapter 6. The diagram shows no dynamics for P, since P is treated as an external parameter, but of course a population needs food to survive, so there most definitely *will* be dynamics for P, giving us the two–dimensional nonlinear dynamic system shown in Figure 7.6b. This latter system is essentially a predator–prey model,

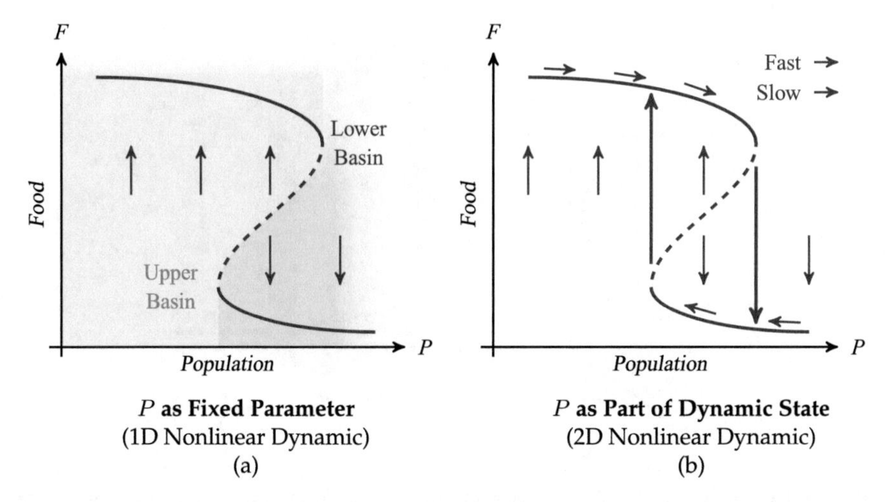

Fig. 7.6 Parameter or state: in systems having a mix of fast and slow dynamics, we might model the slow dynamic as a fixed parameter (left) or as part of the dynamics (right).

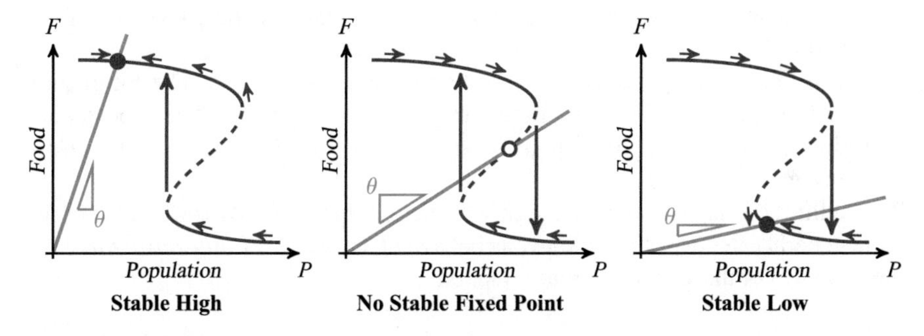

Fig. 7.7 Creating a limit cycle: suppose, as discussed in Figure 7.6b, we treat the population P as dynamic and part of the system state. Let parameter θ represent the food need per individual; the green line therefore depicts the possible fixed points in the F, P domain. For θ large (left) and small (right) we have fixed points in the undergrazed and overgrazed domains, respectively. However for intermediate θ there is no stable state, since the only fixed point in F, P is unstable. Therefore the middle system will cycle, endlessly — a limit cycle.

Prey/food:	Fast food production dynamics,
Predator/consumer:	Slow birth/death population dynamics.

How this resulting two-state system actually behaves depends on where the fixed point ends up in (F, P) space. The required rate of food production $F = \theta P$ will necessarily be proportional to population P, the question is on the degree of proportionality θ, for which several possibilities are sketched in Figure 7.7.

Remarkably, for intermediate values of θ there is no stable fixed point, and the system indefinitely cycles: a limit cycle. Most simple oscillation and limit cycle phenomena are induced in precisely this way — the interaction of two dynamics,

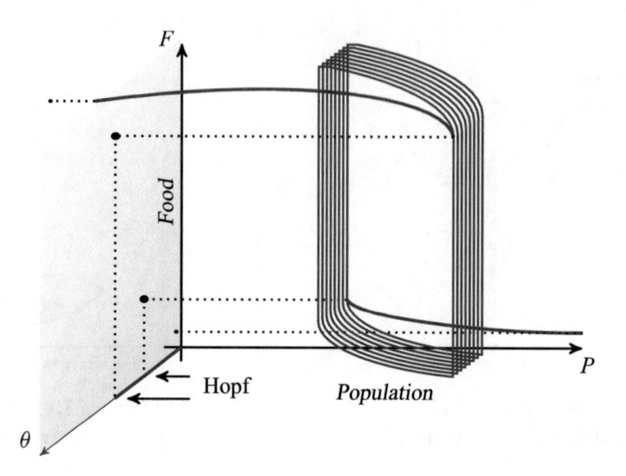

Fig. 7.8 Hopf bifurcation: a Hopf bifurcation involves the sudden appearance or disappearance of a limit cycle. As parameter θ is changed, the system of Figure 7.7 has two such transitions, as shown.

one fast and one slow, where the fixed point is unstable, preventing the system from settling down.

Recall that θ is a parameter for the two–state system (F, P). The bifurcation plot for this system is shown in three dimensions in Figure 7.8. The spontaneous transition from a stable fixed point to/from a limit cycle is known as a *Hopf bifurcation*.

The limit cycling of coupled nonlinear dynamic equations is a special case of the more general category of pseudo-cyclic behaviour known as *strange attractors*. The most famous of these is the Lorenz equation

$$\dot{x}(t) = \sigma\Big(y(t) - x(t)\Big) \quad \dot{y}(t) = x(t)\Big(\rho - z(t)\Big) - y(t) \quad \dot{z}(t) = x(t)y(t) - \beta z(t) \tag{7.13}$$

which was derived as a model of atmospheric convection, a topic to which we shall return in the context of the Navier-Stokes PDE in Chapter 8. The Lorenz equations became famous because of their irregular near-random behaviour, despite being a deterministic differential equation, and their unusually fine dependence on initial conditions, giving rise to the *Butterfly effect* already discussed in Chapter 6.

It is stochastic behaviour, discussed in Section 4.2, together with the initialization sensitivity and chaotic behaviour seen here, which fundamentally limit the predictability and forecasting of nonlinear systems.

7.4 Control and Stabilization

In Section 5.5 we briefly examined the question of system *control*. Control theory is a vast field of study describing the feedback strategies or methodologies required to satisfy or optimize a given objective, itself possibly dynamic, such as the control

Example 7.3: Resilience II — Nonlinear Systems

In our earlier discussion of resilience, in Example 5.3, we encountered the eigenvalue-based definition (5.51) of resilience

$$Resilience(A) = \min_i \left\{ -\text{Real}\big(\lambda_i(A)\big) \right\} \tag{7.14}$$

such that the resilience of a system is measured by the speed with which it returns to the stable fixed point.

However since (7.14) was developed in the context of linear systems, which only have *one* fixed point, (7.14) cannot make reference to *other* fixed points or catastrophic state transitions.

Looking at the phase plots in Figure 7.3 or Example 7.1, or the catastrophic state transitions in Chapter 6, it is clear that resilience must somehow be related to size or shape of a basin. Furthermore not all stable points are "good" (e.g., the snowball earth in Figure 6.19). We are interested in the resilience of a *particular*, *good* fixed point, normally the current state $\underline{z}$ of the system or a desired target.

It would be relatively straightforward to measure the resilience of $\underline{z}$ as the distance to the nearest unstable fixed point

$$Resilience(\underline{z}) = \min_i \left\| \underline{z}, \underline{z}_i \right\| \quad \text{over all } \underline{z}_i \in \mathcal{Z}_U \tag{7.15}$$

where $\mathcal{Z}_U$ is the set of all unstable fixed points. However we're not likely to actually hit an unstable point; it is the distance to the basin boundary that really matters, since crossing a boundary leads to a catastrophic transition:

$$Resilience(\underline{z}) = \min_{\hat{\underline{z}}} \left\| \underline{z}, \hat{\underline{z}} \right\| \quad \text{over all } \hat{\underline{z}} \notin \mathcal{B} \text{ where } \underline{z} \in \mathcal{B} \tag{7.16}$$

which is starting to look a bit more difficult to evaluate.

Our challenges are not yet done. What is this distance $\left\| \underline{z}, \hat{\underline{z}} \right\|$? When we speak of distances, we normally think of a Euclidean distance

$$\underline{z} = \begin{bmatrix} x \\ y \end{bmatrix} \qquad \left\| \underline{z}, \hat{\underline{z}} \right\|_{\text{Euclidean}} = \left\{ (x - \hat{x})^2 + (y - \hat{y})^2 \right\}^{1/2} \tag{7.17}$$

However there is really no reason to think that x and y, the elements of our state $\underline{z}$, should be measured in the same way. If $\underline{z}$ represents a global climate state then is x, the atmospheric CO_2, measured in percent or in parts-per-million? Is the arctic ice extent y measured in km or degrees latitude? The definition of distances in abstract spaces is actually a widely studied question in the topic of *pattern recognition*.

Finally, to what sort of perturbation are we resilient? Should resilience be only a function of the system state $\underline{z}$, or should it also take into account the system parameters $\underline{\theta}$? Although we do not have a dynamic model for the parameters themselves, if the system is subject to outside influences then sensitivity to parameter changes might need to be taken into account.

Further Reading: The subject of resilience is further discussed in Example 10.4. See also [3, 5, 9–11, 13] and the earlier discussion in Examples 5.3.

Example 7.4: Creating and Finding a Limit Cycle

Thinking about motion around the origin, a spiralling motion is most easily articulated in polar coordinates:

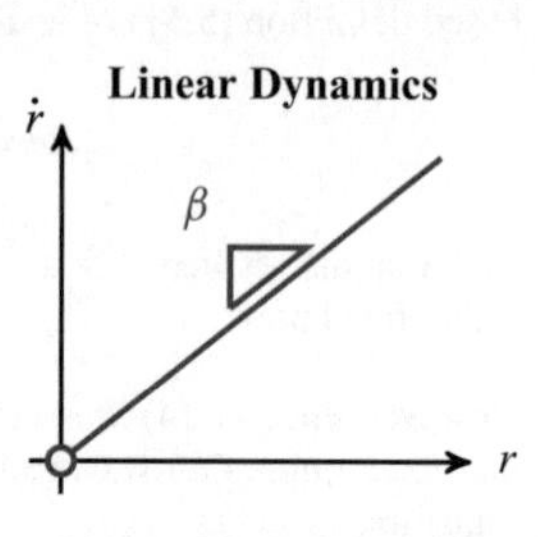

$$\dot{\theta} = \alpha \qquad \dot{r} = \beta r$$

where we have constant angular rotation α and an exponentially growing or shrinking radius. To be confident that we have created a spiral, let's convert to Cartesian coordinates,

$$\dot{x} = \dot{r}\cos(\theta) - r\sin(\theta)\dot{\theta} = \beta x - \alpha y$$

$$\dot{y} = \dot{r}\sin(\theta) + r\cos(\theta)\dot{\theta} = \beta y + \alpha x$$

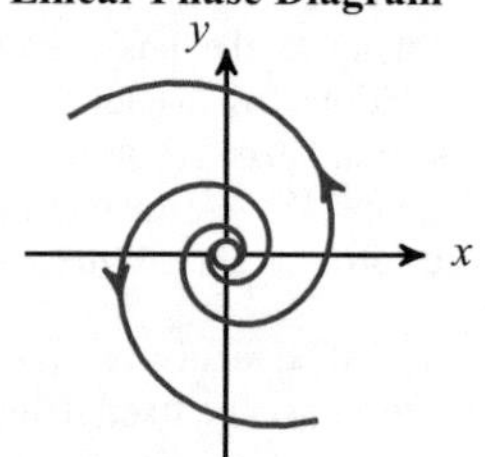

so the dynamics are indeed linear, with a system dynamics matrix

$$A = \begin{bmatrix} \beta & -\alpha \\ \alpha & \beta \end{bmatrix} \quad \longrightarrow \quad \Delta = \beta^2 + \alpha^2,\ \tau = 2\beta$$

Referring back to Figure 7.2 we see indeed that the (τ, Δ) pair corresponds to a spiral, stable if $\beta < 0$, and unstable if $\beta > 0$.

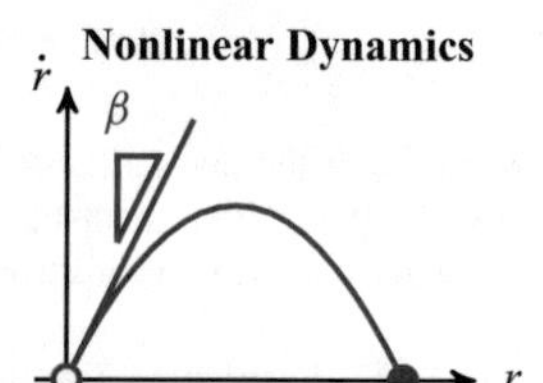

Now let's start with the same idea in polar coordinates, but modified to create a new fixed point in the radial dimension:

$$\dot{\theta} = \alpha \qquad \dot{r} = \beta r(1 - r)$$

We don't actually need to convert to Cartesian coordinates, but as a quick check, we find

$$\dot{x} = \dot{r}\cos(\theta) - r\sin(\theta)\dot{\theta} = \beta x\big(1 - \sqrt{x^2 + y^2}\big) - \alpha y$$

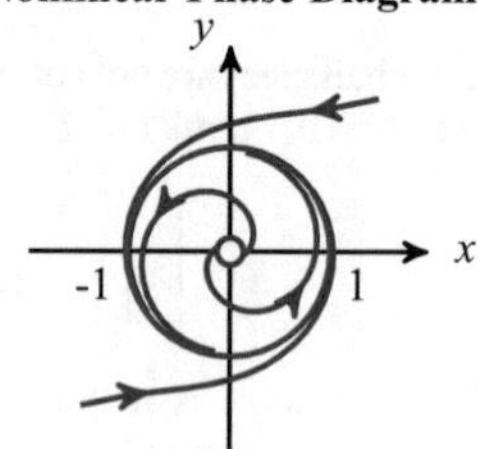

which is clearly nonlinear.

Let's suppose that we now hypothesize the presence of a limit cycle at $r = 1$; we find that

$$r = 1 \quad \longrightarrow \quad \dot{\theta} = 1,\ \dot{r} = 0$$

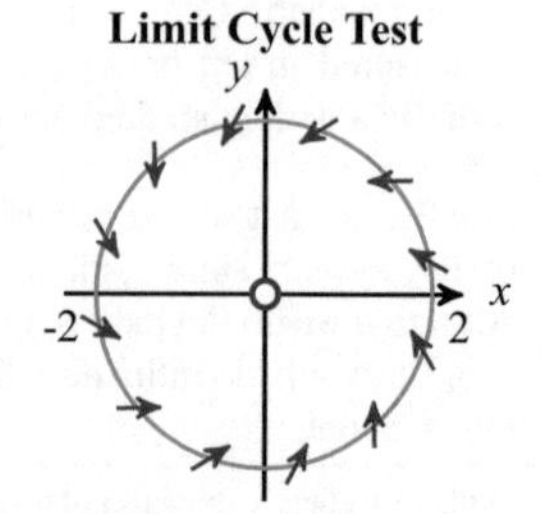

implying indefinite motion on the unit circle, a limit cycle. Instead, we could have hypothesized that there was a limit cycle of some shape and form around the origin. The linearized system at the origin shows an unstable spiral, and testing the dynamics on the green ring at $r = 2$ shows only inward dynamics, so there must be a stable limit cycle somewhere in between.

Example 7.5: An Alarm Bell as Limit Cycle

Any battery-driven ringing alarm or buzzer, whether electronic or mechanical, creates an oscillating signal (ding, ding, ding) from a constant input (the direct current from a battery), therefore we *know* the dynamics must be nonlinear, since linear dynamics cannot spontaneously create oscillations. The analysis of a nonlinear electronic circuit is rather difficult, so let's look at an old-fashioned mechanical alarm bell.

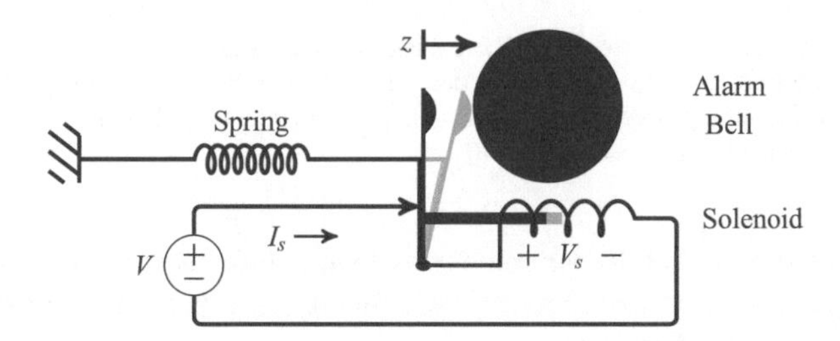

The nonlinearity in the circuit comes from the switching action of the clapper: the solenoid voltage V_s switches discontinuously as a function of clapper position z, leading to an oscillating behaviour. The clapper arm is spring-like, so although its rest position is $z = 0$, it is able to bend slightly to negative z.

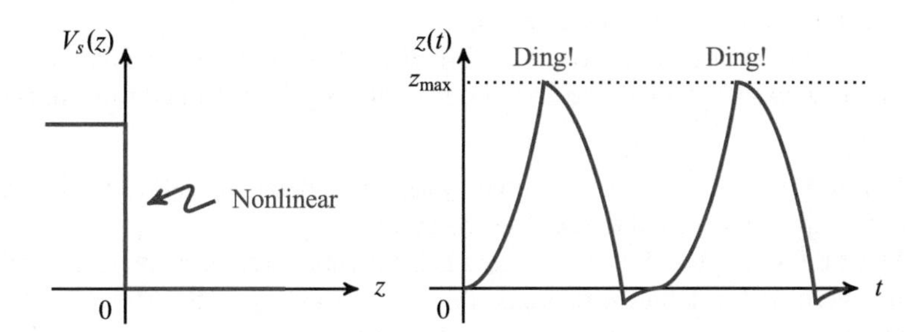

The plotted $z(t)$ is only a sketch; the actual behaviour of $z(t)$ would clearly depend upon details (spring constants etc.) of the electro–mechanical system.

Of course what actually interests us here is to understand the actual limit cycle associated with the mechanism.

The state of the system is described by two coupled variables, mechanical position z and electrical current I_s, plotted here. The voltage V_s is a *response*, not part of the system state.

Note that there is *no* fixed point inside the limit cycle because of the discontinuity in $V_s(z)$.

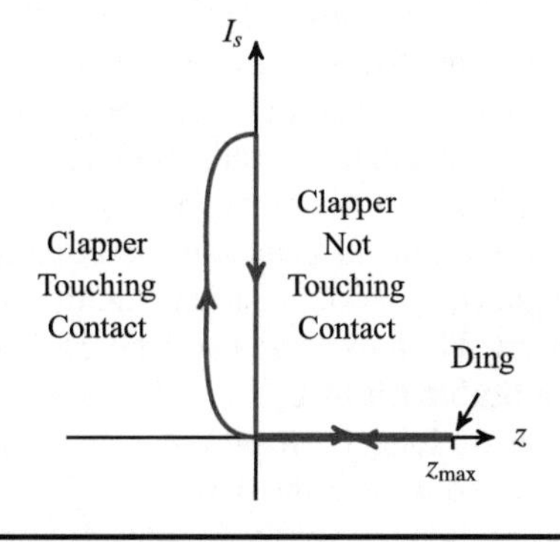

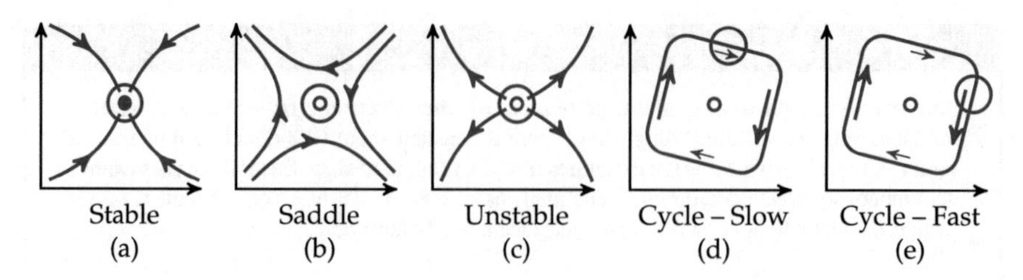

Fig. 7.9 System stabilization: we see here five phase diagrams showing examples of systems to be stabilized at a desired target z_s, indicated by the red circle. The challenge of the feedback control required for stabilization proceeds from trivial (**a**), through small perturbations (**b**), (**c**), to explicit opposing of the dynamics (**d**), (**e**).

of a moving car or rocket. In our context here we will focus more narrowly on how dynamic system properties influence the difficulty of stabilizing a given dynamic system at some desired setpoint.

It turns out that it is a somewhat naïve notion to wish to freeze a system at some point in time. That is, just because a system arrived at some state on its own does *not* make it straightforward to *hold* it there. For example, an economy subject to cycling (Example 7.6) may, at some point in its cycle, exhibit simultaneous high wages and low unemployment; to what extent can a government enact policy in order to hold the system there, as a social objective?

A simple taxonomy of stabilization difficulty is shown in Figure 7.9, where the difficulty is *both* a function of the underlying system dynamics and the desired point of stabilization z_s:

At a Stable Fixed Point the dynamic system is inherently stable, requiring no control input of any kind to stabilize the system.

At a Saddle Point the dynamic system has a combination of convergent (stable) and divergent (unstable) directions. A control strategy is therefore needed to stabilize the divergent directions, which is a slightly simpler problem than at an unstable fixed point, for which *all* directions are divergent.

At an Unstable Fixed Point all directions are divergent, therefore the feedback control needs to be sensitive to and acting upon all of these directions. Nevertheless the control problem remains relatively straightforward, since *at* the fixed point the divergent force is *zero*: as has been pointed out before (Figure 6.7), an unstable fixed point is *not* like a magnet whose repulsive force is increased as the pole is approached, rather the repulsive force shrinks to zero at the unstable fixed point, and grows *away* from the fixed point.

As a result, in principle the control strategy does *not* require large feedback signals, however constant monitoring and a sufficiently fast response time is needed to prevent any significant departure from the fixed point.

On a Stable Limit Cycle, in contrast to the preceding cases, the desired point z_s is no longer a fixed point, therefore the control needs to produce a feedback signal opposing the dynamics. The response time of the feedback control is not so essential, since the limit cycle is stable and its dynamics are known. However

the control may need large inputs, since we now have to explicitly oppose the dynamics.
The magnitude of the control input needed will be a function of the dynamics being opposed. For many limit cycles which are induced by a fast–slow combination (Section 7.3), stabilizing the slow dynamics Figure 7.9d will, in general, be easier than stabilizing in the presence of fast dynamics Figure 7.9e.

The last case is the most interesting. Just because the system is observed to be in target state z at some time t_o does not imply that the system can be stabilized at that point, even though the state arrived there on its own. Particularly in economic policy, where limits cycles are known to be present, we cannot simply choose a point in time and state as an objective the maintaining of that system behaviour, since there are actually underlying dynamics present continuing to push the system and which need to be explicitly opposed.

Finally, there is one problem similar to, and possibly more difficult yet than that of Figure 7.9e. In light of Example 7.8 there may be countries finding themselves caught at an undesirable stable fixed point, such as a poverty trap. In those cases a control strategy is needed to move *away* from a stable fixed point, opposing the dynamics, to cross the basin boundary to some other, desired point.

Case Study 7: Geysers, Earthquakes, and Limit Cycles

There are many forms of cycling behaviour present in the natural world, a number of which were discussed in Case Study 6. Some of these, such as tides or ice ages, are caused by an oscillating input, a forcing function, such as the moon or a variation in the earth's orbit, which produce a varying result.

However other nonlinear systems are inherently oscillatory, even in the absence of any external forcing. There is some evidence that at least three of the hysteresis examples from the global climate discussion in Case Study 6 exhibit cycling behaviour:

Ice Ages:	Rapid warming/cooling cycles
Snowball/Hot Earth:	Ice Dynamics (fast) vs. Atmospheric CO_2 (slow)
Thermo-Haline:	Freshwater (fast) vs. Temperature (slow)

The latter two limit cycles were induced by the pairing of fast and slow dynamics, as discussed in Section 7.3. We know that ice ages are driven by orbital forcing, the Milankovich oscillations shown in Table 6.3, however these forcings have periods of 20,000 or more years, whereas the climate record also shows sub-millennial oscillations which stem from limit cycles in the climate dynamics.

Example 7.6: Limit Cycles in Economics

The *Business Cycle* is one of the most fundamental concepts in macro-economics, describing the large-scale swings observed in economic activity.

Certainly there are repeated cycles in history where people become increasingly confident, undertaking more and more leveraged investments (investing money you don't have, by borrowing); at some point the borrowed money *does* need to be paid back, thus a few people sell, and then more, leading to a cascading effect (catastrophe) whereby many people lose money and have a greatly reduced tolerance for risk. Some of the most famous and extreme events include the seventeenth century Tulip Mania, the eighteenth century South-Sea Bubble, the 1920s stock market, and the turn-of-the-century Dot-Com Bubble.

We do not need to limit our attention to such extreme events to observe cycling behaviour. The *Goodwin Model* is one example of a quantitative economic dynamic model relating unemployment and wages. In its simplest form, the model has two coupled state variables

Employment rate v Wages u

where u represents the fraction of output being returned to workers. The Goodwin Model then takes the form

$$\dot{v} = v\big(\alpha(1-u) - \beta\big) \tag{7.18}$$

$$\dot{u} = u\big(\rho v - \gamma\big) \tag{7.19}$$

where, in a nutshell, (7.18) argues that higher wages u leave less room $(1-u)$ for capital investment, leading to reduced employment. Whereas (7.19) asserts that high employment (low unemployment) leads to an upwards pressure on wages.

However (7.18), (7.19) are, yet again, in the predator-prey class of models, consisting of an unstable fixed point surrounded by a limit cycle. Thus to the extent that the Goodwin model is correct, it suggests that governments face a never-ending battle to stabilize the economic cycle, since the system can be held at the unstable fixed point (or any other point) only with constant control.

To be sure, there are many other models which have been proposed or claimed to describe social and economic cycles ...

Name of cycle	Associated context	Cycle period in years
Kitchin cycle	Inventory	3–5
Juglar cycle	Fixed investment	7–11
Kuznets swing	Infrastructure investment	15–25
Kondratiev wave	Technologically driven	45–60
Elliot wave	Investment psychology	$\ll$ 1–100

A number of these models or claimed behaviours are somewhat controversial, and are really more of an observation of past cycling behaviour than a verifiable model or theory. Nevertheless, the presence of economic cycling is very real, with significant implications in terms of employment, social unrest, policy, and mal-investment.

Further Reading: Predator-prey models are simulated in Problem 7.15. Also see Economic Bubble, Goodwin Model, Business Cycle, Elliott Wave, Kondratiev Wave in *Wikipedia*.

Example 7.7: The Tragedy of the Commons

The tragedy of the commons refers to the effect that individual self–interest leads to the overuse or abuse of shared resources.

In very many cases, the tragedy of the commons stems from the confluence of three factors, each of which examined in this text:

1. *Externalizing*, as discussed in Example 3.2: The tendency to ignore or neglect phenomena outside of our system envelope, whether that of the individual or of a country.
2. *Discounting*, as discussed in Example 9.6: The tendency to ignore the consequences or future impact of current decisions.
3. *Catastrophes*, as discussed in Section 6.3: The inclination for bi-stable nonlinear systems to change abruptly, with little warning, when a fold bifurcation is reached.

The classic explanation of the tragedy comes from *game theory*, in which simple rules are designed characterizing the interaction of two or more players. The most famous of these games is the Prisoner's Dilemma, regarding whether each of two prisoners should confess to a crime or not, a scenario which translates directly to the tragedy of the commons if we similarly consider two farmers, each of whom considers adding another cow (increasing their own profit) to the shared common. If only one farmer adds a cow, that farmer gains additional profits; if *everyone* adds a cow, the common is overgrazed and everyone makes *less* money:

	Farmer B limits (cooperates)	Farmer B increases (self–interest)
Farmer A limits (cooperates)	Farmer A — OK Farmer B — OK	Farmer A — Bad Farmer B — Great
Farmer A increases (self–interest)	Farmer A — Great Farmer B — Bad	Farmer A — Poor Farmer B — Poor

The frustrating outcome of this game is that, invariably, both farmers will act in their own self interest, since *regardless* of Farmer B's decision, in both cases Farmer A is better off (*Great* vs *OK*, *Poor* vs *Bad*) by acting in self-interest, the so-called Nash equilibrium for the game.

Certainly there are many examples, worldwide, of successful shared resources, so the tragedy of the commons is by no means inevitable. In particular, problems are best avoided when people understand their *shared* interest and when there is a measure of trust in the other player, essentially changing the rules of the game. It is therefore no surprise that some of our greatest challenges, both socially and environmentally, involve issues which *cross* borders, where the shared interest is weakest.

Example continues . . .

Example 7.7: The Tragedy of the Commons (continued)

At the same time, there are many examples of failed or failing commons, with the global commons — atmosphere, freshwater, oceans — of particular concern. Even more problematic is combining the game theory context with that of a bi-stable nonlinear system, particularly in those cases where the game theory dynamics push us towards a bifurcation.

Continuing the example of grazing cows on a common, the (oversimplified) growth dynamics of pasture grasses are such that there is

- Very limited growth when fully mature (growth phase is done),
- Very limited growth when cut short (no leaves to photosynthesize),
- Maximum growth when cut back to an intermediate length.

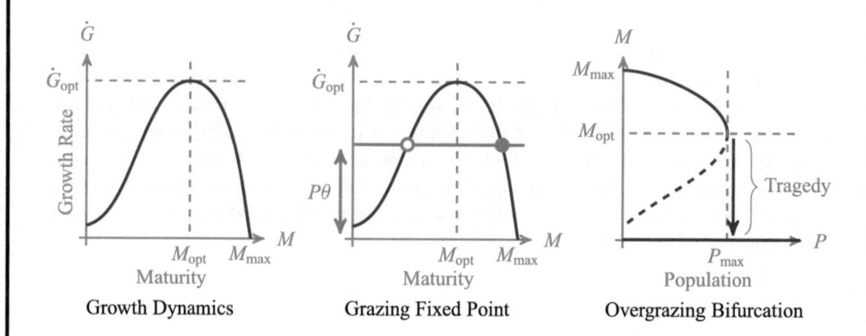

Up to a population of P_{max}, with a food consumption rate per animal of θ, increases in the population make sense, as there are corresponding increases in grass production (middle). However as the bifurcation (right) is approached the game theory dynamics become increasingly dangerous, to the point where the bifurcation is passed, consumption exceeds pasture production, and there is a state transition to an overgrazed pasture (red line) which can now support *zero* animals. The tragedy then lies explicitly in the $\dot{G}_{opt}$ loss in pasture production which could have supported P_{max} animals.

Clearly the challenge is that the *optimum* production point is *also* the bifurcation point. In a well run farm the transition catastrophe does not need to take place, of course: as overgrazing passes the bifurcation point the cows can be kept off the pasture and fed stored hay, allowing the pasture to recover, essentially reducing $P\theta$ to allow the stable fixed point (middle) to be restored.

In general, however, we observe that there are significant challenges and dangers in trying to operate a system near peak efficiency, a topic which will reappear under Complex Systems.

Further Reading: Tragedy of the commons, Prisoner's dilemma, Nash equilibrium in *Wikipedia*.
Hardin, "The Tragedy of the Commons," *Science* (162), 1968
Ostrom, *Governing the Commons,* Cambridge, 2015

Example 7.8: Societal Dynamics

A social system is terribly complex and cannot be boiled down to a simple, low-dimensional state and phase diagram. Nevertheless, similar to that which we saw in Example 7.6, there are indeed patterns in social dynamics which we can understand better from the perspective of nonlinear systems.

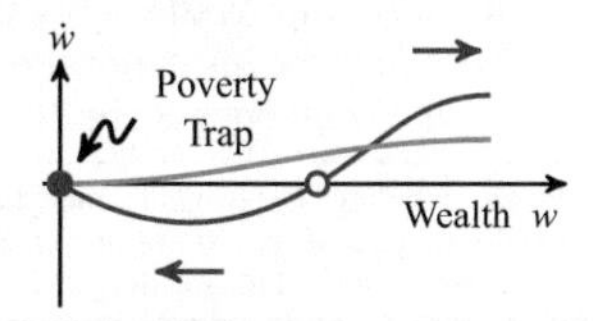

We can start with a relatively simple model of wealth dynamics. For most of the world and for most of history, the wealthy obtain and maintain their wealth through the exploitation, whether overt or implicit, of poorer people. Many poor have unmanageable debt loads, leading to greater poverty over time. The red dynamic, right, illustrates a financial system which has a stable fixed point, a poverty trap. Government policies, such as progressive taxation and tuition support, seek to flatten or raise the dynamic, such as the green line, to either weaken or completely remove the trap at the origin.

A more complex picture is offered by the *Panarchy* model, which is an empirical observation of the cycling seen throughout societies:

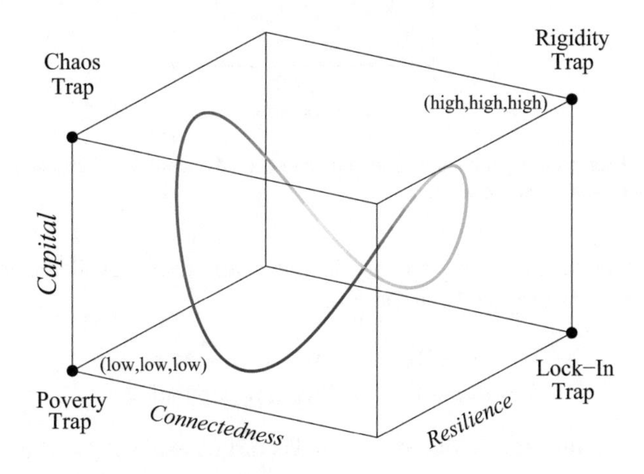

We have all of the familiar elements of nonlinear systems:

- A limit cycle, of typical societal evolution,
- Four fixed points, in which societies often become trapped,
- Attractive basins (not sketched) which lead to the four traps,

Example continues . . .

Example 7.8: Societal Dynamics (continued)

where the four fixed points, the traps, are characterized as [7]

- *Poverty Trap*: Inadequate capital resources to reorganize,
- *Rigidity Trap*: Waste of resources and lack of innovation,
- *Lock-In Trap*: Institutional resilience leads to maladaptive behaviour,
- *Chaos Trap:* Poor connections or resilience, thus sensitive to perturbation.

Further Reading: The panarchy model stems from Gunderson and Holling [5]. Discussions of societal traps are discussed at length in Loring [7], Meadows [8], Scheffer [10], and Wright [14]. Problem 8.12 looks at the question of wealth modelling and dynamics.

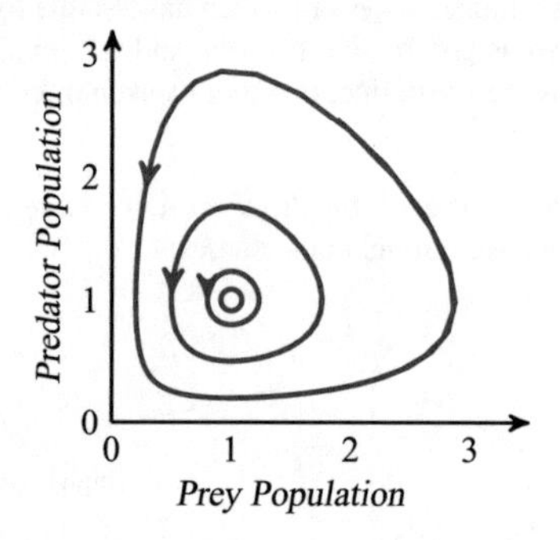

Fig. 7.10 Predator-prey: a simulation of the predator-prey dynamics of (7.20), showing the fixed point (blue) and cyclic (red) behaviour.

In environmental systems, probably the most well-known cyclic behaviour is the *predator-prey* class of models, of the form

$$\begin{aligned} \dot{x} &= (1-y)x \qquad \textit{Prey}\ \text{population}\ x \\ \dot{y} &= a(x-1)y \qquad \textit{Predator}\ \text{population}\ y \end{aligned} \tag{7.20}$$

Model (7.20) is simulated in Figure 7.10 and is further examined in Problem 7.15. It is straightforward to identify the fixed points at $(0, 0)$ and $(1, 1)$ and, following the approach in Section 7.1, by linearization we find the point at $(1, 1)$ to be a centre. The simulated behaviour is clearly cyclic, however there is no radial dynamic, rather we have a periodic cycle for each initial condition.

Building on the hysteresis discussion of Example 6.6, an interesting variation on the predator-prey model is the oscillation in lake turbidity from Scheffer [10]. It has been observed that some lakes oscillate in the spring between being clear and being murky, even though there are no significant driving oscillations in the inputs, such as pollutants (as in the case of Example 6.6). Instead, there are ecological dynamics present here on two different time scales, illustrated in Figure 7.11. We have the growth rate of algae (faster), which cause the water to lose clarity, and that of plankton (slower), which consumes the algae and restores water clarity. It is possible for the cycle to repeat several times, before being interrupted by the appearance of larger predators, such as fish.

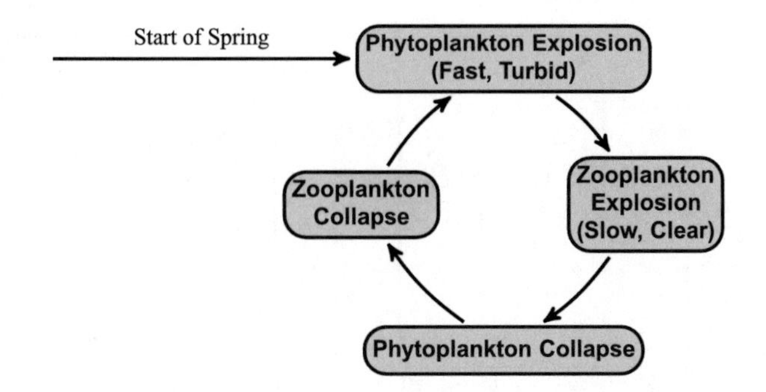

Fig. 7.11 Limit cycle in lake ecology: cold-climate lakes can experience a cyclic alternating between clear (blue states) and murky (red state), because of the different phytoplankton and zooplankton growth rates. The cycle is broken by the arrival of larger predators, such as fish. The arrows indicate state transitions over time.

Based on Scheffer, *Critical Transitions in Nature and Society* [10].

An important class of nonlinear cycling behaviour is that of *Stick-Slip* models, a phenomenon very familiar to anyone who has ever heard a door creak, tires squeal, or fingernails scratched on a blackboard. The simplified model involves a mass m being pulled through a spring at velocity v over a surface subject to friction, as follows:

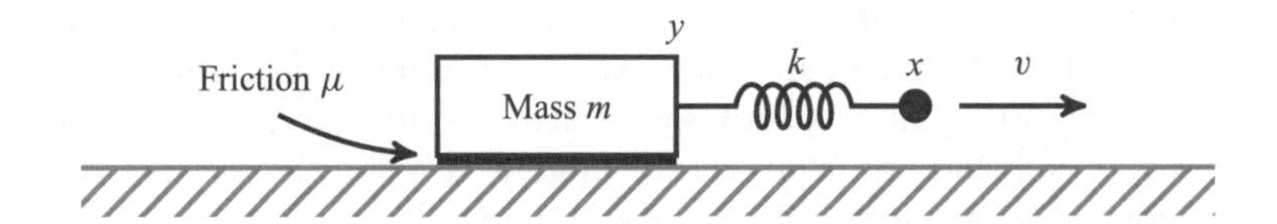

Since static (stopped) friction is normally higher than dynamic (moving) friction, the mass will stay at rest until the spring is sufficiently stretched to produce a force to overcome the static friction, at which point the mass begins to move. As the velocity of the mass accelerates past v, the spring is compressed, producing less force, causing the mass to slow down and stick again, leading to irregular motion even though the spring is being pulled at a constant velocity.

The nonlinearity in the system stems from the velocity–dependent friction term, such that the net force acting upon the mass behaves as

$$\text{Net Force}\, f = \begin{cases} \max\{0, |k(x-y)| - m\mu_s\}\,\text{sign}\big(k(x-y)\big) & \text{if } \dot{y} = 0 \\ k(x-y) - m\mu_d\,\text{sign}(\dot{y}) & \text{if } \dot{y} \neq 0 \end{cases} \tag{7.21}$$

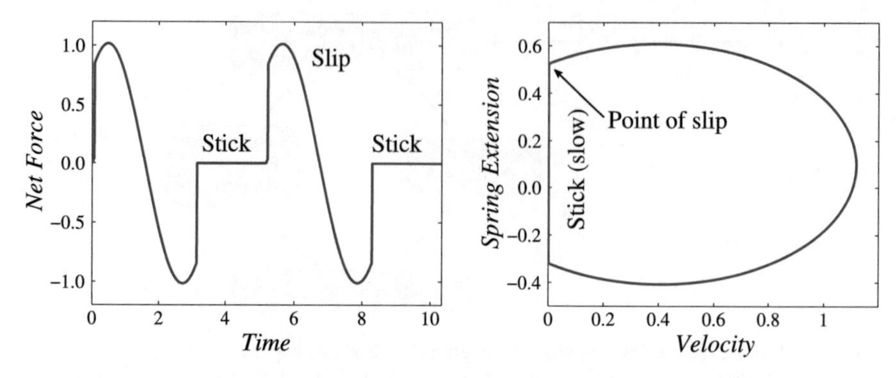

Fig. 7.12 Stick slip: the simulation of a simple stick–slip system, showing the behaviour over time (left) and the corresponding limit cycle (right).

This is a relatively simple approximation, modelling the dynamic friction $m\mu_d$ as constant, whereas in practice it may have a more complex dependence on the mass velocity $\dot{y}$.

Figure 7.12 shows a time–simulation based on the friction model of (7.21) where we can observe the corresponding limit cycle. There is actually a Hopf bifurcation present here, such that a higher velocity v or stiffer spring k prevents the cycling, which is precisely why a door creaks when opened slowly, but not at all when opened quickly. Oiling a hinge also prevents creaking, because oil sufficiently reduces the static friction to stabilize the fixed point, eliminating cycling.

One of the most significant stick-slip systems in the natural world is that of earthquakes. Continental drift pushes the crust at a relatively constant velocity, with the crust able to deform (like a spring) to a point, with very sudden "slips" (earthquakes) when the accumulated force exceeds the static friction, so to speak. However rather than a single point of friction, two tectonic plates have a sequence of contact points along a fault, so a spatially-distributed stick-slip model is required, as will be seen in Example 10.2.

The stick–slip concept is analogous to the limit cycles of Section 7.3, whereby a system is stable (sticking), but being pushed along the stable dynamic (increasing force) to the point of bifurcation and catastrophe (slip). A wonderful and familiar illustration of exactly this dynamic is that of geysers, as shown in Figure 7.13.

Suppose we have a deep shaft, Figure 7.13b, surrounded by a hot rock formation. The temperature and pressure of the water will clearly be a function of depth, but we will define the state to be the temperature T and pressure P near the bottom of the shaft.

Most intuitively, we would expect the geyser to settle into steady state, boiling continuously like a kettle. Why, then, does it erupt? From this chapter you know that the geyser system can settle down only if there is a stable fixed point; however the Geyser system has a Hopf bifurcation, thus under certain circumstances there is no stable fixed point, only a stable limit cycle (leading to periodic eruptions).

(a) Geysir, Iceland

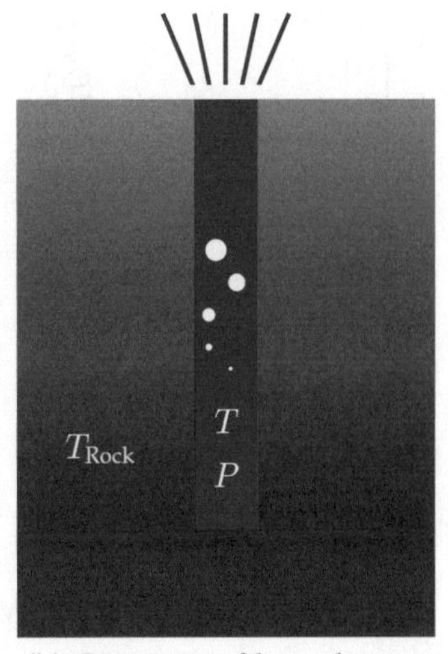

(b) Geyser profile and state

Fig. 7.13 Geysers: the Geysir geyser, left, in Iceland erupts every few minutes. The geyser is formed as a deep channel, right, into which groundwater flows and is heated to the boiling point. The associated limit cycle is shown in Figure 7.14.

The boiling temperature of water increases with increased pressure, thus the pressure of the water column implies a local boiling point, at the bottom of the column, above 100 °C. At some point this temperature is reached, and the column begins to boil; the resulting air bubbles push some of the water out of the top of the column, *reducing* the pressure, meaning that the water column is now super-heated and boils explosively. The geyser eruption can be interpreted as a catastrophic state transition, as shown in the limit cycle in Figure 7.14.

Stick-slip, earthquakes and geysers are examples of bounded systems subject to a constant forcing function. Since all natural systems are bounded, and in most cases driven by external inputs, this scenario is relatively common and important to understand. The resulting behaviour depends on what happens at the nonlinear bound:

Catastrophic State Transition: A significant state transition occurs, from which it takes *time* for the system to recover

⟶ Limit Cycle

Settling Back to Stability: No significant state transition occurs, rather the system settles slightly to back away from the bounding limit. However each settling is slightly different, leading to complex/chaotic behaviour.

⟶ Self-Organized Criticality

The latter case does not lead to cycling behaviour because there is no significant state jump from which the system needs to recover; however such systems are very interesting and will be further explored in Chapter 10.

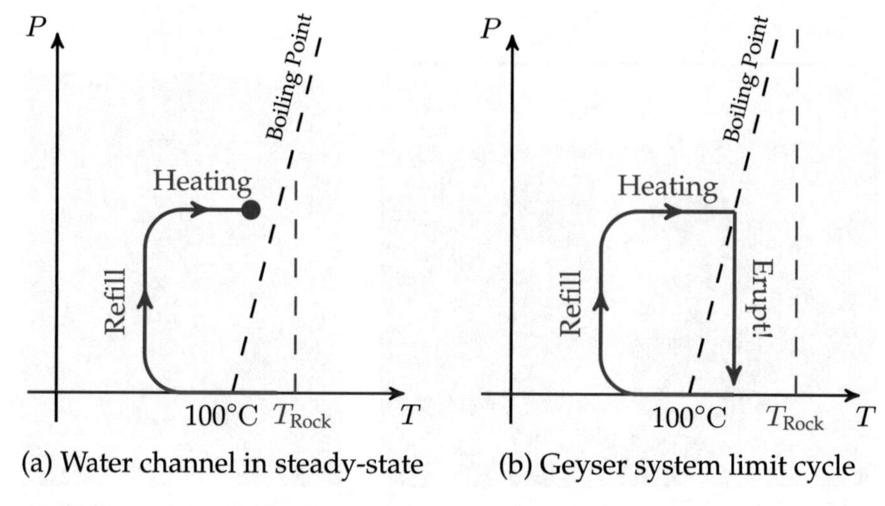

Fig. 7.14 Geyser limit cycle: the periodic eruption of a geyser stems from a limit cycle in the temperature — pressure state of Figure 7.14. If the rate of heat flow into the water channel equals the heat flow lost by convection, then we have a stable fixed point and the system sits in steady state, *left*. However if the rock is sufficiently hot such that the local boiling point is exceeded, then we have a Hopf bifurcation to a limit cycle, right, leading to periodic eruptions.

Further Reading

The references may be found at the end of each chapter. Also note that the textbook further reading page maintains updated references and links.

Wikipedia Links — Nonlinear Systems: Nonlinear System, Hopf bifurcation, Limit cycle, Chaos theory

Wikipedia Links — Examples: El Niño, La Niña, Lotka-Volterra Equation, Newton's Method, Attractor, Fractal, Goodwin Model, Business Cycle, Elliott Wave, Kondratiev Wave, Stick-Slip phenomenon

The suggestions for further reading here parallel those from the previous chapter, on page 128. In particular, the books by Scheffer [10] and by Strogatz [12] are both highly recommended as next steps for the reader.

To follow up on the connection between limit cycles and ice-ages, the paper by Crucifix [2] is an accessible review. Resilience references were suggested in Chapter 6; probably the most accessible material is that of Meadows [8] and Scheffer et al. [10, 11]. The issue of distance metrics and pattern recognition was discussed in the context of resilience in Example 7.3; a great many pattern recognition texts have been written, however I would humbly suggest that of Fieguth and Wong [4] as an accessible introduction.

Sample Problems

Problem 7.1: Short Questions

Provide brief definitions (in perhaps 2–3 sentences and/or a quick sketch) of each of the following:

(a) Degeneracy
(b) A Hopf bifurcation
(c) A limit cycle

Problem 7.2: Analytical — Linearization

For each of the following dynamic matrices ...

$$\begin{bmatrix} 3 & 2 \\ 0 & 0 \end{bmatrix} \quad \begin{bmatrix} 1 & 0 \\ 0 & 1 \end{bmatrix} \quad \begin{bmatrix} 1 & 1 \\ -1 & 0 \end{bmatrix} \quad \begin{bmatrix} 2 & 5 \\ 1 & -2 \end{bmatrix} \quad \begin{bmatrix} -3 & 2 \\ 1 & -1 \end{bmatrix}$$

(a) Identify the behaviour type.
(b) If the matrix had come from the linearization of a 2D nonlinear system, what would you be able to conclude regarding the nonlinear system?

Problem 7.3: Analytical — Linearization

For each of the following dynamic matrices ...

$$\begin{bmatrix} -2 & -2 \\ -3 & -3 \end{bmatrix} \quad \begin{bmatrix} -2 & 0 \\ 0 & -2 \end{bmatrix} \quad \begin{bmatrix} 1 & 2 \\ -2 & -1 \end{bmatrix} \quad \begin{bmatrix} 1 & 1 \\ 3 & 2 \end{bmatrix} \quad \begin{bmatrix} 0 & 2 \\ 2 & 4 \end{bmatrix}$$

(a) Draw a $\tau - \Delta$ diagram and show the locations of the five matrices
(b) Name the behaviour type
(c) Draw a rough sketch of the corresponding phase diagram

Problem 7.4: Analytical — System Characterization

In (7.8) we saw that a complex conjugate pair of eigenvalues necessarily implies that

$$\tau^2 - 4\Delta < 0 \tag{7.22}$$

in the $\tau - \Delta$ diagram of Figure 7.2b.

Now show that the converse is true, that *any* unstable node, stable node, or saddle must correspond to

$$\tau^2 - 4\Delta > 0, \tag{7.23}$$

and hence lie on the other side of the parabola.

Problem 7.5: Conceptual — Limit Cycles

Suppose we identify a boundary such that the dynamics along the boundary always point "inwards", as shown:

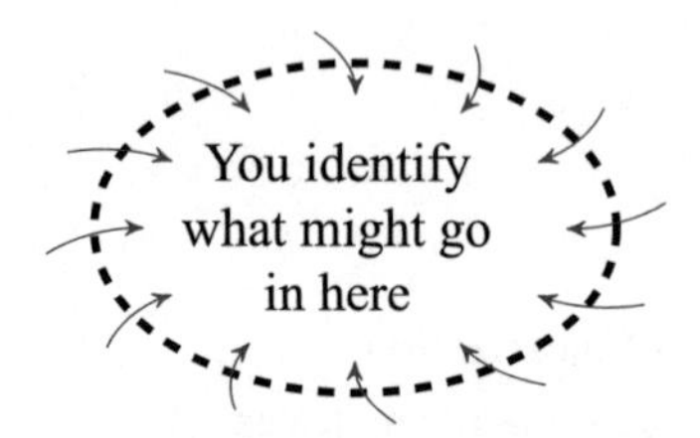

Using combinations of fixed point(s) and/or limit cycle(s), draw illustrations (no math, just sketches) of *three different* systems consistent with the given figure.

Problem 7.6: Conceptual — Limit Cycles

For each of the four cases in Figure 7.5 where a limit cycle is identified as possibly being present,

(a) Sketch a phase plot for a solution which includes a *half-stable* cycle;
(b) Sketch a phase plot for a solution which includes an *unstable* cycle.

Problem 7.7: Conceptual — Limit Cycles

The alarm bell of Example 7.5 clearly has a stable limit cycle. No matter what state the bell is in, if you turn the electricity on the bell rings.

Yet the limit cycle of the bell contains no fixed point, whereas Figures 7.4 and 7.5 both clearly showed stable limit cycles containing an unstable fixed point.

(a) Explain why there is no fixed point in the limit cycle of Example 7.5.
(b) Suppose the $V_s(z)$ relationship in Example 7.5 was steeply sloped rather than discontinuous. Where does the fixed point appear? Argue persuasively why this fixed point is unstable.

Problem 7.8: Conceptual — Stick-Slip Models

Once you understand the Stick-Slip model in Case Study 7, you can begin to see examples everywhere. For each of the following contexts,

(a) squealing brakes on a train
(b) fingernails on a blackboard
(c) blowing through pursed lips

qualitatively describe under what circumstances we have a limit cycle and when not.

Suggest three other phenomena, of any kind, that are similarly subject to a stick-slip behaviour.

Problem 7.9: Analytical — Nonlinear Dynamics

Given the following nonlinear system:

$$\dot{x} = x - y^2$$

$$\dot{y} = 3y - 4x + xy$$

(a) Identify the fixed points of the system.
(b) At each fixed point, find the linearized system and the type of dynamic behaviour.
(c) Draw the phase plot.

Problem 7.10: Analytical — Nonlinear Dynamics

You will examine the following nonlinear system numerically in Problem 7.15:

$$\dot{x} = (3 - x - 2y)x \qquad \dot{y} = (2 - x - y)y$$

(a) Solve for the fixed points of this system.
(b) Linearize the system at each of the fixed points and identify the type of behaviour of the linearized system.
(c) Draw a sketch of the 2D phase plot, clearly showing the fixed points.

Problem 7.11: Analytical — Nonlinear Dynamics (Challenging)

Suppose we have the nonlinear dynamic system

$$\dot{x} = x^2 + xy - 2x$$

$$\dot{y} = xy^2 + 3y$$

(a) Solve for the fixed points
(b) Describe the system behaviour at each fixed point

The next two parts are somewhat challenging:

(c) Draw the phase diagram corresponding to this system. Show all of your work.
(d) Suppose the system dynamics are generalized to

$$\dot{x} = x^2 + xy - 2x$$

$$\dot{y} = xy^2 + \theta y$$

Describe the fixed point behaviour as a function of θ, and identify the type(s) of any bifurcations and corresponding value(s) of θ. Draw a sketch clearly showing what is happening.

Problem 7.12: Analytical — Limit Cycles (Challenging)

Suppose we consider the coupled nonlinear system

$$\dot{x} = y \qquad \dot{y} = -\left(x + y^3 - y\right)$$

We would like to understand the behaviour of this system.

(a) Confirm that there is only one fixed point and that the linearized dynamics correspond to an unstable spiral. The presence of an unstable spiral certainly makes it possible that there is a limit cycle around the fixed point.

(b) Sketch the behaviour of $\dot{y}$ versus y for $x = 0$. Are the roots in this sketch necessarily fixed points, or what is the sketch telling you about a possible limit cycle?

(c) Now sketch the direction vectors $(\dot{x}, \dot{y})$ in the (x, y) plane. It is simplest to sketch the vectors only along the x and y axes.

(d) We are going to hypothesize that we have a stable limit cycle at a radius of approximately $r = 1$. So consider an envelope at $r = 2$; derive the dynamic for $\dot{r}$ on all points on the circle. Show that $\dot{r}$ is not always negative, meaning that the circular envelope does *not* allow us to establish the presence of an enclosed limit cycle.

(e) It *is*, however, possible to select an appropriate envelope. It isn't obvious, but a careful observation of your two-dimensional vector sketch in (b) will reveal a suitable envelope. Identify this envelope, and argue that the dynamics must always be pointing inward along this envelope, proving the presence of a limit cycle.

Problem 7.13: Numerical/Computational — Fractals

We encountered fractals in Example 7.2; although not directly related to environmental or social systems, fractals are very simple to generate and do serve to reveal some of the complexity and astonishing structure present in a great many nonlinear dynamic systems.

By far the most famous fractal is the Mandelbrot set, which is easily generated. At spatial coordinate (x, y), let $z_0 = 0$ and then iterate

$$z_{n+1} = z_n^2 + (x + iy)$$

up to a maximum of M iterates. The value of the fractal $F(x, y)$ is then given by the maximum value of n for which $|z_n| < 4$.

Evaluate F over the domain $-2 < x < 0.5$, $-1.25 < y < 1.25$.

Much of the structure which you see in F is related to the period-doubling bifurcation of Problem 8.7.

Problem 7.14: Numerical/Computational — Strange Attractors

The Lorenz equations of (7.13) are given by

$$\dot{x}(t) = \sigma\Big(y(t) - x(t)\Big) \quad \dot{y}(t) = x(t)\Big(\rho - z(t)\Big) - y(t) \quad \dot{z}(t) = x(t)y(t) - \beta z(t) \tag{7.24}$$

This system is very simple to simulate. We will assume initial conditions and parameters as follows:

$$x(0) = 1 \quad y(0) = 1 \quad z(0) = 1 \qquad \sigma = 10 \quad \beta = 2.66 \quad \rho = 28 \tag{7.25}$$

Use a Forward-Euler time discretization [see (8.22)] with a time step of $\delta_t = 0.001$:

(a) Numerically simulate the Lorenz system of (7.24) over $0 \leq t \leq 100$. There are three state variables, so you will need to plot in 3D.

(b) Test the initialization sensitivity of the system by changing the initial condition of (7.25) to $\hat{x}(0) = 1 + \epsilon$. We will assume the perturbed system to have diverged at time t when

$$\Big(x(t) - \hat{x}(t)\Big)^2 + \Big(y(t) - \hat{y}(t)\Big)^2 + \Big(z(t) - \hat{z}(t)\Big)^2 > 1$$

Plot the time to divergence as a function of ϵ and discuss.

(c) The Lorenz system fixed points are stable for smaller values of ρ. Identify the value of ρ corresponding to the bifurcation between stable attractor and unstable/chaotic behaviour.

Problem 7.15: Numerical/Computational — Predator–Prey, Limit Cycles

We will study three two-dimensional nonlinear systems. In each case, discretize the system using Forward Euler [see (8.22)] with a time step of $\delta_t = 0.1$:

(a) **Competing Dynamics Model**

$$\dot{x} = (3 - x - 2y)x$$
$$\dot{y} = (2 - y - x)y$$

Select 100 random starting points,

$$0 < x < 3 \qquad\qquad 0 < y < 2$$

and superimpose the state evolutions over time, plotting on x, y axes, essentially producing a phase diagram. What do you observe? Compare your numerical results, here, with your analytical ones in Problem 7.10.

(b) **Classic Predator–Prey Model**

$$\dot{x} = (1 - y)x$$
$$\dot{y} = \theta(x - 1)y \qquad \theta > 0$$

Analytically find the locations of the fixed points. Determine the behaviour of the system as a function of θ at each fixed point by linearization.

Let $\theta = 2$. As before, select 100 random starting points,

$$0 < x < 1 \qquad\qquad 0 < y < 1$$

and superimpose the state evolutions over time, plotting on x, y axes. State your observations.

(c) **Selkov Model**

$$\dot{x} = -x + 0.1y + x^2y$$
$$\dot{y} = \theta - 0.1y - x^2y$$

The behaviour here is a bit more complex.

(i) The equations look a bit tricky, but finding the fixed point is actually not hard. Find the fixed point as a function of θ.
(ii) Find the linearized system at the fixed point. Numerically evaluate τ and Δ as a function of θ, for $0 \leq \theta \leq 1$, and plot on a $\tau - \Delta$ plot.
(iii) Based on the $\tau - \Delta$ plot, specify the ranges of θ corresponding to unique types of behaviour from the system.
(iv) Finally, select only five random points and plot the state evolution. Produce one such plot for each of $\theta = 0.1, 0.5, 1$. Discuss your observations.

Problem 7.16: Reading — Resilience

The question of resilience is examined in Examples 5.3, 7.3, and 10.4.

I would suggest reading[2] *Chapter 1: What is Resilience?* from [3], however you can choose to read a section from any of [5, 9, 10, 13].

Summarize the challenges in attempting to even formulate a definition of system resilience, and itemize the most common resilience definitions currently in use.

[2]Excerpts of most books can be found online; links are available from the textbook reading questions page.

Problem 7.17: Policy — Limit Cycles in Economics

In general, it is intuitive to believe that most systems have some point of balance or equilibrium, whether that system be an economic, ecological, or social one. Therefore issues such as recessions or the yearly swings in the price of oil must be due to incompetent governments or evil speculators.

This "intuitive" picture is, of course, not correct.[3] One of the greatest challenges to our intuitive understanding of how systems work is to really understand that many nonlinear systems, such as the Goodwin model of the business cycle discussed in Example 7.6, have no stable fixed point at all and that the *only* stable behaviour is a limit cycle.

Using Example 7.6 as a starting point, do some reading and prepare a brief summary on one of two directions:

1. *Business Cycles:* Read further on the subject of economic cycling. What methods are currently popular in modelling cycling? Are certain economic systems (energy prices, employment, ...) more susceptible to cycling than others?
2. *Policy Response:* Not all limit cycles are equal — the oscillation may have large or small amplitudes, and large or small periods. What are some policy tools or economic factors which can affect the size and frequency of the business cycle?

References

1. K. Alligood, T. Sauer, and J. Yorke. *Chaos: An Introduction to Dynamical Systems*. Springer, 2000.
2. M. Crucifix. Oscillators and relaxation phenomena in pleistocene climate theory. *Philosophical Transactions of the Royal Society A*, 370, 2012.
3. G. Deffuant and N. Gilbert (ed.s). *Viability and Resilience of Complex Systems: Concepts, Methods and Case Studies from Ecology and Society*. Springer, 2011.
4. P. Fieguth and A. Wong. *Introduction to Pattern Recognition*. Springer. (in preparation).
5. L. Gunderson and C. Holling. *Panarchy: Understanding Transformations in Human and Natural Systems*. Island Press, 2012.
6. R. Hilborn. *Chaos and Nonlinear Dynamics: An Introduction for Scientists and Engineers*. Oxford, 2000.
7. P. Loring. The most resilient show on earth. *Ecology and Society*, 12, 2007.
8. D. Meadows. *Thinking in Systems: A Primer*. Chelsea Green, 2008.
9. D. Meadows, J. Randers, and D. Meadows. *Limits to Growth: The 30-Year Update*. Chelsea Green, 2004.
10. M. Scheffer. *Critical transitions in nature and society*. Princeton University Press, 2009.
11. M. Scheffer, S. Carpenter, V. Dakos, and E. van Nes. *Generic Indicators of Ecological Resilience: Inferring the Chance of a Critical Transition*. Annual Reviews, 2015.

[3] Well ... some governments *are* incompetent, and much of financial speculation and high-frequency trading is indeed parasitic on society.

12. S. Strogatz. *Nonlinear Dynamics and Chaos*. Westview Press, 2014.
13. B. Walker and D. Salt. *Resilience Thinking: Sustaining Ecosystems and People in a Changing World*. Island Press, 2006.
14. R. Wright. *A Short History of Progress*. House of Anansi Press, 2004.

Chapter 8
Spatial Systems

In early April 2014, people across England woke up to a very visible layer of Saharan desert dust covering their vehicles, an astonishing transport of mass from one continent to another.

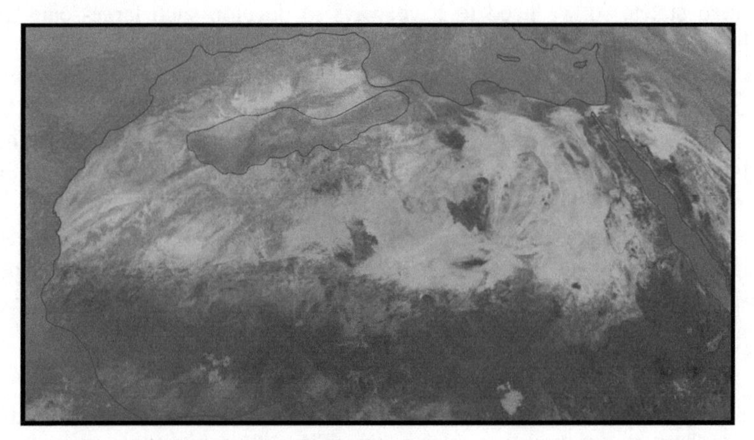

Data from the SEVIRI imager of EUMETSAT

Indeed, there is a staggering world-wide transport of dust, approximately 2 billion tons per year, most of it originating from deserts, and about half of this total from the Sahara alone [8]. Outlined in red in the infrared figure is a huge dust storm from 2006, approximately 1500 km in width; a fraction of this dust is thrown sufficiently high into the atmosphere to travel right across the Atlantic, 8000 km to Florida, even to the Pacific ocean past California, well over 10,000 km away. In addition to

P. Fieguth, *An Introduction to Complex Systems*,
DOI 10.1007/978-3-319-44606-6_8

being a respiratory irritant and a source of air pollution, dust and other aerosols exert significant influences:

Weather: Dust in the air blocks sunlight; over the ocean, dust clouds can reduce the sea surface temperatures by 1°C, which can have a major impact on reducing hurricane formation.

Soil Nutrients: About 40 million tons of dust are transported, annually, from the Sahara desert to the Amazon basin in South America. The Saharan dust is therefore one of the primary mineral sources fertilizing the Amazon basin, an interesting observation in the context of Case Study 3.

Climate Change: In many parts of the ocean life is iron-limited, whereas Saharan dust is 4 % iron by weight, so such iron deposition is a major factor in certain ocean ecosystems. The dust can induce the growth of huge algae blooms (red tides), responsible for the deaths of fish and other marine life, however at the same time the accelerated growth absorbs carbon-dioxide and is a non-trivial factor in global climate.

Dust, aerosols, and other transport phenomena are all multi-dimensional spatial processes; indeed, *most* ecological and social systems have many-element spatially distributed states, so we need to have ways to develop such large-scale 2D/3D models.

Up to this point our dynamic systems have had state vectors consisting of only a few elements. However representing a global time-varying map of ocean temperature, ocean currents, atmospheric pressure, or more subtle ideas such as global aerosol and dust transport, will require states of millions or billions of elements

Table 8.1 offers an overview of the main categories of spatial models: the state may consist of elements which are irregularly or regularly gridded, representing a process in some number of dimensions, for which the state elements themselves are continuous or discrete in value.

Table 8.1 Spatial states: we may have systems with discrete-valued or continuous-valued states, and where the states are continuously, discretely, or irregularly distributed over one or more dimensions. Simple dynamic systems we have studied earlier in this text; the green systems are those of interest to us in this chapter.

	System state	
Spatial arrangement	Discrete-valued	Continuous-valued
Non-spatial, small	Markov chains	Simple dynamic systems
Discrete — 1D	1D cellular automata	Array processing
Discrete — 2D	2D cellular automata	Video processing
Discrete — 2D/3D	Classification	Discretized PDEs
Continuous — 2D/3D	Labelling	PDEs
Moving/abstract	Particle/agent-based models	

8.1 PDEs

Our dynamic models thus far have been ordinary differential equations,

$$\frac{du}{dt} = f(u). \tag{8.1}$$

However if $T(x, y, z, t)$ represents a world map of temperature, for example, then we need some way of describing the space–time behaviour of

$$\frac{dT}{dt} = f\big(T(x, y, z, t)\big). \tag{8.2}$$

The most fundamental concept is one of *conservation*. In physical systems there are many spatially-distributed quantities that can move around but are not created or destroyed. For example, all of

water, salt, energy, linear momentum, angular momentum, heat

are conserved. Let $u(x, y, t)$ describe the spatial *density* of some conserved quantity, such as salt, with flows $\underline{j}^T = \begin{bmatrix} j_x \; j_y \end{bmatrix}$, as shown in Figure 8.1. Because of conservation, within some small domain

$$\text{(Increase over time)} = \text{(Input across boundaries)} + \text{(Sources in box)} \tag{8.3}$$

Since u is a spatial density, let U be the total integrated amount

$$U = \iint_{\text{Box}} u \, dx \, dy, \tag{8.4}$$

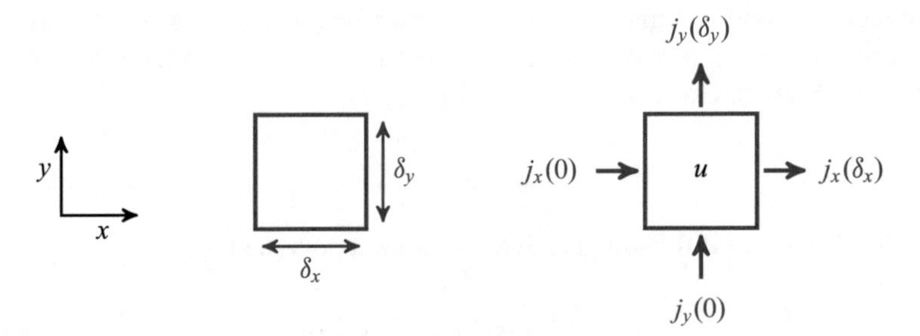

Fig. 8.1 Flow: we wish to imagine the flow of some quantity u, such as water or heat, through a small differential element of width δ_x and height δ_y. The flows are described by a vector field $\underline{j}$.

then if u is a continuous function and the box is sufficiently small,

$$\text{(Increase of } U \text{ over time } \delta_t) = \delta_x \delta_y \cdot \big(u(\delta_t) - u(0)\big) \tag{8.5}$$

Similarly for a given rate of flow, the total change in U depends on the width of the flow and the period of time of the flow:

$$\text{(Input of } U \text{ across boundaries}) = \delta_y \delta_t \cdot \big(j_x(0) - j_x(\delta_x)\big) + \delta_x \delta_t \cdot \big(j_y(0) - j_y(\delta_y)\big) \tag{8.6}$$

Finally, if S is the density of source or sink flows, then

$$\text{(Sources of } U \text{ within box}) = \delta_x \delta_y \delta_t \cdot S. \tag{8.7}$$

Dividing through by $\delta_x \delta_y \delta_t$, (8.3) becomes

$$\frac{u(\delta_t) - u(0)}{\delta_t} + \frac{j_x(\delta_x) - j_x(0)}{\delta_x} + \frac{j_y(\delta_y) - j_y(0)}{\delta_y} - S = 0 \tag{8.8}$$

In the limit as the box size and time interval reduce to zero, we have

$$\frac{\partial u}{\partial t} + \frac{\partial j_x}{\partial x} + \frac{\partial j_y}{\partial y} - S = 0. \tag{8.9}$$

That is, the conservation of u with respect to flow $\underline{j}$ and sources S is given by

$$\frac{\partial u}{\partial t} + \nabla \cdot \underline{j} - S = 0, \tag{8.10}$$

also known as the classic advection–diffusion equation. There are basically two common types of flow $\underline{j}$:

1. **Advective Flow:** A movement or transport due to some carrier, for example salt or heat being moved in the ocean from the tropics to northern latitudes via the Gulf Stream. Given a velocity field $\underline{v}(x, y, t)$,

$$\underline{j}_{\text{advection}} = \underline{v} u, \tag{8.11}$$

in the absence of diffusion the advective flow is governed by

$$\frac{\partial u}{\partial t} + \nabla \cdot (\underline{v} u) - S = 0 \tag{8.12}$$

2. **Diffusive Flow:** A spontaneous flow down gradients (as discussed in Chapter 3), such as the movement of heat through rock to drive a geyser in Case

Study 7. The flow is proportional to the gradient

$$\underline{j}_{\text{diffusion}} = -k\nabla u \tag{8.13}$$

for conductivity coefficient k. Thus in the absence of advection or sources/sinks ($S = 0$), (8.10) becomes

$$\frac{\partial u}{\partial t} = k\nabla^2 u \tag{8.14}$$

which is known as the heat or diffusion equation.

Let us derive the Navier–Stokes equation, which describes the motion $\underline{v}$ of air and water, and which is therefore fundamental to modelling the atmosphere and oceans. We begin by asserting conservation of mass; if we let ρ be the spatial mass density, then because there are no sources or sinks of mass (8.10) becomes

$$\frac{\partial \rho}{\partial t} + \nabla \cdot (\underline{v}\rho) = 0 \tag{8.15}$$

Next, the momentum density ($\rho\underline{v}$) is, like mass, *also* conserved; therefore again (8.10) becomes

$$\frac{\partial \rho \underline{v}}{\partial t} + \nabla \cdot \big(\underline{v} \otimes (\rho\underline{v})\big) = S \tag{8.16}$$

Expanding the terms[1] in (8.16) gives us

$$\underline{v}\frac{\partial \rho}{\partial t} + \rho\frac{\partial \underline{v}}{\partial t} + \underline{v}\big(\nabla \cdot (\rho\underline{v})\big) + \rho\underline{v} \cdot \nabla\underline{v} = S \tag{8.17}$$

thus

$$\underbrace{\underline{v}\left(\frac{\partial \rho}{\partial t} + \big(\nabla \cdot (\rho\underline{v})\big)\right)}_{=0} + \rho\left(\frac{\partial \underline{v}}{\partial t} + \underline{v} \cdot \nabla\underline{v}\right) = S \tag{8.18}$$

where the first term on the left goes to zero by conservation of mass, from (8.15). As a result of the $\underline{v} \cdot \nabla\underline{v}$ term in (8.18), the Navier–Stokes equation is clearly nonlinear, with all of the related issues of multiple fixed points, hysteresis, and

[1]The details of vector manipulations and differential equations are well beyond the scope of this text. The focus here is not on the details of the derivation, rather on understanding the high-level form of the Navier–Stokes equation [23].

the inapplicability of superposition. The nonlinearity also makes (8.18) impossible to solve analytically, significantly motivating numerical simulations, discussed in Section 8.3.

Since ρ is a mass density and $\partial \underline{v}/\partial t$ an acceleration, by Newton's $F = ma$ we interpret the sinks and sources S in (8.18) to represent forces. Since water and air are fairly complex fluids, in principle there is a wide variety of possible forces to model and take into account, which partly explains the bewildering range of forms of the Navier–Stokes equation. For our purposes, three of the most important forces are that air and water move under pressure, have a certain viscosity which resists movement, and are present on a rotating planet leading to Coriolis forces:

$$\rho\left(\frac{\partial \underline{v}}{\partial t} + \underline{v} \cdot \nabla \underline{v}\right) = \sum \text{Forces} = \underbrace{-\nabla p}_{\text{Pressure}} + \underbrace{\mu \nabla^2 \underline{v}}_{\text{Viscosity}} \underbrace{- 2\rho \underline{\Omega} \times \underline{v}}_{\text{Coriolis}} \tag{8.19}$$

8.2 PDEs and Earth Systems

We are interested in knowing the climate-scale consequences stemming from the PDEs developed in the previous section. Certainly the advection–diffusion and Navier–Stokes PDEs apply to a broad range of spatial earth systems:

Diffusion:	underground movement of heat, pollutants
Advection:	oceanic/atmospheric transport of heat, salt, nutrients
Navier–Stokes:	fundamental model for water (oceans), air (atmosphere)

There are a great many Navier–Stokes variations, one of the most common simplifications being an incompressibility assumption for water, but preserving compressibility for air.

On local scales, turbulence in both air and water is due to the nonlinearity in Navier–Stokes (8.19), giving rise to eddies being shed from major ocean currents, such as the Gulf Stream.

On global scales, one of the key terms in (8.19) is the final term, representing the Coriolis effect, caused by the fact that we live on a rotating planet.

Our earth rotates about its axis, as illustrated in Figure 8.2, so that a parcel of water or air at the equator has a significantly higher velocity than a similar parcel at higher latitudes. As a result, a parcel of water or air moving away from the equator would tend to move eastward. A bit more subtly, an eastward motion causes a centripetal force away from the axis of rotation, leading to a deflection towards the equator.

More formally, at latitude ψ and ignoring upward motion (since the atmosphere and ocean are relatively thin, in height, relative to their spatial extents), for a planet with angular rotational velocity ω the Coriolis acceleration is given by

$$\underline{a}^{\text{Cor}} = \begin{bmatrix} a_{\text{East}}^{\text{Cor}} \\ a_{\text{North}}^{\text{Cor}} \end{bmatrix} = \begin{bmatrix} v_{\text{North}} \\ -v_{\text{East}} \end{bmatrix} \cdot 2\omega \sin \psi \tag{8.20}$$

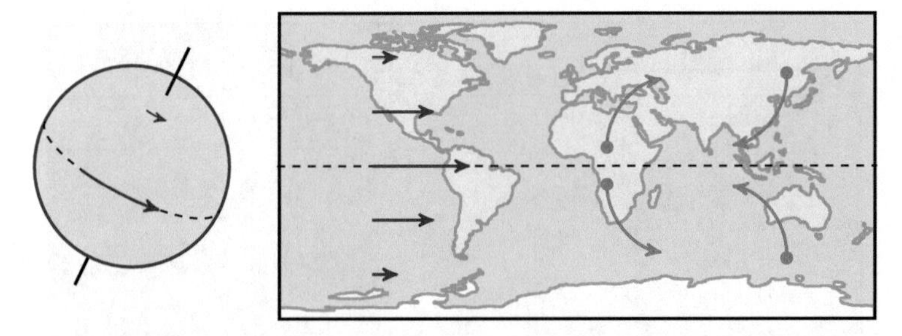

Fig. 8.2 The Coriolis effect: the Coriolis effect has a major influence on the movement of water and air around the planet. Because the earth is rotating about an axis, at the equator (dashed) a parcel of air or water moves faster than near the poles, as shown by the red arrows in both panels. Consequently *moving* air or water experiences a deflection, as shown by the green arrows, *right*.

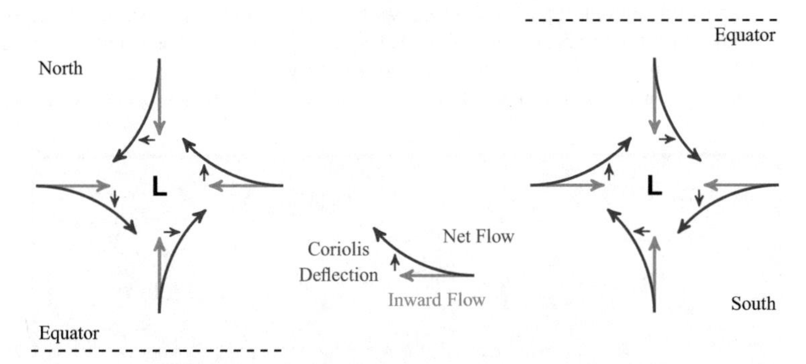

Fig. 8.3 Coriolis and storm rotation: an atmospheric low-pressure area induces inward flows upon which the Coriolis effect introduces a deflection, leading to the overall curved airflow. Therefore hurricanes and other storms rotate counter-clockwise in the northern hemisphere, left, and clockwise in the southern hemisphere, right.

Therefore, in the northern hemisphere ($\sin \psi > 0$) a northward velocity leads to an eastward acceleration, and an eastward velocity leads to a southward acceleration. Indeed, in general the Coriolis acceleration is rotated by 90° with respect to the velocity.

The Coriolis force is rather weak, and therefore does *not* influence the water in your sink or bathtub, but becomes increasingly significant at larger scales. Most storms are associated with atmospheric low-pressure areas, for which the Coriolis force induces a characteristic direction of rotation, as sketched in Figure 8.3.

Indeed, very large storms such as hurricanes require two ingredients for formation:

1. Hurricanes derive their energy from warm sea-surface temperatures, and therefore must form relatively close to the equator.
2. Hurricanes rely on rotation as a way of concentrating energy, and therefore cannot form at the equator, where the Coriolis force is zero.

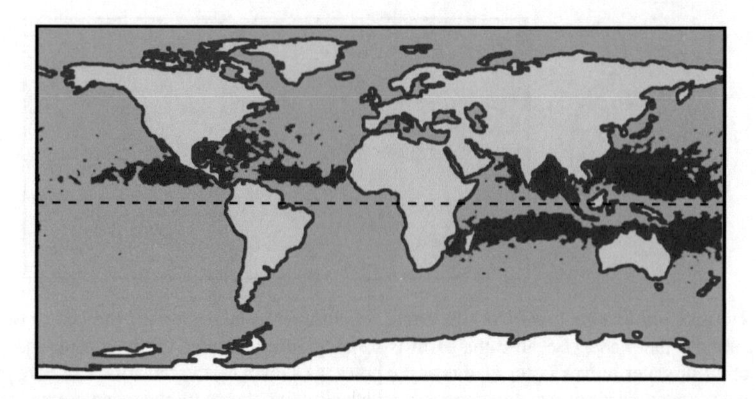

Fig. 8.4 Hurricane formation: hurricanes require warm sea-surface temperatures, which peak at the equator, and sufficient energy from Coriolis spin, which is zero at the equator, therefore we expect hurricanes primarily in bands away from but near to the equator. The actual historical data of hurricane formation, red, shows this behaviour very clearly.

Adapted from M. Pidwirny, "Tropical Weather and Hurricanes," *Encyclopedia of Earth*, www.eoarth.org.

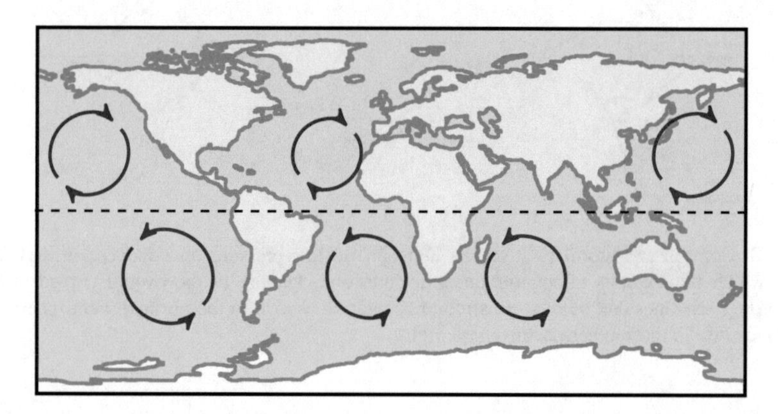

Fig. 8.5 Ocean gyres: there is a *gyre*, an ocean-basin scale rotating current, in each of the major ocean basins. The gyres are wind driven and curved by Coriolis forces.

With the above two constraints in mind, it is particularly satisfying to observe the worldwide map of hurricane formation regions, sketched in Figure 8.4, showing a striking gap along the equator where the absence of Coriolis rotation precludes hurricane formation.

At even larger scales, in each ocean basin there is a *gyre*, a circulating current of water, circulating clockwise in the northern hemisphere and counter-clockwise in the southern, as shown in Figure 8.5. The curvature and shape of the gyres is controlled by Coriolis forces, with the currents themselves driven by winds. However the persistent global winds driving these gyres stem from atmospheric cells, circulations which are driven by warm air at the equator and cold at the poles,

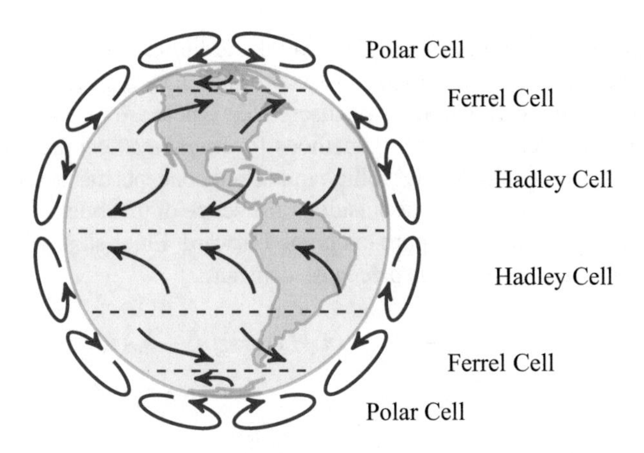

Fig. 8.6 Atmospheric cells: there are six latitudinally separated cells, whose northern or southern surface flows (blue) are deflected by the Coriolis force to give rise to prevailing winds (red).

as sketched in Figure 8.6. However, yet again, it is the Coriolis force which causes the north–south air movements to lead to easterly winds (Hadley, Polar cells) and westerly winds (Ferrel cells).

The conclusion is that (8.19), a relatively simple PDE (that is, simple to write down, *not* that it is simple to solve) leads relatively quickly to interpreting broad, global-scale phenomena.

8.3 Discretization

High-dimensional nonlinear systems, such as the Navier–Stokes equation in (8.19), are impossible to solve analytically. Admittedly we were able, in Section 8.2, to understand the large-scale behaviour of (8.19) in a qualitative way. However far more challenging to understand or interpret are discontinuous models, particularly those models incorporating a discrete-valued state, such as the state of a forest (extant versus clear-cut), where the discontinuity leads to spontaneous jumps of behaviour.

As a result, we are drawn to numerically simulating dynamic models, whether linear or nonlinear, continuous or discontinuous. Such a simulation requires a spatio-temporal problem to be quantized into discrete steps in both time and space.

8.3.1 Discretization in Time

We have focused almost exclusively on continuous-time systems, such as

$$\dot{\underline{z}}(t) = A\underline{z}(t), \tag{8.21}$$

since almost all physical systems in the natural and human worlds evolve continuously over time. However if we wish to study or simulate the system numerically, in a computer, then the system must be discretized. There are many strategies here, including higher-order Runge–Kutta methods for improved accuracy, and implicit methods for improved numerical stability, important concepts discussed at length in numerical methods texts [6, 9] but outside of the scope of this book.

The conceptually simplest approach is the Forward–Euler step, which approximates the continuous-time $\underline{z}(t)$ as piecewise constant:

$$\dot{\underline{z}}(t) = A\underline{z}(t) \qquad \longrightarrow \qquad \underline{z}(n+1) = \underline{z}(n) + \delta_t A\underline{z}(n) \tag{8.22}$$

$$= \big(I + \delta_t A\big)\underline{z}(n) \tag{8.23}$$

where continuous-time t is related to the discrete-time index n as $t = n\delta_t$. From Chapter 5 we know that a linear system is primarily characterized in terms of its fixed points and its eigendecomposition, so we would like to investigate their sensitivity to discretization. For a continuous-time fixed point $\bar{\underline{z}}$,

$$z(n) = \bar{\underline{z}} \qquad \longrightarrow \qquad \underline{z}(n+1) = \bar{\underline{z}} + \delta_t A\bar{\underline{z}} = \bar{\underline{z}} + \underline{0} = \bar{\underline{z}}, \tag{8.24}$$

so a fixed point in continuous time remains a fixed point in discrete time. Next, suppose we are given the modes

$$A\underline{v}_i = \lambda_i \underline{v}_i \tag{8.25}$$

for the continuous-time system; then in discrete time, from (8.23),

$$\big(I + \delta_t A\big)\underline{v}_i = \underline{v}_i + \delta_t \lambda_i \underline{v}_i = \big(1 + \delta_t \lambda_i\big)\underline{v}_i. \tag{8.26}$$

That is, the mode *directions*, the eigenvectors $\underline{v}_i$, are unaffected by discretization, however the eigenvalues themselves are definitely changed.

The conditions for stability [16] are sketched in Figure 8.7:

$$\text{Continuous Time: } \text{Real}(\lambda) < 0$$
$$\text{Discrete Time: } |1 + \delta_t \lambda| < 1$$

From the figure it is clear that stable eigenvalues (green) may become *unstable* (red) when time-discretized, and certainly that the eigenvalues change as a function of δ_t.

In general we have a tradeoff between accuracy and computational complexity: as δ_t is decreased we have more accurate simulations, but requiring more computations to reflect the greater number of time steps. Particularly challenging are systems having both fast and slow dynamics, since

- The fastest eigenvalue $\lambda_{\max}$ will limit the step size: $\delta_t < 2/\lambda_{\max}$
- The slowest eigenvalue $\lambda_{\min}$ will set the simulation time: $T > 3/\lambda_{\min}$.

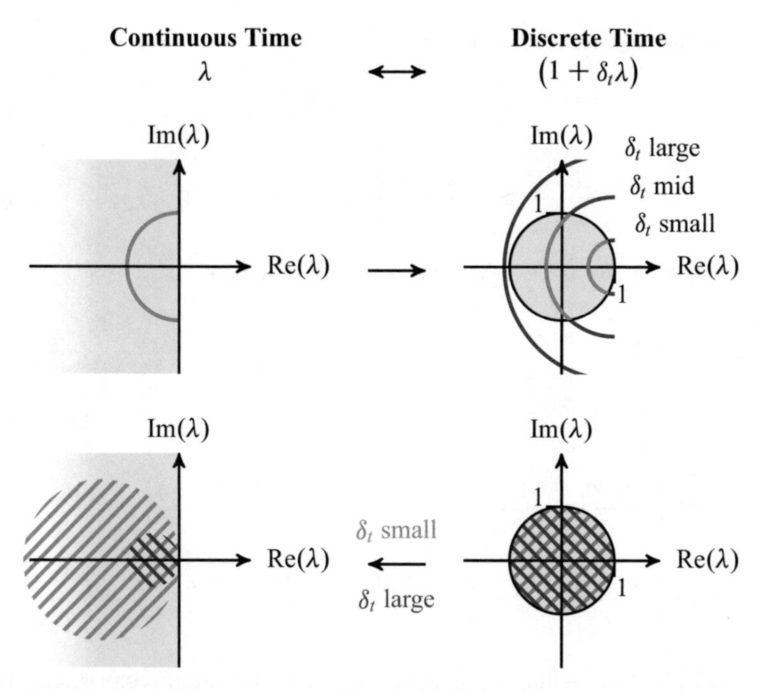

Fig. 8.7 Time discretization: the domain of stable eigenvalues is shown shaded, the left half plane in continuous time and inside the unit circle in discrete time. Stable eigenvalues in continuous time (top left) may remain stable (green) or may become *unstable* (red) when time-discretized with Forward Euler, and more eigenvalues become unstable as δ_t is increased. Under Forward Euler, any system that is stable in discrete time (bottom right) must have been stable in continuous time (bottom left), regardless of δ_t.

Therefore the total *number* of simulation steps is controlled by the ratio

$$\frac{T}{\delta_t} > \frac{|\mathrm{Re}(\lambda_{\max})|}{|\mathrm{Re}(\lambda_{\min})|} = \kappa, \tag{8.27}$$

measuring the numerical *stiffness* or challenge, a subject we will encounter again in Chapter 11 in the context of problem conditioning.

8.3.2 Discretization in Space

Spatial discretization is, in general, a more complex topic than time discretization: time resides, by definition, on a single dimension; spatial components, on the other hand, need to be discretized simultaneously in two or more dimensions.

Certainly the most straightforward approach to solving spatial PDEs is via *finite differences*, to assert a constant spatial step size δ_x, δ_y etc. and then to discretize all of the derivatives present in the PDE:

$$\frac{\partial z(x,y)}{\partial x} \simeq \frac{z(x+\delta_x,y)-z(x,y)}{\delta_x} = z_x(x,y) \tag{8.28}$$

$$\frac{\partial^2 z(x,y)}{\partial x^2} \simeq \frac{z_x(x,y)-z_x(x-\delta_x,y)}{\delta_x} = z_{xx}(x,y) \tag{8.29}$$

etc.[2] For example, for the two-dimensional wave equation

$$\frac{\partial^2 z}{\partial t^2} = c^2\nabla^2 z \tag{8.30}$$

$$= c^2\left\{\frac{\partial^2 z}{\partial x^2} + \frac{\partial^2 z}{\partial y^2}\right\} \tag{8.31}$$

$$= c^2\left\{\frac{z_x(x,y)-z_x(x-\delta_x,y)}{\delta_x} + \frac{z_y(x,y)-z_y(x,y-\delta_y)}{\delta_y}\right\} \tag{8.32}$$

$$= c^2\left\{\frac{z(x+\delta_x,y)-2z(x,y)+z(x-\delta_x,y)}{\delta_x^2} + \frac{\textit{same idea}\ldots}{\delta_y^2}\right\} \tag{8.33}$$

If we stack all of the sampled points of $z(x,y)$ into a column vector $\underline{z}$, then (8.33) can be written as a matrix–vector product

$$\frac{\partial^2 \underline{z}}{\partial t^2} \simeq A\underline{z} \tag{8.34}$$

where A is a sparse matrix of coefficients. This product representation of (8.34) can accommodate any linear partial differential equation. However the wave equation of (8.30) is spatially stationary, meaning that the nature of the dynamics does not vary as a function of location, therefore (8.33) can be expressed far more naturally and succinctly via a convolution as

$$\frac{\partial^2 z}{\partial t^2} = \frac{c^2}{\delta_x^2}\begin{bmatrix} 1 & \textcircled{-2} & 1 \end{bmatrix} * z + \frac{c^2}{\delta_y^2}\begin{bmatrix} 1 \\ \textcircled{-2} \\ 1 \end{bmatrix} * z \tag{8.35}$$

where $*$ is the two-dimensional convolution operation and where the circled element indicates the location of the origin in the convolutional kernel.

The limitation of the above finite difference approach is that we are forced onto a uniform grid. In considering the spatial discretization of $z(x,y)$, ideally the discretization grid points would *not* be uniformly spaced, rather would place

[2] The careful reader may observe an apparent inconsistency in shifting by $+\delta_x$ in (8.28) and by $-\delta_x$ in (8.29). This change is deliberate and ensures the proper centering of the second derivative; see Problem 8.5 to work out the details.

more attention (more points) where the function is complex, and fewer where it is smooth, leading to a uniform degree of error. Indeed, the state of the art in spatial discretization is a *finite element* strategy which discretizes space *non*-uniformly, normally onto a triangular mesh, allowing for variable discretization steps and complex geometries. However spatial derivatives are not naturally represented on a triangle, so the original problem needs to be transformed, a more complex process than we can discuss here.

As in time-discretization, where the choice of discretization step δ_t had implications on eigenvalue stability, the choice of spatial discretization step $\delta_{x,y}$ similarly can have significant impacts on the resulting solution, particularly for nonlinear systems. In particular, suppose we have dynamics A_F defined at a fine scale, and let S be a linear subsampling operator, a matrix transforming a system from fine to coarse:

$$\underline{z}_{\text{Coarse}} = S\,\underline{z}_{\text{Fine}} \tag{8.36}$$

If A_F is *linear* it is possible to define coarse-scale dynamics A_C, such that the dynamics and subsampling *commute*, meaning that we can switch their order:

$$\text{if} \quad A_C = SA_FS^+ \qquad \text{then} \qquad SA_F\,\underline{z}_{\text{Fine}} = A_C\,S\underline{z}_{\text{Fine}} = A_C\underline{z}_{\text{Coarse}} \tag{8.37}$$

where S^+ is the pseudoinverse[3] of S. In other words, the numerical solution of a coarse-scale problem is equivalent to subsampling the numerical solution of a fine-scale problem. This may seem obvious, however it is not: for *nonlinear* dynamics, the numerical solution at a fine-scale solution may lead to large-scale phenomena which simply do not appear in the coarse-scale solution.

One of the best-known examples is that of eddies in Navier–Stokes simulations [13]. Ocean eddies can be relatively large and represent a large fraction of ocean kinetic energy, however they will simply not appear, not even as low-resolution blurs, until a spatial resolution of approximately 1/3 degree, and become realistic in size and distribution at around 1/10 degree.

8.4 Spatial Continuous-State Models

Our phenomena of interest are almost always *continuously*-valued functions (temperature, pressure, salinity, population density) over space, either two-dimensionally over a surface or three-dimensionally in a volume, and all of the PDEs in which we are interested fall into this context.

[3] Pseudoinverses are essentially a generalization of the concept of matrix inverses but applied to rectangular matrices. Given a rectangular matrix S with more columns than rows, it may be possible to find S^+ such that $S \cdot S^+ = I$, however $S^+ \cdot S \neq I$.

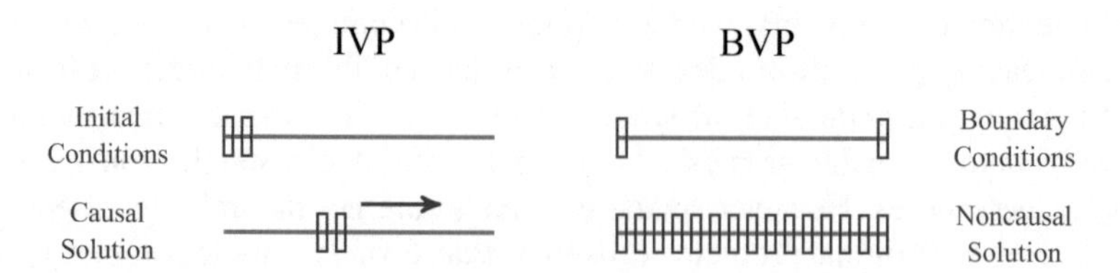

Fig. 8.8 IVP and BVP: the IVP and BVP may both solve the same underlying differential equation, but constrain it in very different ways. The IVP is a *causal* problem, with sufficient constraints to allow for a unique solution specified at the beginning of the domain, making for a relatively simple step-by-step computational solution. In contrast, the BVP is *noncausal*, with constraints at both ends of the domain; as a result there is no one place to begin solving the problem, and the entire problem needs to be tackled simultaneously.

Given a differential equation, whether ordinary or partial, the equation alone does not specify a unique solution, and the problem needs to be constrained in some way. Fundamentally, differential equation solutions fall into two categories, as sketched in Figure 8.8, depending upon what constraints are asserted:

Initial Value Problem (IVP):

Typically encountered in *temporal* problems, the solution is made unique by asserting initial conditions at the *beginning* of the problem domain.

These constraints can be expressed in continuous time by specifying the values of the solution $z(t)$ and some number of derivatives $z(0), \dot{z}(0), \ddot{z}(0), \ldots$ or, equivalently, as we saw in (8.29), each derivative may be represented as some number of points in discrete time $z(0), z(1), z(2), \ldots$

Boundary Value Problem (BVP):

Typically encountered in *spatial* problems, the solution is made unique by asserting boundary conditions *around* the problem domain.

The constraints around the boundary are significantly more flexible than the initial constraints in IVPs, in that the type of condition can vary from place to place along the boundary. For example, it would be possible to specify only the value z along one side, only the gradient along another, and both value and gradient elsewhere, as shown in Figure 8.9.

Although it is convenient to think about gradients and derivatives in the initial/boundary conditions, in practice the problems will be temporally and spatially discretized, so that the actual assertion of any derivative will be accomplished by specifying certain differences between values of z.

In addition to the conceptual differences just described, Figures 8.8 and 8.10 also illustrate the very significant computational differences between IVPs and BVPs. In particular, since the IVP is fully specified by its initial conditions, the solution to the IVP can march forward, in a causal manner, from beginning to end; at any given point in time the number of elements in our state vector is equal to the order of the underlying differential equation, normally relatively small. In contrast, since

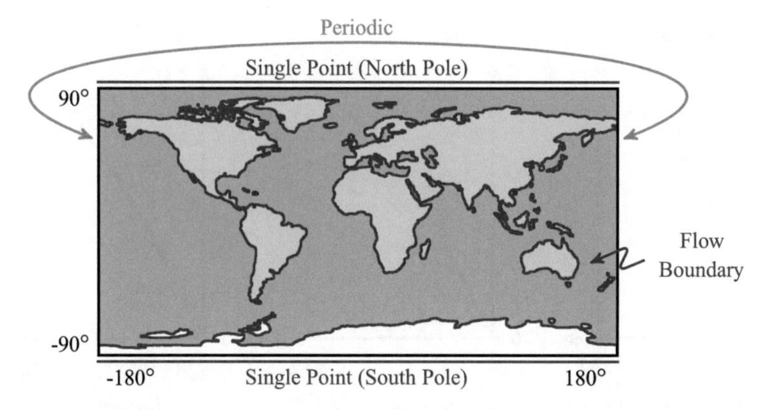

Fig. 8.9 Boundary conditions: consider a model for the ocean on a rectangular grid. We know the ocean doesn't *end* at the east and west sides, so we have periodic conditions there. Similarly all of the pixels at the poles, at the north and south ends of the domain, actually represent the same point, and so need to be coupled. Finally, currents can flow *along* a land boundary but not *into* it, therefore there are boundary conditions on flow along all coastlines.

the BVP has constraints specified along the boundary, in various places, we have a *non*-causal problem, as in the example in Figure 8.11. Where do we start? Actually, there is no *location* to start; we need to store the *entire* discretized problem, possibly very large, and somehow solve the whole problem simultaneously.

Therefore for an nth order problem to be solved on domain $[0, T]$ with step size δ_t,

The IVP requires T/δ_t steps with n variables
The BVP requires an *unknown* number of iterations with T/δ_t variables

as illustrated in Figure 8.10. So if BVPs require so much more memory and computational effort, what is their appeal? In a nutshell, nearly *all* spatial problems *are* BVP — it would be highly unusual to specify the behaviour of a forest, lake, or ocean from only one side.

Now suppose that we are given a spatial problem $z(x, y)$, characterized by a linear PDE which we spatially discretize and stack into a vector of unknowns, per (8.34), such that

$$\frac{\partial z}{\partial t} = A\underline{z}(t). \tag{8.38}$$

As described in Example 8.1, the boundary conditions assert constraints on $\underline{z}$,

$$B\underline{z}(t) = \underline{b}(t) \tag{8.39}$$

where, in principle, the constraints $\underline{b}$ could change over time.

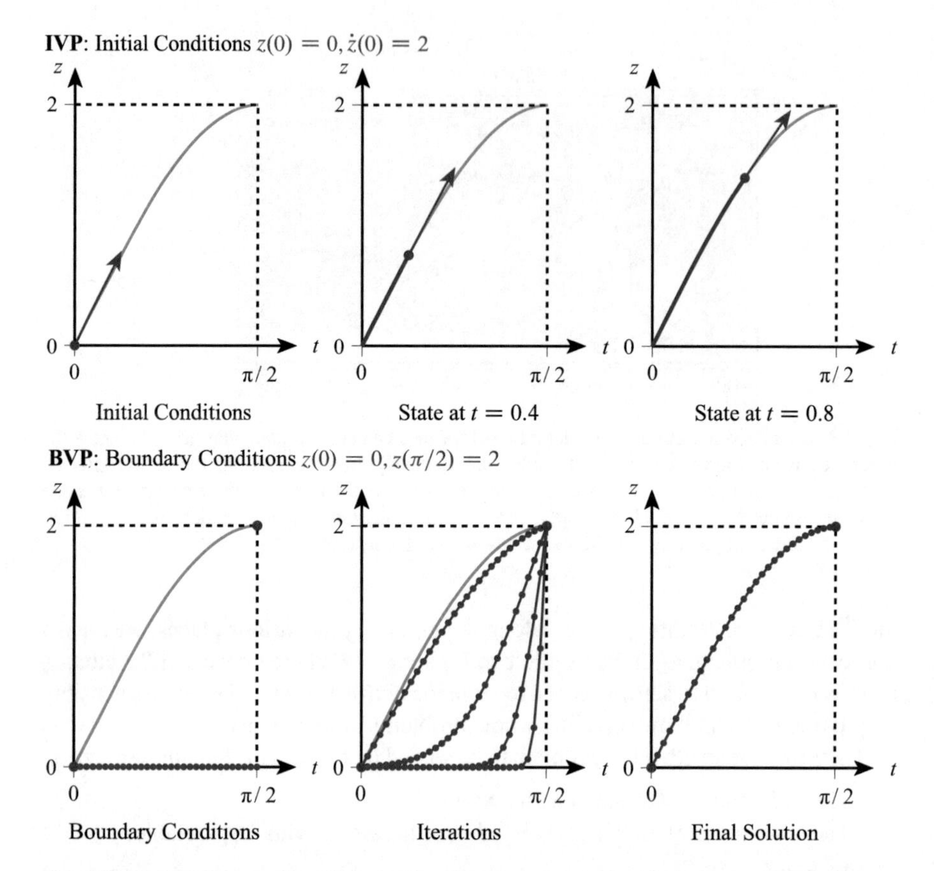

Fig. 8.10 Suppose we are given $\ddot{z}(t) + z(t) = 0$. The differential equation in no way specifies the kind of problem: it can be solved either as an IVP (top) or as a BVP (bottom), depending on the asserted conditions (red). The BVP solutions (blue) are shown iteratively converging to the exact solution (green), however non-iterative approaches are possible.

We then have four contexts of increasing complexity:

1. Steady-State Solution:

We are looking for the state of the system after it settles into steady state, meaning that it is no longer changing over time. Therefore[4]

$$\frac{\partial z}{\partial t} = 0 \qquad \longrightarrow \qquad A\underline{z} = \underline{0}. \tag{8.47}$$

[4]Although the notation may look a bit odd, it is important mathematically to distinguish between the digit 0 and the zero vector $\underline{0}$, *both* of which appear in (8.47).

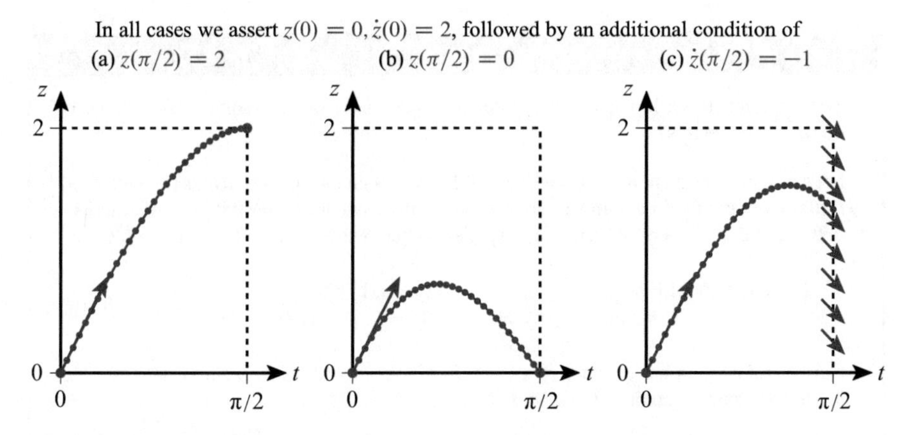

Fig. 8.11 Boundary conditions: a BVP allows for any number of constraints along the boundary. All three panels are based on the same differential equation as in Figure 8.10, but with different constraints (red), leading to different solutions (blue).

Together with the boundary conditions, which in the steady-state context are unchanging over time, we have a large linear system of equations to solve:

$$A\underline{z} = \underline{0}, \quad B\underline{z} = \underline{b} \qquad \longrightarrow \qquad \begin{bmatrix} A \\ B \end{bmatrix} \underline{z} = \begin{bmatrix} \underline{0} \\ \underline{b} \end{bmatrix} \tag{8.48}$$

2. BVP Over Time:

Many systems do not settle into steady-state; certainly most social and ecological systems are constantly subject to change, so there are many cases in which we will be interested in the system behaviour over time. The spatially discretized time-dependent problem is

$$\frac{\partial z}{\partial t} = A\underline{z}(t) \qquad Bz(t) = \underline{b}(t) \tag{8.49}$$

That is, we have a large number of coupled ordinary differential equations over time, one equation at every spatial location, essentially a mixed IVP-BVP (IVP in time, BVP in space). If we discretize (8.49) over time, then

$$\underline{z}(n) = \underline{z}(n-1) + \delta_t A\underline{z}(n-1) \qquad B\underline{z}(n) = \underline{b}(n) \tag{8.50}$$

In order to be able to cast the temporal and boundary conditions into a common linear system we need to employ *Backward Euler* in the time step, thus

$$\begin{aligned} \underline{z}(n) - \delta_t A\underline{z}(n) &= \underline{z}(n-1) \\ B\underline{z}(n) &= \underline{b}(n) \end{aligned} \qquad \longrightarrow \qquad \begin{bmatrix} I - \delta_t A \\ B \end{bmatrix} \underline{z}(n) = \begin{bmatrix} \underline{z}(n-1) \\ \underline{b}(n) \end{bmatrix} \tag{8.51}$$

Thus we have a large linear system to solve at each discretized time step.

Example 8.1: Spatial Discretization

The text refers to setting up a linear system of equations in order to solve a BVP. So how do we *really* do this?

Let's re-create the right panel from Figure 8.11. Although the horizontal axis is labelled as t (time), really t is just a dummy variable, and in the context of a BVP it is more intuitive to think of the independent variable as spatial, so here we will use x. Thus our problem is

$$\begin{array}{ll} \text{Differential Equation:} & \ddot{z}(x) + z(x) = 0 \\ \text{Boundary Conditions:} & z(0) = 0, \quad \dot{z}(0) = 2, \quad \dot{z}(\pi/2) = -1 \end{array} \tag{8.40}$$

Our spatial domain has extent $0 \le x \le \pi/2$, which we will discretize into steps of size δ_x, thus we have a vector of unknowns

$$\underline{z}^T = \begin{bmatrix} z(0) & z(\delta_x) & z(2\delta_x) & \dots & z(\pi/2) \end{bmatrix} \tag{8.41}$$

From (8.29)–(8.33) we know how to discretize the second derivative $\ddot{z}$,

$$\ddot{z} + z = \left\{ \frac{z(x+\delta_x, y) - 2z(x,y) + z(x-\delta_x, y)}{\delta_x^2} \right\} + z \tag{8.42}$$

therefore $\ddot{z}(x) + z(x) = 0$ expressed in the first three elements of $\underline{z}$ is

$$\begin{bmatrix} 1/\delta_x^2 & (1 - 2/\delta_x^2) & 1/\delta_x^2 & 0 & 0 & \dots \end{bmatrix} \cdot \underline{z} = 0 \tag{8.43}$$

We can begin to see the linear system of (8.48) taking shape, so that

$$A\underline{z} = \frac{1}{\delta_x^2} \begin{bmatrix} 1 & (\delta_x^2 - 2) & 1 & 0 & 0 & 0 & 0 \dots \\ 0 & 1 & (\delta_x^2 - 2) & 1 & 0 & 0 & 0 \dots \\ 0 & 0 & 1 & (\delta_x^2 - 2) & 1 & 0 & 0 \dots \\ & \vdots & \vdots & \vdots & \vdots & & \end{bmatrix} \cdot \underline{z} = \underline{0} \tag{8.44}$$

We also need to assert the boundary conditions $Bz = \underline{b}$. Just like the preceding construction, each condition will correspond to one row in B:

$$B\underline{z} = \begin{bmatrix} 1 & 0 & 0 \dots 0 & 0 & 0 \\ -1/\delta_x & 1/\delta_x & 0 \dots 0 & 0 & 0 \\ 0 & 0 & 0 \dots 0 & -1/\delta_x & 1/\delta_x \end{bmatrix} \cdot \underline{z} = \begin{bmatrix} 0 \\ 2 \\ -1 \end{bmatrix} = \underline{b} \qquad \begin{array}{r} z(0) = 0 \\ \dot{z}(0) = 2 \\ \dot{z}(\pi/2) = -1 \end{array} \tag{8.45}$$

One final detail — the combined linear system of A and B is overconstrained (more equations than unknowns), so the linear systems solver cannot satisfy (8.44) and (8.45) with equality. We knew this: we *want* the differential equation to accommodate the asserted boundary conditions, but we do *not* want the boundary conditions to accommodate the differential equation. The simplest approach is to assert the boundary conditions more strongly; that is, to solve

$$Q\underline{z} = \begin{bmatrix} A \\ 1000 \cdot B \end{bmatrix} \underline{z} = \begin{bmatrix} \underline{0} \\ 1000 \cdot \underline{b} \end{bmatrix} = \underline{q} \quad \longrightarrow \quad \begin{array}{c} \text{In Matlab} \\ \texttt{z = Q \textbackslash{} q;} \end{array} \tag{8.46}$$

It is precisely (8.46) which was used to generate Figure 8.11c.

Example 8.2: The Spatial Dynamics of Human Impact

A highway built through a forest might occupy much less than 1 % of the surface area of the forest, however that fractional coverage may be a highly misleading assessment of impact.

Any human impact, whether a water well, a pollutant, or a road, acts as a source or sink on the surrounding natural system, a source or sink which appears as a spatial boundary condition in the associated dynamic system.

In the case of a road, local living creatures such as worms, snakes, raccoons and moose all experience significant death rates from car traffic. We could model such a system in detail, as an ensemble of spatial agents (Section 8.6), or in PDE form, like a diffusion model.

For a given natural pre-existing, unimpacted animal population density P_{Pre}, the impacted population density P will be affected by three basic parameters:

- The roadkill model, a probability of death with every road crossing, introducing a constraint on P *at* the road.
- The mobility model, a measure of movement speed or territory size, controlling the degree of spatial coupling in P.
- The reproduction model, controlling the rate over time at which P increases towards P_{Pre}.

The impacted population density P will be affected over some spatial extent τ, an extent which will increase with higher mobility and lower reproduction rate, leading to steady-state solutions

Example continues ...

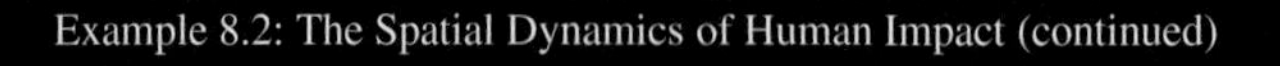
Example 8.2: The Spatial Dynamics of Human Impact (continued)

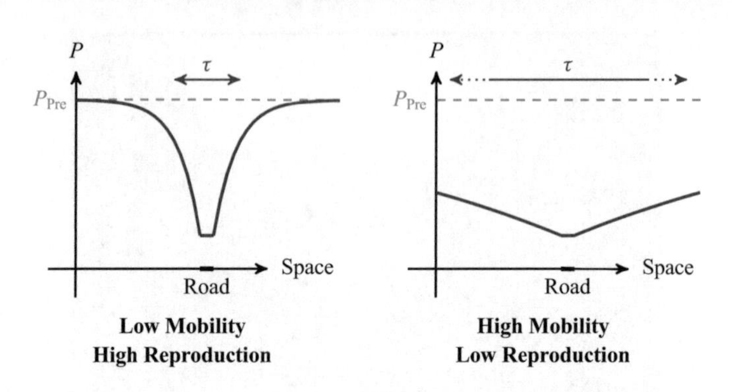

Therefore human impact can extend *far* from a given disturbance, particularly on sensitive animals having long lifespans, low reproductive rates, and large territories, such as moose or wolves.

3. BVP Over Time, Plus Other Constraints:
Why should we have constraints only at the boundaries of the domain? We have satellites continuously measuring the whole planet, thus we have constraints all over the atmosphere and oceans, for example, and not just at boundaries.
In principle, the measurements $\underline{m}$ are just another piece of information regarding the unknown field $\underline{z}$, mathematically no different from a boundary condition, so we just introduce additional constraints

$$\frac{\partial \underline{z}}{\partial t} = A\underline{z}(t) \qquad B\underline{z}(t) = \underline{b}(t) \qquad C\underline{z}(t) = \underline{m}(t) \tag{8.52}$$

which could all be combined, in principle, as in (8.51). In practice, (8.51) will almost certainly have either more or fewer equations (constraints) than unknowns (state elements), and any solution will need to be stochastic.

4. Stochastic BVP Over Time, Plus Other Constraints:
The framework of (8.52) is deterministic, however most of the world is not! As discussed in Section 4.2, all physical systems are subject to random influences and fluctuations, thus our dynamic model, boundary conditions, and associated measurements are all subject to some degree of uncertainty:

$$\frac{\partial \underline{z}}{\partial t} = A\underline{z}(t) + noise \qquad B\underline{z}(t) = \underline{b}(t) + noise \qquad C\underline{z}(t) = \underline{m}(t) + noise, \tag{8.53}$$

no longer a deterministic linear system.

Even in the absence of an explicitly stochastic model we will rarely be solving a deterministic problem, since the linear system of (8.51), (8.52) will almost certainly not be invertible, and will require a stochastic approach.

What (8.52), (8.53) describe is an *inverse problem*, to be discussed further in Section 11.4.

The discussion throughout this section has focused entirely on spatially-discretized differential equations. However there are at least three reasons to want to consider alternative spatial models:

1. *The PDE is too challenging to solve.* There are many large-scale spatio-temporal problems which would require an exceptionally fine spatial discretization in order for certain phenomena of interest to be manifested, such as the appearance of ocean eddies, discussed on page 181, or the modelling of precipitation.
2. *The PDE may obscure insight.* The PDE may be small enough to be numerically solvable, however that does not necessarily make the results easy to interpret. To acquire an understanding of system behaviour we may prefer a relatively high-level model that reflects certain important aspects of the system, but does not try to simulate the underlying physics in detail.
3. *We may not be able to write down the PDE.* Even if we understand the physics, there may be so many unknown parameters that we are unable to specify a particular PDE. Both clouds and arctic ice are examples whereby we understand very well the physics of water condensation and freezing, but complicating factors of aerosols, droplet formation, arctic heat transfer etc. make modelling exceptionally challenging.

Lumped-parameter models are one approach to spatial modelling, whereby a problem is significantly simplified by representing higher-level behaviour directly, rather than expecting it to emerge from a fine-scale PDE. We already encountered such aggregated, simplified models in the global climate representation of Figure 6.19 and the societal Panarchy model of Example 7.8. In this chapter, Case Study 8 will discuss this strategy in the context of climate models, for which a tenth-degree spatial resolution is not computationally possible, so that the size, frequency, and distribution of ocean eddies is explicitly parameterized into the model.

A key benefit of the lumped approach is that the reduction in state size makes more likely the ability to analyze and understand the system behaviour. Particularly if the system is nonlinear, the methods of Chapter 7 are applicable only to rather low-dimensional states and not to PDEs. See Example 8.3 for one such illustration, whereby we can infer bifurcation behaviour in a spatial environmental problem.

Example 8.3: Two-Box Lumped-Parameter Model

In Case Study 6 we saw a variety of environmental systems which exhibit bifurcation and hysteresis, including the Thermo–Haline Circulation on page 126. Simulating the Thermo–Haline Circulation is a daunting task, requiring us to model precipitation, evaporation, arctic freshwater runoff, coastlines, and ocean-bottom bathymetry. In light of this complexity, a broad system-level lumped-parameter model might be more instructive [11].

What actually *drives* flows in the ocean? Wind dominates the surface flows of Section 8.2, however deeper flows are driven by differences in density ρ, caused primarily by differences in temperature T and salinity S. Consider the so-called "two box" model:

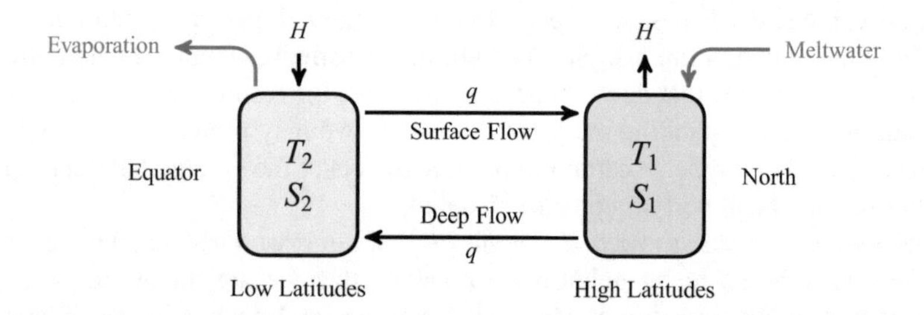

There is a net evaporation of water at low latitudes, and a net input of meltwater and rain at high latitudes. The two-box model mimics these effects by creating an input H of salt at low latitudes (as water evaporates the salinity goes up) and an output H of salt at high latitudes (as freshwater arrives the salinity goes down). The flow q is driven by the difference in density ρ, where density increases with salinity but decreases with temperature:

$$q = k(\rho_1 - \rho_2) = k(\beta(S_1 - S_2) - \alpha(T_1 - T_2)) = k(\beta \Delta_S - \alpha \Delta_T) \tag{8.54}$$

It is interesting to note that temperature and salinity drive the flow in opposite directions: the low latitudes are warmer ($q > 0$) but also more saline ($q < 0$).

We will focus on salinity, assuming that temperature equilibrates relatively quickly on the time scales of interest and that Δ_T is constant. Therefore

$$\frac{d\Delta_S}{dt} = -2H - 2|q|\Delta_S = -2H - 2\left|k(\beta\Delta_S - \alpha\Delta_T)\right|\Delta_S \tag{8.55}$$

With a change of variable and decluttering the notation slightly, we have

$$x = \frac{\beta\Delta_S}{\alpha\Delta_T} \quad \longrightarrow \quad \frac{dx}{dt} = a - b\,|x-1|\,x \tag{8.56}$$

That is, we have a one-dimensional nonlinear system in two parameters, for which the fixed points and bifurcations can be found (see Problem 8.4).

Further Reading: H. Stommel, "Thermohaline Convection . . . ," *Tellus* (13) #2, 1961
H. Kaper, H. Engler, *Mathematics & Climate*, SIAM, 2013 [11]

Example 8.4: Global Flows II

Global flows were briefly looked at, from a systems perspective, in Example 3.7. Although each of the water, carbon, nitrogen, aerosol, and heat cycles are global in scope, we may or may not choose to explicitly construct a spatial model. That is, we have three modelling alternatives:

Spatial Model: The model is a spatially-discretized PDE, with large-scale structure emerging *implicitly* from the fine-scale model. Such models include most weather forecasting systems and the dust/aerosol flow modelling from page 169.

Lumped Model: Although the model represents a spatial phenomenon (flow around the globe), the model itself is not spatial, instead representing large-scale structure *explicitly*. Many systems-level simulations fall into this category, such as the primitive carbon-cycle diagram on page 6, the nutrient-cycle models of Case Study 3, and the substantial population/resource modelling of the *Limits to Growth* work [14].

Mixed Model: Most models will not be purely abstract lumps or strictly a discretized PDE. There is significant rationale in having low-resolution spatial models, able to differentiate between arctic, temperate, and tropical regions, for example, but with significant explicitly modelled behaviour (ocean eddies, aerosol sources, cloud formation etc.).

All five cycles present spatial modelling challenges and represent highly active research areas, however the carbon and heat cycles are of particular interest, both because of concerns regarding human–driven climate change, but also because of intriguing uncertainties.

That is, to some extent we do not know where significant quantities of carbon are actually going. We know quite accurately the extent of fossil fuel consumption and the carbon levels in the atmosphere, however there remain unexplained mismatches.

Similarly there are imbalances in global heat/energy flows. Outside of the atmosphere it is possible for satellites to measure incoming and outgoing radiation quite precisely, however heat fluxes within and below the atmosphere are far more difficult to measure.

In particular, many global flows are very large, for example solar energy input of 10^{17} W or carbon flows of 10^{17} g/y, and are subject to modest measurement errors. These errors are *much* larger, however, relative to the small *difference between flows* (incoming minus outgoing), which represents the accumulated carbon or heat of interest to us.

To be sure there is enormous interest in spatially mapping carbon and radiation. The NASA *Orbiting Carbon Observatory-2 (OCO-2)* mission, launched in 2014, focuses on spatial carbon flows. Similarly the NASA *Clouds and the Earth's Radiant Energy System (CERES)* project is dedicated to refining our understanding of spatial heat flows.

Further Reading: See Water cycle, Nitrogen cycle, Carbon cycle, Earth's Energy Budget in *Wikipedia*. Link to NASA OCO-2 page at the textbook further reading page.

8.5 Spatial Discrete-State Models

The question of discretization is always one of *degree*: from Section 8.3 we know that a sufficiently fine temporal and spatial discretization leads to a discrete-time pixellated model which has very nearly the same behaviour as the original PDE.

Similarly the numeric representation of the problem state $\underline{z}$ is, in principle, discrete in value, since it is stored in a fixed number of bits in a computer. However modern double-precision floating-point representations offer 16 digits of accuracy and are, therefore, awfully close to continuous in value.

In contrast, when we talk about discrete-state models, we mean a model in which each state element is limited to a relatively small number of distinct values, such as

$$z \in \{0, 1\} \qquad z \in \{A, \ldots, Z\} \qquad z \in \{Land, Water\}. \tag{8.57}$$

Discrete-state fields are most commonly encountered in label maps, where the state represents a category:

Maps of arctic ice distribution $z \in \{Water, Ice\}$
Application: Ocean shipping, Climate change

Maps of urban areas $z \in \{Urban, Natural\}$
Application: Studies of urban sprawl, land use, farm policy

Maps of forest state $z \in \{Disturbed, Undisturbed\}$
Application: Tropical deforestation, ecology

Discrete-state dynamic models are, by definition, nonlinear, since any linear dynamic function $f()$ must be continuous and cannot be limited to a discrete set of state values. A *spatial* discrete-state model is, therefore, a coupled set of nonlinear dynamics, and is therefore nearly impossible to analyze mathematically.[5] Consequently our emphasis is now very much on model simulation. We are still interested in analysis, but now on the analysis of simulation results, rather than an analysis of the model fixed points etc.

The simplest possible spatial dynamic model is a one-dimensional sequence of binary states having local evolution rules governing their dynamic behaviour, as shown in Figure 8.12. That is, the state $z_i^n \in \{0, 1\}$ at location i and time iteration n evolves as

$$z_i^{n+1} = f\left(z_{i-1}^n, z_i^n, z_{i+1}^n\right) \tag{8.58}$$

Because all of the states are binary, there are only $8 = 2 \times 2 \times 2$ possible unique input combinations, and for each of these eight possibilities f can assign to z_i^{n+1} either zero or one, meaning that there are $2^8 = 256$ possible rules, the so-called Wolfram Rules 0 through 255 [24].

[5] A famous exception being the analytical solution to the Ising model of a ferromagnet, discussed later in this section.

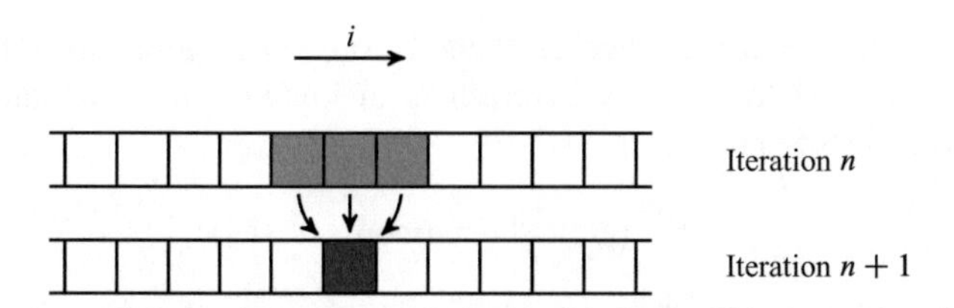

Fig. 8.12 1D automaton: a one-dimensional cellular automaton allows a local neighbourhood of values (green) at iteration n to determine the value of the state z_i^{n+1} (red) at iteration $n+1$.

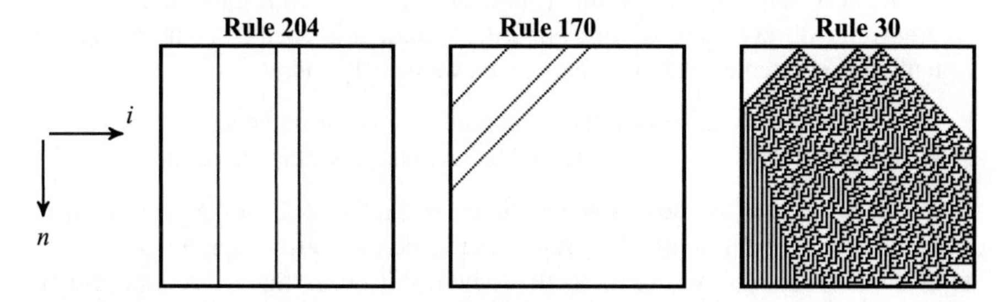

Fig. 8.13 1D automata: three examples of one-dimensional binary automata, per Figure 8.12. Such automata predictably lead to simple behaviour, like copying cells from one row to the next (left) or from one column over (middle), however the *same* class of models also includes very unusual, nearly random behaviour (right). Each example was initialized in the top row with all zero (white) except for three values of one (black).

Most of the rules give rise to relatively simple behaviour, such as Rule 204 ...

$$\begin{array}{r} \text{Current state values } \left(z_{i-1}^n, z_i^n, z_{i+1}^n\right) \\ 111\ 110\ 101\ 100\ 011\ 010\ 001\ 000 \\ \downarrow \quad \downarrow \quad \downarrow \quad \downarrow \quad \downarrow \quad \downarrow \quad \downarrow \quad \downarrow \\ \text{Wolfram Rule } 204_{10} = (\ 1 \quad 1 \quad 0 \quad 0 \quad 1 \quad 1 \quad 0 \quad 0\ )_2 \\ \text{Next state value } z_i^{n+1} \end{array} \tag{8.59}$$

which just copies the state elements from iteration n into iteration $n+1$, or

$$\begin{array}{r} \text{Current state values } \left(z_{i-1}^n, z_i^n, z_{i+1}^n\right) \\ 111\ 110\ 101\ 100\ 011\ 010\ 001\ 000 \\ \downarrow \quad \downarrow \quad \downarrow \quad \downarrow \quad \downarrow \quad \downarrow \quad \downarrow \quad \downarrow \\ \text{Wolfram Rule } 170_{10} = (\ 1 \quad 0 \quad 1 \quad 0 \quad 1 \quad 0 \quad 1 \quad 0\ )_2 \\ \text{Next state value } z_i^{n+1} \end{array} \tag{8.60}$$

which copies the value of the next state element over, leading to a shifting behaviour over time, both of which are shown in Figure 8.13.

Nevertheless Rule 30, also shown in Figure 8.13, demonstrates a degree of irregularity or chaos that seems hard to fathom could arise from such an exceedingly simple, deterministic, local rule. We will have more to say about Rule 30 in the context of complex systems in Chapter 10.

The one-dimensional sequence of states z_i^n obviously generalizes to the two-dimensional case, where we now have a map or image of discrete states $z_{i,j}^n$. The local dynamic of (8.58) becomes

$$z_{i,j}^{n+1} = f\big(z_{i,j}^n \text{ and neighbours of } (i,j)\big), \tag{8.61}$$

as sketched in Figure 8.14. The most famous of the two-dimensional automata is Conway's *Game of Life*, as shown in Figure 8.15, a simple rule set that leads to intriguingly complex behaviour.

We need to move on: the cellular automata we have seen thus far are fun to program and observe, but of limited use in modelling social or environmental systems. Two ingredients are missing, which we need to address:

1. We need stochastic models, rather than purely deterministic ones.
2. We need cell rules driven by observed behaviour, rather than made up.

The need for stochastic models was outlined in Section 4.2. In our context here, in simulating a given automaton, representing perhaps urban sprawl or tropical deforestation, we don't want *exactly* the same results each time we run the model, for the same reason that not all sprawl or deforestation patterns are the same. That is, we need a certain degree of irregularity or unpredictability.

The most famous of the stochastic cellular automata is the *Ising* model, which is a simple model of ferromagnetism. Each iron atom is permitted a binary state —

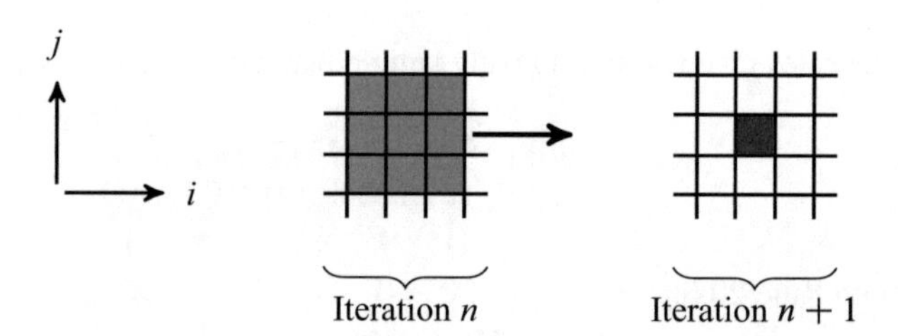

Fig. 8.14 2D automaton: a simple generalization of Figure 8.12, a *two*-dimensional cellular automaton allows a local neighbourhood of values (green) at n to determine the value of the state z_{ij}^{n+1} (red) at iteration $n+1$.

Evolution Rule

# Neighbours of z_{ij}^n which are "on"	z_{ij}^{n+1}
0 – 1	off
2	z_{ij}^n
3	on
4 – 8	off

Iteration 0

Iteration 125

Fig. 8.15 The game of life: Conway's *Game of Life* is a two-dimensional cellular automaton, with dynamic rules as listed in the table. These simple, local rules can give rise to quite striking behaviour, similar to Wolfram's Rule 30 in Figure 8.13.

spin-up or spin-down, as sketched in Figure 8.16. The tendency for an atom to align with its neighbours, to form larger magnetized crystals, is controlled by a coupling or inverse-temperature parameter β. Thus when very hot ($\beta \approx 0$) the atoms flip their state more or less at random, whereas when very cold ($\beta > 1$) the atoms are tightly coupled, essentially frozen or crystallized together; both of these examples are plotted in Figure 8.17. We shall return to the Ising model in Chapter 10.

In contrast to the deterministic model of Figure 8.15, the iterative dynamic for the Ising model is now stochastic (that is, containing a random component), as shown in Table 8.2. Similarly, most of the spatial models of interest to us are stochastic, three examples of which are listed in Table 8.3.

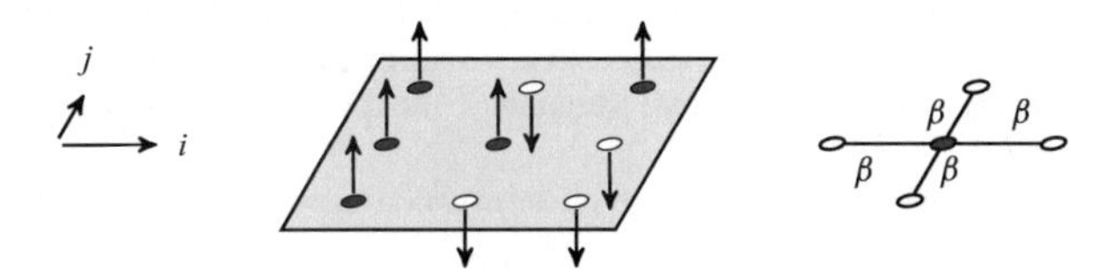

Fig. 8.16 The ising model: we have a grid of iron atoms, where the state of each atom corresponds to a magnetization of up or down. A given atom is coupled with a coupling strength β to its nearest four neighbours. Simulated results are shown in Figure 8.17.

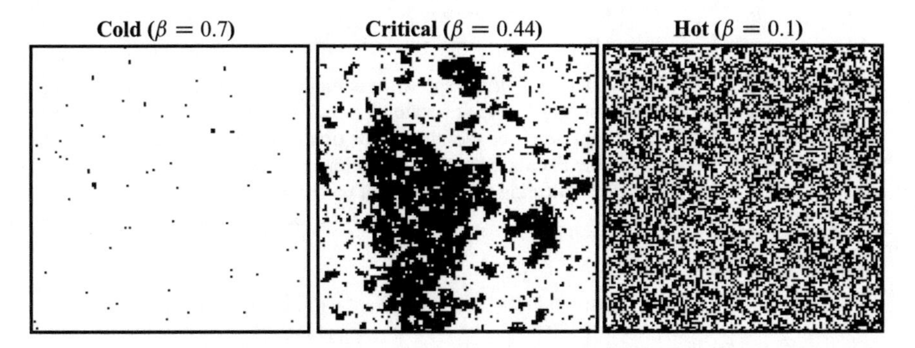

Fig. 8.17 The ising model: we can generate random samples from the Ising model of Figure 8.16; samples are shown here for three different degrees of coupling, controlled by β, where the colour of each pixel (black/white) indicates its spin (up/down). At hot temperatures the behaviour is nearly random; at cold temperatures the state is nearly "frozen" into a single direction. We shall return to the critical transition, in between the two extremes, in Chapter 10.

Table 8.2 The ising model: the Ising model is governed by a stochastic model, just like the deterministic rule in Figure 8.15, but now stated probabilistically. If $\beta = 0$ or $q = 0$ then z_{ij}^n is chosen at random; as β increases it becomes more likely that z_{ij}^n is chosen to equal the majority of its neighbours.

How many *Up* minus *Down* states in the neighbours of z_{ij}^n	$\Pr\left(z_{ij}^{n+1} = Down\right)$	$\Pr\left(z_{ij}^{n+1} = Up\right)$
q	$\dfrac{\exp(-q\beta)}{\exp(q\beta) + \exp(-q\beta)}$	$\dfrac{\exp(q\beta)}{\exp(q\beta) + \exp(-q\beta)}$

Table 8.3 Three examples of stochastic automata: these models are relatively primitive, however they do begin to illustrate the range of spatial phenomena which we can model as stochastic cellular automata; that is, automata with some degree of randomness or uncertainly, more closely reflecting the real world.

Forest-Fire Model:

State of z_{ij}^n	# Burning neighbours		Probability that z_{ij}^{n+1} has state: Ash	Tree	Burning
Tree	0	$\longrightarrow$	0	$1-f$	f
Tree	>0	$\longrightarrow$	0	0	1
Burning	$\times$	$\longrightarrow$	1	0	0
Ash	$\times$	$\longrightarrow$	$1-p$	p	0

This simple stochastic spatial cellular automaton models the spread of forest fires. In the table, $\times$ means "doesn't matter;" that is, if a tree is already burning, it doesn't matter how many of its neighbours are burning. The model has two parameters: the forest regeneration rate p and the rate of lightning strikes f.

See Drossel, Schwabl, "Self-organized critical forest-fire model.", *Phys. Rev. Lett. 69*, 1992

Urban Development Model:

Current z_{ij}^n	State: t_{ij}^n	Number of developed neighbours		Next state at $n+1$: $\Pr(z_{ij}^{n+1}=U)$	$\Pr(z_{ij}^{n+1}=D)$	t_{ij}^{n+1}
D	$\times$	$\times$	$\longrightarrow$	0	1	$\times$
U	q	0	$\longrightarrow$	1	0	q
U	q	>0	$\longrightarrow$	$1-p^q$	p^q	$q+1$

A simple model of city development and urban sprawl, this model is a step more complex than previous ones in that each cell now has a vector state (z,t), where z reflects developed (D) or undeveloped (U) land, and t keeps track of the number of times that development failed. The model has a single parameter p, the likelihood of parcel development. Over time a given parcel of land is less likely to be developed, reflecting the presence of undeveloped parcels in cities.

See Batty, *The New Science of Cities*, MIT Press, 2013 [3]

Invasive Plant Species Model:

State Transition Probability from Plant p to Plant q

$$\Pr\big(z_{ij}^n = p \longrightarrow z_{ij}^{n+1} = q\big) = P_{p\to q} \cdot \frac{\text{\# Neighbours of type } q}{4}$$

This model goes to some effort to be ecologically realistic by incorporating observed invasive behaviours. Six plant species were simulated, thus $P_{p\to q}$ is a 6×6 matrix of invasion probabilities. The probabilities are scaled by the number of the four immediate neighbours having a plant species of competing type q.

See Silvertown et al., "Cellular Automaton Models of Interspecific Competition ...," *J. Ecology* (80) #3, 1992

8.6 Agent Models

Most of the spatial models discussed earlier in this chapter have focused on planet-wide environmental phenomena, however there are many other examples of spatially-distributed systems, particularly social systems, consisting of interacting and mobile individuals. Such systems are poorly represented as PDEs, which are best suited for continuous flows, or as cellular automata, which have no ability for cells to move around, in the way that people, animals, insects, and bacteria would do. Agent models are the generalization of cellular automata, in which we have some number of states

$$\underline{z}_i^n = \begin{bmatrix} x_i^n \\ y_i^n \\ \underline{\phi}_i^n \end{bmatrix} \qquad 1 \le i \le S^n \qquad \text{or} \qquad \underline{z}_i^n = \begin{bmatrix} \underline{\phi}_i^n \end{bmatrix} \qquad 1 \le i \le S^n \tag{8.62}$$

for spatial and non-spatial models, respectively, where the number of state elements S^n can change as a function of time n, what are known as birth–death processes. For spatial models the location (x, y) associated with each state can change over time, where the location could be discrete (on a grid, like a cellular automaton) or continuous. For that matter, the non-spatial state elements in $\underline{\phi}$ can also be a mix of continuous and discrete values.

The dynamic system model represents an update rule on the state:

$$\left\{\underline{z}_1^{n+1}, \ldots, \underline{z}_{S^{n+1}}^{n+1}, S^{n+1}\right\} = f\left(\underline{z}_1^n, \ldots, \underline{z}_{S^n}^n, S^n, \underline{\theta}\right\}, \tag{8.63}$$

where the state interaction and model evolution are controlled by learned parameter vector $\underline{\theta}$.

The philosophy underlying agent-based modelling is that we have many entities/agents. We assert micro-level interactions between individual agents, and then wish to observe the macro-level, large-scale emergent behaviour. In many cases, the most interesting cases involve self-organizing behaviours of structure and order:

- Can evolutionary agents develop large-scale structures such as differentiated cultures or strata?
- Can elements of game theory spontaneously arise, such as coalitions or collective actions, without such groupings having been asserted via some sort of top-down control?

Like the discrete-state models of the previous section, agent models are nonlinear. The whole rationale is to examine models and rule-sets too complex to tackle analytically, analyzing the simulated results as a function of rule or parameter changes. Certainly quite unusual and life-like flocking/swarming/herding behaviours can appear in agent models, precisely one of the aspects which make them intriguing; we will come back to some of these in our discussion of complex systems in Chapter 10.

One of the classic agent models in the literature is a simple model of segregation, for example to study the roots of racial segregation in cities. The agent rule is exceptionally simple:

$$\textit{if}\ (\%\ \text{neighbours not like me} > \rho)\ \textit{then}\ (\text{move at random to empty location}) \tag{8.64}$$

The iteration continues as long as there are agents not satisfied with their neighbourhood.

We therefore have an agent state $\underline{z}$ with a constant number S of agents living on a discrete spatial grid, where the state value associated with a grid location is zero, if empty, or racial type $1, \ldots, T$. The following parameters would need to be specified:

- The number of types T
- The size of the spatial grid
- The number of agents, or the fraction of empty grid locations
- The preference parameter ρ
- The definition of "neighbour"

Simulated results are shown in Figure 8.18 as a function of ρ.

A vast range of agent-based models has been developed, studying human behaviour (altruism, collective action), economics (poverty, unemployment, taxation), ecology (spread of insects or invasive species) and many more. Indeed, in principle we could take *any* collection of individuals (people, insects etc.), define interactions, and run a simulation.

It is important to be aware, however, of the limitations of such modelling: Producing agent-model simulation results certainly does not imply that the results are meaningful or have predictive power. In very many models the agent interfaces are trivialized representations of human behaviour. As a consequence the results may represent self-fulfilling prophecies in that the model simulation gives rise to precisely those actions specified in the program. Furthermore, since the agents live in an artificial world, the conclusions can be challenging to generalize or apply in practice. As with any model, whether agent–based or not, a key step is validation:

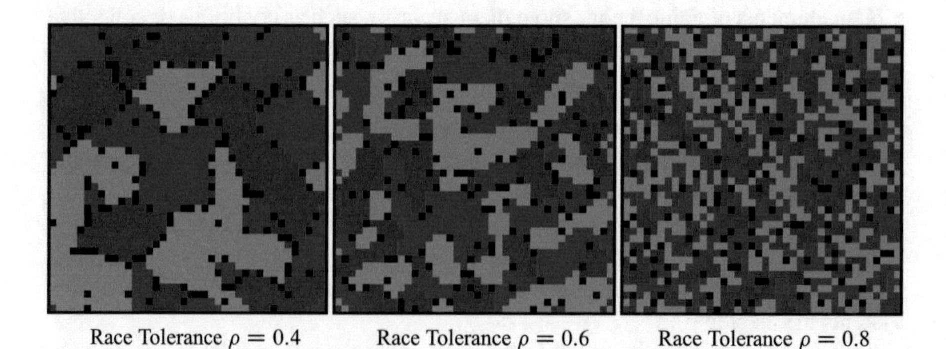

Fig. 8.18 Segregation model: What degree of racial tolerance gives rise to spatial segregation? Three simulations of Schelling's agent model are shown, based on the preference rule in (8.64). A $T = 3$ simulation is shown on a 40×40 grid; black squares represent unoccupied locations. In the middle panel people are wiling to have a *majority* of their neighbours different from themselves, nevertheless relatively large homogeneous regions still form.

Model from Schelling, "Dynamic Models of Segregation," *J. Mathematical Sociology*, 1971.

1. The model f in (8.63) was learned, indirectly, by the explicit learning of parameter vector $\underline{\theta}$ from a given set of training data $\mathcal{D}_1$. A first step in validation, the *in–sample validation*, is to test the extent to which the resulting model f is consistent with the training data.
2. It is easy, in principle, for f to exhibit excellent consistency with the training data by *overfitting*, essentially by memorizing $\mathcal{D}_1$. What we actually desire, however, is *predictive* power, the *out–of–sample validation*, such that model f is consistent with a training data set, $\mathcal{D}_2$, which was *not* presented during the learning of $\underline{\theta}$.

The goal of the validation is to ensure the meaningfulness of the model output. That is, we wish the agent–based model to reflect the *persistent*, higher level aspects of the system being studied, rather than being an overfitting to *temporary* aspects of the system as appearing in the training data.

Case Study 8: Global Circulation Models

Global circulation models attempt to simulate the ocean–atmosphere system of the planet, taking into account the physical topography of the earth (mountain ranges, oceans etc.) and simulating physical processes (evaporation, clouds, ice formation etc.). There are two key motivations driving the development of such algorithms:

- Weather forecasting;
- Understanding global climate, particularly climate change.

How might we create such a model? Presumably we would start with Navier–Stokes in compressible (atmosphere) and incompressible (ocean) forms and spatially discretize these onto a tenth-degree map to be eddy-resolving, with some number of layers in depth, typically 40. At every spatial location we would need a representation of the essential model state:

Atmosphere:	Air velocity (wind), temperature, humidity
Ocean:	Water velocity (current), temperature, salinity

Clearly additional variables are important, including clouds, soil moisture, and ice thickness, to name a few.

It is time, however, for a quick reality check; the ocean state consists of

$$\underbrace{3600}_{\text{Longitude}} \cdot \underbrace{1800}_{\text{Latitude}} \cdot \underbrace{40}_{\text{Layers}} \cdot \underbrace{4}_{\text{State}} = 1 \text{ Billion variables!} \tag{8.65}$$

So at double-precision accuracy the state vector for the ocean, at just *one* time step, requires 8GB of RAM. Next, for the time discretization, we need a time step sufficiently fast to capture the time scale of oceanic and atmospheric dynamics (minutes), and then to run the simulation on time scales over which climate change

Example 8.5: Spatial Nonlinear Systems and Traffic Jams

Automobile traffic is a fascinating problem: it has great socio-economic significance, can be studied using PDEs, nonlinear methods, and agent-based models, and everybody has an opinion on it!

Per lane of highway, the flow f of cars (number per unit time) increases with speed, but only to a point, since higher speeds lead to greater car separation. Starting with an empty highway, as demand (flow) increases the speed is unaffected (green arrow), but then speed begins to decrease (red arrow) to accommodate additional cars.

However the connection between measured traffic and speed can be unclear, since some people choose to drive slowly, below the speed limit. Instead, the state of congestion is the *density* of cars on the road. For $f < f_{\max}$ we have a fixed point (lower curve) of free flow; at $f_{\max}$ we have a bifurcation, and attempts to exceed $f_{\max}$ lead to a sudden state transition from free to congested flow, and f is forcibly throttled down (upper curve).

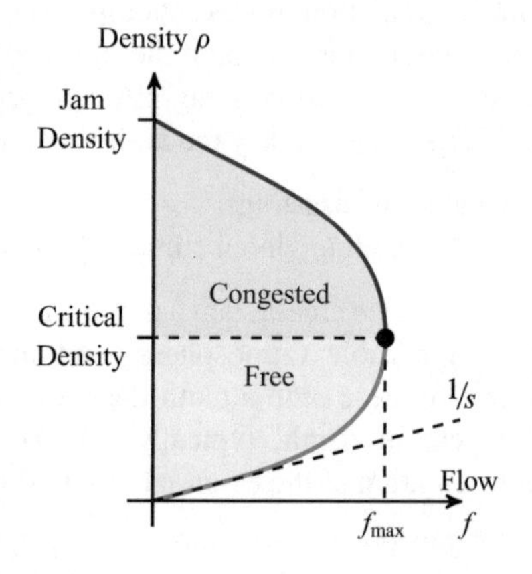

In the congested state, traffic leads to shock-wave behaviour that can be studied using PDEs. These shock waves travel upstream (against traffic), typically at a speed around 20 km/h.

Traffic congestion is fantastically well-suited to agent-based modelling, since individual agents are easy to define (drivers who are lane-changers, tailgaters, aggressive, rule-abiding, distracted, slow etc.), with associated driving parameters of reaction time, rate of acceleration, preferred separation to next vehicle, lane-change likelihood etc. Furthermore, computer simulations allow for more complex geometries than a single lane, such as lane reductions, highway ramps, and speed reductions in tunnels.

A wide variety of sophisticated traffic models has been developed, many of them open-source and available for experimentation. Some of the better known packages include ARCADY, ARCHISIM, CORSIM, and MATSim.

Further Reading: See Traffic flow, Traffic simulation, Three-phase traffic theory in *Wikipedia*;
Banks et al., *Discrete-Event System Simulation*, Prentice-Hall, 2009.
Vanderbilt, *Traffic*, Knopf, 2008.
Links to ARCADY, ARCHISIM, CORSIM, and MATSim at the textbook further reading page.

might be observed (centuries):

$$\left.\begin{array}{rcl}\text{Time Scale of Dynamics} & \rightarrow & \text{Minutes} \\ \text{Time Scale of Climate Change} & \rightarrow & \text{Centuries}\end{array}\right\} 30 \cdot 10^6 \text{ time steps!} \qquad (8.66)$$

Hopefully it is becoming clear that global circulation or global climate models are among the most challenging numerical simulation problems world-wide, such that many of the largest supercomputers in the world are dedicated to challenges of weather and climate forecasting.

Since the idealized spatial/temporal discretization scenario, just outlined, is computationally unfeasible, by definition global climate simulations are always starved for computer resources. The overall complexity is more or less the product of

- Spatial discretization,
- Temporal discretization,
- Simulation period, and
- State/lumped model complexity,

therefore there is necessarily some tradeoff present between these four goals. Table 8.4 gives some discretization parameters for three models, where we can observe a stark contrast between the high-resolution/tiny time-step/short period model for weather forecasting (WRF), which is intended to simulate over a period of a week or two, and the much coarser models for climate change (CM3/GEM2), which need to simulate over a much longer period of centuries to millennia.

To be sure, there is a great deal more to comparing models than just the number of atmospheric and oceanic state elements. At the very least we require extensive boundary conditions:

- Ocean bathymetry, characterizing the shape of the ocean floor;
- Continental and island coastlines;
- Sea surface temperature as the boundary between oceanic and atmospheric models;
- Radiation in/out, surface albedo (surface reflectivities of ice, forest, desert, cities etc.).

Table 8.4 Example parameter settings for climate models HadCM3, HadGEM2 and weather forecasting model ES-WRF.

Model	Atmosphere				Ocean			
	Resolution		Vertical	Time	Resolution		Vertical	Time
	Latitude	Longitude	layers	step	Latitude	Longitude	layers	step
HadCM3	2.50°	3.75°	19	30 min	1.25°	1.25°	20	1 h
HadGEM2	1.25°	1.88°	38–60	30 min	1.00°	1.00°	40	1 h
ES-WRF	0.05°	0.05°	101	6 s				

Next, following on the discussion in Section 8.4, we know that a resolution significantly smaller than 1° is required to resolve eddies, a resolution which is not provided by the state-of-the-art HadGEM2 model. However from Section 8.4 we also know that we can incorporate aggregate behaviour in the form of a lumped model, rather than limiting the model to being purely the discretization of a single PDE. Indeed most climate models are very much lumped models, such that certain dynamics which do not implicitly appear from the discretized model (such as ocean eddies in a low-resolution model) can be explicitly included in some way. For example, very much like the two-box model of Example 8.3, the HadCM2 model which preceded HadCM3 required so-called flux adjustments, forced additions or removals of heat and freshwater in order to produce realistic simulations.

Similar to the concerns regarding the limited prediction power of agent models, one concern with lumped models for global climate is that the model behaviour is built in. That is, the ideal case is to create a simulated earth with oceans, atmosphere, and boundary conditions, and then to run the Navier–Stokes PDE and watch a simulated world appear. However because our spatial discretization is too coarse it is known that ocean eddies, for example, will *not* appear spontaneously, so their presence needs to be explicitly pre-planned and programmed into the model. As more and more empirically observed phenomena are parameterized into the model, the more the outcome of the model becomes a self-fulfilling prophesy. Particularly in the case of climate change this distinction is important, since we wish to believe that the model predictions are based on a simulation of physics, and not the consequence of the phenomena which a scientist chose to include or not include in a lumped model. These modelled phenomena can include ocean eddies, ice formation, clouds, ocean-atmosphere coupling, vegetation, ocean biology and atmospheric chemistry.

Further Reading

The references may be found at the end of each chapter. Also note that the textbook further reading page maintains updated references and links.

Wikipedia Links — Partial Differential Equations: Partial Differential Equation, Initial value problem, Boundary value problem, Discretization, Finite element method

Wikipedia Links — Earth Systems: Advection, Navier–Stokes, Coriolis effect, Low-pressure area, Ocean gyre, Buys Ballot's law

Wikipedia Links — Cellular Automata and Agents: Cellular automaton, Conway's Game of Life, Forest Fire model, Ising model, Agent-Based model

There are a variety of good books explaining connections between mathematics and nature/climate, including [1, 2, 4, 11]. For the more specific topic of partial differential equations there are many accessible texts, such as [6, 21, 23]. A more advanced but very complete treatment can be found in the text by Evans [5].

In terms of the process of discretizing continuous-time or -space signals, the reader is referred to books on signal processing [16], control theory [7], and numerical methods [9]. For discrete-state models, one of the best-known texts [24] is written by Stephen Wolfram, a highly enthusiastic advocate for cellular automata. Other texts include that by Mitchell [15] and a more mathematical book by Sipser [20].

There are nearly endless programs available online for developing and simulating models, particularly agent-based. These range from Java implementations, which can run within a web browser, to full programming environments. Some of the most active agent-based communities are based on *GAMA*, *NetLogo*, and *StarLogo*.

In terms of the dust and aerosol example at the beginning of this chapter the reader is referred to [8, 12].

Sample Problems

Problem 8.1: Short Questions — PDEs

Provide a short answer for each of the following:

(a) What aspects of the continuous-time problem are preserved, and which are changed, in temporally discretizing a given PDE?
(b) We do we mean by a mixed initial-value/boundary-value problem?
(c) What makes solving a BVP problem numerically/computationally more challenging than solving an IVP?
(d) What conservation properties are asserted in deriving the Navier–Stokes PDE?

Problem 8.2: Short Questions — Automata and Agents

Provide a short answer for each of the following:

(a) What makes a discrete-state model inherently nonlinear?
(b) What are the tradeoffs between selecting a large or small neighbourhood?
(c) What are the qualitative differences between a cellular automaton and an agent model?
(d) Briefly summarize the strengths and weaknesses of agent-based modelling of social systems.

Problem 8.3: Analytical — PDEs

Briefly discuss the issues surrounding the choices of spatial and temporal discretization for oceanic and atmospheric climate models.

Problem 8.4: Analytical — Lumped Parameter

The *Two Box* lumped-parameter model for the Thermo-Haline circulation was presented in Example 8.3, where we derived a model

$$\frac{d\Delta_S}{dt} = -2H - 2|q|\Delta_S = -2H - 2\big|k(\beta\Delta_S - \alpha\Delta_T)\big|\Delta_S \tag{8.67}$$

and a simplified form

$$x = \frac{\beta\Delta_S}{\alpha\Delta_T} \quad \longrightarrow \quad \frac{dx}{dt} = a - b\,|x-1|\,x \tag{8.68}$$

(a) Find the fixed points of (8.68) as a function of $\gamma = a/b$.
(b) Sketch and describe the bifurcation plot for this system.
(c) H is the salt flux which drives the system. If $H = 0$, describe both mathematically and in earth-system terms what is happening.

Problem 8.5: Analytical — Spatial Discretization

In (8.28), (8.29) we saw the basic equations for discretizing along a spatial dimension. In this problem we would like to explore the difference between the two equations, in terms of the appearance of δ_x:

(a) Suppose that we made the equations consistent:

$$z_x(x,y) = \frac{z(x+\delta_x,y) - z(x,y)}{\delta_x} \qquad z_{xx}(x,y) = \frac{z_x(x+\delta_x,y) - z_x(x,y)}{\delta_x}$$

Substitute the expression for z_x into z_{xx}, to write z_{xx} explicitly in terms of z. Around which value of x is the second derivative effectively being computed?
(b) Write the effective second-derivative operator from (a) in terms of the convolutional representation of (8.35), being deliberate to show where the origin is located.
(c) The chosen first-derivative operator

$$z_x(x,y) = \frac{z(x+\delta_x,y) - z(x,y)}{\delta_x}$$

is itself asymmetric. As in (b), write this operator in convolutional form. Can you suggest a different definition of a discretized first derivative which is symmetric?

Problem 8.6: Analytical/Computational — Time Discretization

We would like to study the stability of a numerical simulation as a function of the time discretization.

Suppose we have a stable spiral. We know from Chapter 5 that a stable spiral is caused by a conjugate pair of eigenvalues with negative real parts:

$$V = \begin{bmatrix} 1 & 1 \\ -i & i \end{bmatrix} \qquad \lambda_1 = -1 + \alpha i \quad \lambda_2 = -1 - \alpha i$$

If we multiply this out, we deduce the system matrix

$$A = V \Lambda V^{-1} = \begin{bmatrix} -1 & \alpha \\ -\alpha & -1 \end{bmatrix}$$

for which it is easy to see that $\tau < 0, \Delta > 0, \tau^2 - 4\Delta < 0$, so from Figure 7.2 we observe that we do, indeed, have a stable spiral.

(a) Let the time-discretization step be δ. Determine the location of the eigenvalues λ_δ of the time-discretized system.
(b) For the discrete-time system to be stable, we know that discrete-time eigenvalues must satisfy $|\lambda_\delta| < 1$. Compute analytically the upper limit on δ, as a function of α, such that the discretized iteration remains stable. That is, find $\delta_{\max}(\alpha)$ such that the iterated system is stable for $0 < \delta < \delta_{\max}(\alpha)$.
(c) The discrete-time system is very easy to simulate:

$$\text{Given } A \text{ we find } A_\delta = I + \delta A \qquad \text{then} \qquad \underline{z}_{n+1} = A_\delta \underline{z}_n \qquad \underline{z}_0 = \begin{bmatrix} 1 \\ 1 \end{bmatrix}$$

Validate $\delta_{\max}(\alpha)$ which you derived by simulating the discrete-time system over 1000 time steps, and seeing whether

$$|\underline{z}_{1000}| \longrightarrow 0 \text{ for } \delta < \delta_{\max}(\alpha), \quad |\underline{z}_{1000}| \longrightarrow \infty \text{ for } \delta > \delta_{\max}(\alpha).$$

Problem 8.7: Analytical/Computational — Time Discretization

Discrete-time nonlinear systems have some quite fascinating properties, of which we saw a hint in Example 7.2. In general, for a one-dimensional iterated (i.e., discrete-time) system we write the dynamic as

$$z_{n+1} = f(z_n, \theta)$$

for system state z and parameter θ. The fixed points therefore occur at

$$\bar{z} = f(\bar{z}, \theta)$$

that is, where function f intersects the line $y = x$. A saddle bifurcation therefore occurs for some value of θ where f *just* touches $y = x$. The stability of each fixed point is controlled by the slope of f:

$$|f'(\bar{z})| < 1 \rightarrow \bar{z} \text{ is Stable} \qquad |f'(\bar{z})| > 1 \rightarrow \bar{z} \text{ is Unstable}$$

Let us consider the nonlinear iteration

$$z_{n+1} = f\big(z_n, \theta\big) = z_n^2 + \theta \tag{8.69}$$

(a) Derive the location and stability of the fixed point(s) as a function of θ.
(b) For what value of θ is there a saddle bifurcation? Draw a sketch to explain.
(c) Simulate the system over $-2 \le \theta \le 0.5$. For each value of θ,

  (i) Let $z_0 = 1$
  (ii) Iterate (8.69) 100 times to allow the system to settle down
  (iii) Continue iterating (8.69) another 100 times and plot z_n (y-axis) at each iteration versus θ (x-axis)

(d) State your observations. As much as possible, draw connections between the empirical simulation and your analysis.

There is clearly something interesting happening at $\theta = -3/4$. This is a *period–doubling bifurcation*, such that $\bar{z}$ is not a fixed point of $f(z)$, but *is* a stable fixed of $f\big(f(z)\big) \equiv f^{(2)}(z)$. In general, $\bar{z}$ lies on a stable cycle of order n if

$$f^{(n)}(\bar{z}) = \bar{z} \qquad \text{and} \qquad \text{abs}\left(\left.\frac{\partial f^{(n)}(z)}{\partial z}\right|_{z=\bar{z}}\right) < 1$$

In your simulation you will see many such bifurcations.

What is quite intriguing is how (8.69), "just" a parabola, gives rise to such a complex structure. Indeed, (8.69) is actually the nonlinear equation underlying the Mandelbrot Set of Problem 7.13; a quick online search or a bit of reading [17, 19, 22] will convince you that this iteration has quite remarkable structure.

Problem 8.8: Numerical/Computational — Time Discretization

For linear systems we have well defined eigenvalues, as in Problem 8.6, such that the effects of time discretization can be computed exactly.

In nonlinear systems the discretization requirements are far more difficult to analyze, but no less relevant. Let us consider the nonlinear limit cycle from Problem 7.12:

$$\dot{x} = y \qquad \dot{y} = -\beta\big(x + y^3 - y\big)$$

where β controls the relative dynamic rates in the x and y directions.

Implement this system numerically and test a variety of time discretization steps $0.01 < \delta_t < 1$ for both $\beta = 1$ and $\beta = 10$. State your observations.

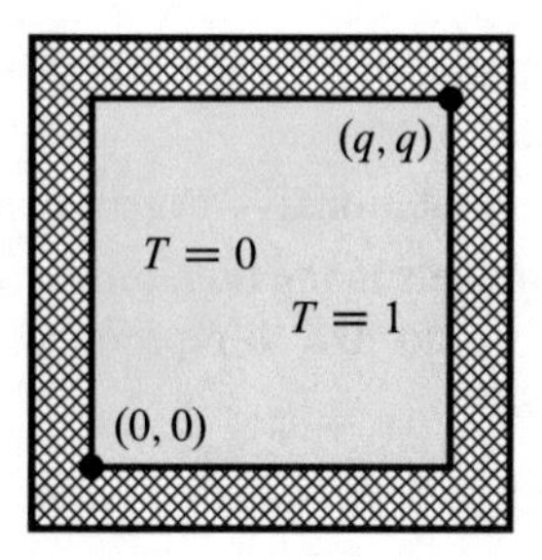

Fig. 8.19 The well insulated domain of Problem 8.9, of size $q \times q$, with the spatial initial temperatures as shown.

Problem 8.9: Numerical/Computational — Heat Flow

The heat equation is given by

$$\frac{\partial u}{\partial t} = c^2 \nabla^2 u.$$

Let's set $c = 1$. We want to simulate a well insulated square domain with initial temperatures as shown in Figure 8.19.

We wish to spatially discretize the domain to $q = 60$ where, for convenience, $\delta_x = \delta_y = 1$:

(a) Describe your method for computing $\nabla^2 u$.
(b) What are the boundary conditions for the discretized domain?
(c) For $\delta_t = 0.1$ and using Forward–Euler from (8.22), show the resulting temperature map after 100 time steps; that is, show $u(x, y, 10)$ (use one of `mesh, surf, image, imagesc`).
(d) For $\delta_t = 0.2$, superimpose the line plots (see the Matlab `hold` command) of $u(x, q/2, t)$ for $t = 0, 10, 20, \ldots, 400$ and discuss your observations.

Problem 8.10: Numerical/Computational — Spatial Models

We would like to implement a the forest-fire model of Table 8.3. At each point in time, each cell can be one of three states: { *tree, burning, ash* }.

1. Start with a 200 by 200 grid of *ash* as the state Z_0 at iteration 0.
2. Given the state map Z_n, then at iteration $n + 1$ ...

- Any *tree* in Z_n next to *burning* in Z_n is *burning* in Z_{n+1}
- Any *burning* in Z_n becomes *ash* in Z_{n+1}
- Each *ash* in Z_n becomes *tree* at random with probability p
- Each *tree* in Z_n becomes *burning* at random with probability f

So p is a measure of forest growth rate, and f is a measure of the likelihood of lightning strikes. The key parameter controlling the scale of the resulting burns is the ratio f/p.

Set $p = 0.01$. Run simulations for $f = 0.0001$, $f = 0.001$, and $f = 0.01$. Discuss your observations.

Problem 8.11: Numerical/Computational — Agent Based Model

One of the classic agent models in the literature is Shelling's segregation model, described in (8.64). We would like to reproduce and expand upon the model results shown in Figure 8.18.

(a) Start by reproducing Figure 8.18, based on the following parameters:

- Number of types $T = 3$
- Size s of the spatial grid is 40×40
- Fraction f of empty grid locations is 10 %
- Preference parameter $\rho = 0.4, 0.6, 0.8$
- Each pixel has eight "neighbours" (those directly and diagonally adjacent).

Iterate over all pixels (agents) until all agents are satisfied with their neighbourhood, or up to some maximum iteration, at which point we declare the simulation to have failed to converge.

Next, we wish to explore the behaviour of the model:

(b) The number of iterations to converge is a strong function of the empty fraction f; for $\rho = 0.6$, produce a plot demonstrating this relationship.
(c) The size of the resulting regions is clearly a function of ρ; produce a plot showing average region size versus ρ (see Matlab `bwlabel`).
(d) As $\rho \longrightarrow 0$, each element tolerates no neighbour of a different type. For $\rho = 0$ deduce, analytically, what the minimum value of f is as a function of T and s. Next, numerically, with $\rho = 0$ and keeping s fixed at 40×40, what is the minimum value of f for which your code successfully converges for each value of $T = 1, \ldots, 5$?

Problem 8.12: Numerical/Computational — Agent Models

Poverty is world-wide, a trouble affecting billions of people. The origins of poverty are many, and we will not pretend to have a realistic world simulation here.

However it *is* possible to use an agent-based simulation to understand some of the reasons for persistent wealth inequality, connected to the notion of a *poverty trap*, as discussed in Example 7.8.

In an *American Scientist* article [10], Brian Hayes described a simple "yard-sale" model, whereby people meet and trade (goods, money), but the trade is not necessarily equitable, and can favour one person or the other:

1. Initialize $w_i^0 = 100$, where w_i describes the wealth of person $i = 1, \ldots, n$ in a group of n people
2. Select a random pair of people $j \neq k$
3. The two people put in an equal amount of wealth, set to $\alpha = \min\{w_j, w_k\}$
4. We select a uniform random number $0 \leq p \leq 1$

5. The money 2α is divided up

$$w_j^{n+1} = w_j^n + (2p-1)\alpha \qquad w_k^{n+1} = w_k^n + (1-2p)\alpha$$

6. Return to step 2.

This is a distributed, non-spatial agent-based model, as described in Section 8.6, modelling the financial interaction of pairs of people.

(a) Implement the algorithm, run it repeatedly, and comment on the repeatability of your observations.
(b) A huge number of variations can be imagined:

 (i) The role of welfare: Taxation and redistribution
 (ii) The role of taxation: Comparing progressive and regressive taxation
 (iii) The role of interest: Players having to borrow money, at interest, from other players or from a bank

 Experiment with one or more of these scenarios, or propose a different scheme.

Problem 8.13: Reading Question — Waves

Most of the PDE related discussion in this chapter focused on phenomena related to advection–diffusion, however most definitely one other phenomenon worth understanding is that of waves:

(a) Look up the derivation of the *wave equation*

$$\frac{\partial^2 u}{\partial t^2} = c^2 \nabla^2 u$$

 and offer a succinct summary of the derivation.
(b) There are many wave–like phenomena present on the earth, from ripples on the water surface of a pond, shock waves in automobile traffic, to large–scale inertial waves, gravity[6] waves, and Rossby waves, to name a few. Look up one or more large–scale wave phenomena and offer a qualitative summary of their role in or impact on climate.

Problem 8.14: Reading Question — Cellular Automata

A great deal has been written on understanding nature through the lens of cellular automata and related models. Two recommended texts include

(a) P. Bak, *How Nature Works* [2], particular chapter 11, "On Economics and Traffic Jams"
(b) S. Wolfram, *A New Kind of Science* [24], particularly one of

[6]Not to be confused with the recently detected *gravitational* waves. Gravitational waves come from outer space; gravity waves are an earth ocean/cloud phenomenon.

- Chapter 7, “Mechanisms in Programs and Nature”, or
- Chapter 8, “Implications for Everyday Systems”

Based on one of these chapters, or another of your choice, give some illustrations of the types of problems examined via cellular automata.

Problem 8.15: Policy — Agent Models

Agent models are highly attractive in the study of those social systems for which analytical models are difficult to apply.

Look up recent work on using agent-based models as some component of government policy. Certainly as of the writing of this book (2016), most such work involves economic models, however other applications (energy, health care, environment) can also be found.

Look up one or two models and briefly summarize the modelling strategy (what sorts of models, what parameters, how many agents, spatial or non-spatial), the intended application, and to what extent the model sheds light on policy alternatives.

References

1. J. Adam. *Mathematics in Nature: Modeling Patterns in the Natural World.* Princeton, 2006.
2. P. Bak. *How Nature Works.* Copernicus, 1996.
3. M. Batty. *The New Science of Cities.* MIT Press, 2013.
4. H. Dijkstra. *Nonlinear Climate Dynamics.* Cambridge, 2013.
5. L. Evans. *Partial Differential Equations.* American Mathematical Society, 2010.
6. S. Farlow. *Partial Differential Equations for Scientists and Engineers.* Dover Books, 1993.
7. B. Friedland. *Control System Design.* Dover, 2005.
8. J. Giles. Climate science: The dustiest place on earth. *Nature*, 434, 2005.
9. R. Hamming. *Numerical Methods for Scientists and Engineers.* Dover, 1987.
10. B. Hayes. Follow the money. *American Scientist*, 90(5), 2002.
11. H. Kaper and H. Engler. *Mathematics & Climate.* SIAM, 2013.
12. I. Koren et al. The Bodl depression. *Environmental Research Letters*, 1, 2006.
13. R. Malone, R. Smith, M. Maltrud, and M. Hecht. *Eddy–Resolving Ocean Model.* Los Alamos Science, 2003.
14. D. Meadows, J. Randers, and D. Meadows. *Limits to Growth: The 30-Year Update.* Chelsea Green, 2004.
15. M. Mitchell. *Complexity: A Guided Tour.* Oxford, 2009.
16. A. Oppenheim, A. Willsky, and H. Nawab. *Signals & Systems.* Prentice Hall, 1997.
17. H. Peitgen, H. Jürgens, and D. Saupe. *Chaos and Fractals: New Frontiers of Science.* Springer, 2004.
18. B. Saltzman. *Dynamical Paleoclimatology: Generalized Theory of Global Climate Change.* Academic Press, 2001.
19. M. Schroeder. *Fractals, Chaos, Power Laws: Minutes from an Infinite Paradise.* Dover, 2009.
20. M. Sipser. *Introduction to the Theory of Computation.* Cengage Learning, 2012.
21. W. Strauss. *Partial Differential Equations: An Introduction.* Wiley, 2007.
22. S. Strogatz. Exploring complex networks. *Nature*, 410, 2001.
23. F. White. *Fluid Mechanics.* McGraw Hill, 2010.
24. S. Wolfram. *A New Kind of Science.* Wolfram Media, 2002.

Chapter 9
Power Laws and Non-Gaussian Systems

In 1993 the *Long-Term Capital Management* (LTCM) hedge fund was created, most notably including Myron S. Scholes and Robert C. Merton who would later win the Nobel Memorial Prize (the so-called Nobel Prize in Economics).

Scholes, Merton, and Black were instrumental in the theory behind derivative pricing, deducing the mathematics that described the time-evolving fair price of an option.[1] The key to the hedge fund was *arbitrage*, meaning finding inconsistencies in pricing, either across different markets, or between the value of a stock and its associated options. The mathematical theory was still relatively new, and certainly not widely implemented by trading houses, so there were certainly many arbitrage opportunities available. The challenge with arbitrage, however, is that the inconsistencies are very small; a glaring inconsistency would, after all, be spotted by others in the markets. So to make money off tiny inconsistencies, you need to undertake very large trades, *very* large.

You therefore need to buy an enormous amount of stock x and sell a huge amount of option y. At this point you are subjected to *risk*, a statistical distribution of possible outcomes, and the assessment and management of risk requires some model of this distribution. The LTCM models suggested that a loss of 50 % percent of its portfolio was impossibly low, not to occur in 10^{28} *years, billions of times the life of the universe*. More "conservative" risk models suggested such a loss not to occur in 10^7 years.

Unfortunately their models, even their "conservative" ones, had not accounted for an Asian financial crisis (1997) nor Russian default (1998), and the hedge fund collapsed, spectacularly, not in 10^{28} years, nor in 10^7 years, but in 4 years. How could a statistical model possibly be so catastrophically wrong?

[1]An option is the right to buy or sell stocks or some other financial instrument at some price at some point in time.

P. Fieguth, *An Introduction to Complex Systems*,
DOI 10.1007/978-3-319-44606-6_9

Whereas nonlinear systems challenge our assumptions regarding superposition, distributions known as *power laws* challenge our assumptions regarding averages and learning from experience. Far from occurring only in esoteric branches of finance, power laws are actually remarkably commonplace, even if not widely understood.

Although there are a great many distributions that play important roles in statistical analyses of data, particularly χ^2 tests, broadly speaking there are three families of distributions that describe an overwhelming fraction of observed phenomena:

I. Gaussian (Normal) Distribution

No matter how unusual the statistics of a single object (coin, molecule, person), the Central Limit Theorem tells us that the distribution of the sum or average becomes ever more Gaussian as the number of independent objects increases.

Typical Context: Sums or averages of random events
Examples: Brownian motion, census data, student grades

II. Uniform/Exponential/Poisson Distributions

You may be surprised to see the uniform, exponential, and Poisson distributions lumped together, since they look very different. However all three distributions do appear in the same context; the particular distribution depends on the type of question which you ask. So, for example, given the arrival times of unconnected or unrelated people walking into a bank,

- The distribution of arrival *times* is *uniform*
- The distribution of the time *between arrivals* is *exponential*
- The distribution of the *number of people* over some time interval is *Poisson*

Once we start asking about large numbers of arrivals, the Central Limit Theorem makes an appearance and the Poisson distribution becomes ever more Gaussian.

Typical Context: The statistics of individual independent events
Examples: Radioactive decay, cars on a highway

III. Power Laws/Heavy-Tailed Distributions

The previous distributions relied on independence or unrelatedness of objects or events. However complex systems are *highly* related — avalanches are not billions of snowflakes which *coincidentally* move in the same direction, rather the snow mass is a coupled system with strong relationships, leading to very large events, frequently in the form of a power law.

Typical Context: Rare events, complex systems, bounded systems
Examples: Earthquakes, asteroid impacts, personal wealth

We will now examine each of these in turn.

9.1 The Gaussian Distribution

This chapter's title on *non*-Gaussian systems does not imply that Gaussian distributions are, in general, incorrect or invalid. To the contrary, Gaussian distributions do indeed show up very frequently:

The Central Limit Theorem (Figure 9.1) says that the mean of independent, finite-variance random variables converges to a Gaussian. Specifically,

$$\text{Given } x_i \sim (\mu, \sigma^2) \quad \text{then} \quad \frac{1}{n}\sum_{i=1}^{n} x_i \rightarrow \mathcal{N}(\mu, \sigma^2/n) \tag{9.1}$$

where[2] *the Gaussian probability distribution $p(x)$ is given as*

$$x \sim \mathcal{N}(\mu, \sigma^2) \quad \text{if} \quad p(x) = \frac{1}{\sigma\sqrt{2\pi}} \exp\left(-\frac{(x-\mu)^2}{2\sigma^2}\right) \tag{9.2}$$

for mean μ and variance σ^2.

The reason why we need to emphasize *non*-Gaussianity is precisely because the Gaussian distribution is so widespread and convenient:

- Any linear function of Gaussian random variables is still Gaussian;
- There is an analytical, closed-form solution to the estimation of unknowns in linear, Gaussian problems;
- The estimation of unknowns is a *linear* function of the measurements.

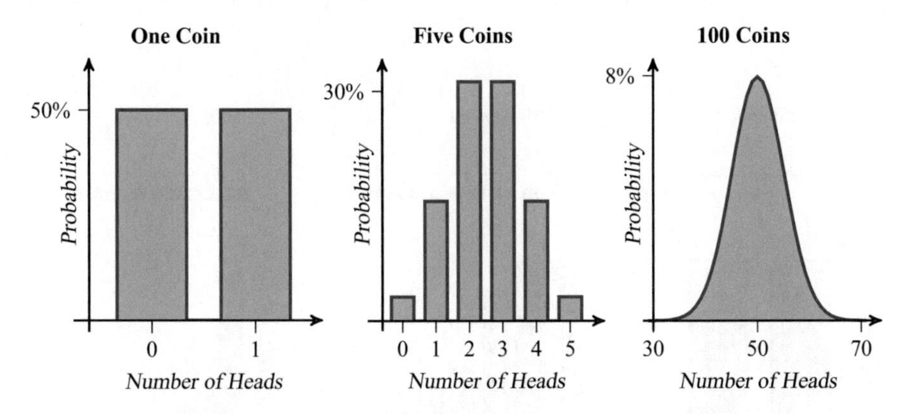

Fig. 9.1 The central limit theorem: tossing one coin, left, we have a 50–50 chance of seeing heads or tails, a binary distribution which looks not at all Gaussian. However as we toss more coins, the statistics of heads start to average over many coins, and we obtain a Gaussian distribution, right.

[2]See Appendix C for a clarification of $x \sim \mathcal{N}()$ notation.

9.2 The Exponential Distribution

The exponential distribution is commonly associated with independent events over time, as shown in Figure 9.2. Events could include raindrops falling on your head, leaves falling from a tree, or people arriving at a grocery store. It is essential that the events be memoryless, or unconnected with each other, as discussed in Example 9.1; connections between events can lead to cascading effects and very *non*-exponential behaviour, as we shall see in Section 10.2.

Given a one-dimensional domain $0 \leq t \leq T$ containing N arrivals, the *uniform distribution* $\mathcal{U}$ describes the distribution of any single point:

$$p_{\mathcal{U}}(t) = \begin{cases} 1/T & 0 \leq t \leq T \\ 0 & \text{Otherwise} \end{cases} \tag{9.3}$$

the *exponential distribution* $\mathcal{E}$ describes the distribution of inter-point spacing:

$$p_{\mathcal{E}}(\Delta) = \begin{cases} \frac{N}{T} \exp\left(-\frac{N}{T}\Delta\right) & \Delta \geq 0 \\ 0 & \Delta < 0 \end{cases} \tag{9.4}$$

and the *Poisson distribution* describes the probability of finding k points in a small interval:

$$P(k \text{ points in interval } \delta) = \frac{\left(\frac{N\delta}{T}\right)^k \exp\left(-\frac{N\delta}{T}\right)}{k!} \qquad \delta \ll T \tag{9.5}$$

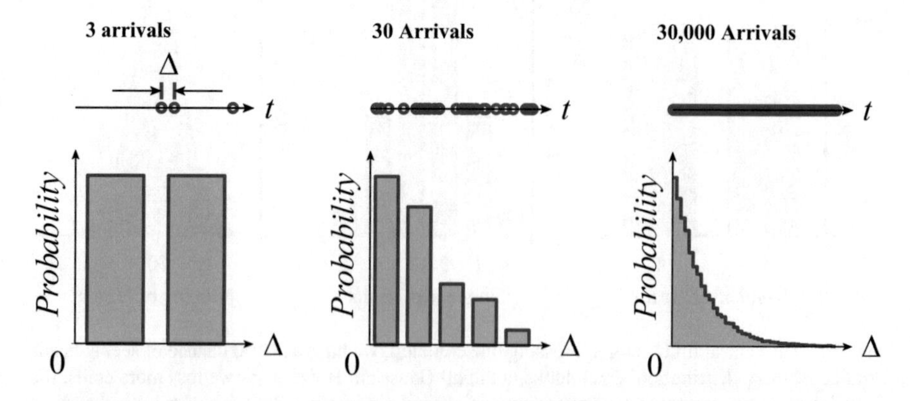

Fig. 9.2 Independent arrivals: three random arrivals, left, arriving over time t, can be characterized by the inter-arrival time Δ between them. With more data points, middle, we can see the random, irregular spacing of the points, with most gaps small but a few larger. With many points, right, we observe that the time between successive arrivals of random events is exponentially distributed.

Example 9.1: Memory and Exponentials

Suppose we have a sample of radioactive material, such as uranium. If we hold the sample next to an old-fashioned Geiger counter, which emits a click with every measured decay, then the time between clicks will be exponentially distributed. What can uranium atoms possibly "know" about exponentials?

Indeed, the time interval between *any* set of independent arrivals — (cars on the highway, people arriving at a bank machine) — must be exponentially distributed. Why is this so? Suppose we have a set of arrivals at times $t_1, t_2, \ldots$. Starting at an arbitrary point in time, let random variable t with distribution $p(t)$ represent the time until the next arrival. The sketch is deliberately *unlike* an exponential, to avoid biasing our thinking, for now. Suppose there is no arrival up to time τ. Here is the key concept:

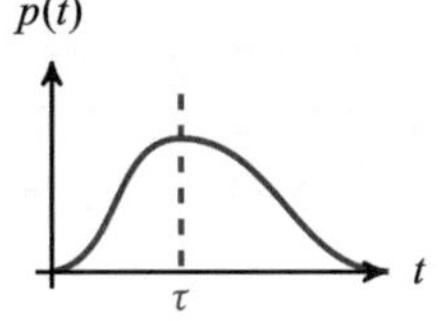

Since the arrivals are independent and know nothing about each other, the distribution looking ahead from $t = \tau$ must be the same as the original distribution looking ahead from $t = 0$.

In other words, if nobody has come to the bank machine for the last 5 min, it does *not* make it more likely that someone will come soon, since the next person coming *doesn't know* that no-one has come. Stated mathematically, the conditional distribution

$$p(t \mid \text{No arrival to } \tau) = \begin{cases} 0 & t < \tau \\ p(t) / \int_\tau^\infty p(\alpha)\, d\alpha & t \geq \tau \end{cases} \tag{9.6}$$

must just be a shifted version of the original distribution $p(t)$:

$$p(t) = p(t + \tau \mid \text{No arrival to } \tau) = p(t+\tau) \Big/ \int_\tau^\infty p(\alpha)\, d\alpha \tag{9.7}$$

thus

$$p(t) \cdot \int_\tau^\infty p(\alpha)\, d\alpha = p(t + \tau) \tag{9.8}$$

Taking the derivative by τ of each side,

$$p(t) \cdot -p(\tau) = \dot{p}(t + \tau) \cdot 1 \tag{9.9}$$

Now (9.9) must be valid for all $\tau \geq 0$, so we'll pick the simplest value of τ, which is to assert $\tau = 0$:

$$\dot{p}(t) = -p(t) \cdot p(0) \quad \longrightarrow \quad p(t) = p(0)\, e^{-p(0) \cdot t} \tag{9.10}$$

Well, look at that! A simple, first-order differential equation whose solution *must* be the exponential function. Thus if the time between arrivals is *not* exponential, then the events cannot be memoryless.

Further Reading: Books on stochastic processes, such as [4, 7], or Poisson Process in *Wikipedia*.

9.3 Heavy Tailed Distributions

Most distributions, including those in Figure 9.3, consist of some events, which are relatively likely, and other events, the "tails," where the unlikely, extreme events lie. The study of heavy-tailed and power-law distributions is a characterization of these rare, extreme events.

In many cases our statistics may only be approximately normal or exponential, but that distinction may be a small detail relative to the more fundamental question of whether the problem variances are finite or infinite.

As long as the data $\{x_i\}$ have modest *variance*, then the sample mean and variance of (B.22) converge, given sufficient data, to the true mean and variance:

$$\begin{array}{c}\text{Sample}\\ \text{Mean}\end{array} \quad \mu_N = \frac{1}{N}\sum_{i=1}^{N} x_i \quad \xrightarrow{N \to \infty} \quad \begin{array}{c}\text{True}\\ \text{Mean}\end{array} \quad \mu \tag{9.11}$$

$$\begin{array}{c}\text{Sample}\\ \text{Variance}\end{array} \quad \sigma_N^2 = \frac{1}{N-1}\sum_{i=1}^{N} \left(x_i - \mu_N\right)^2 \quad \xrightarrow{N \to \infty} \quad \begin{array}{c}\text{True}\\ \text{Variance}\end{array} \quad \sigma^2 \tag{9.12}$$

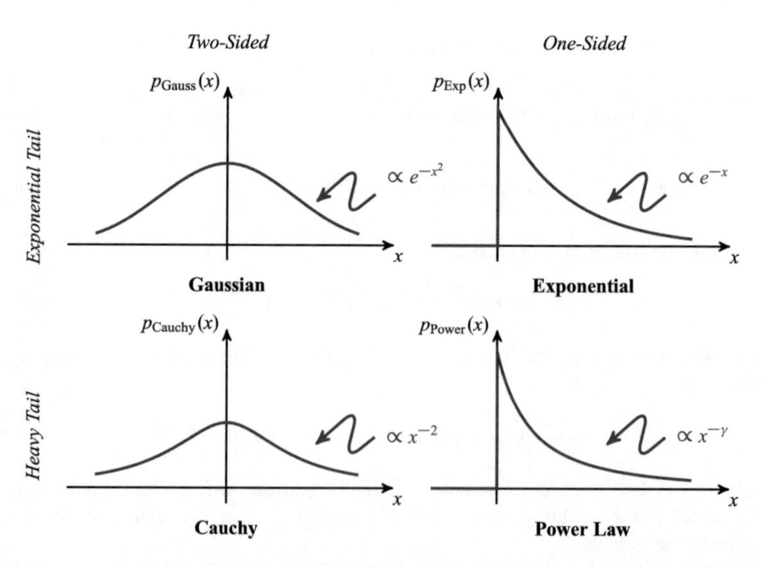

Fig. 9.3 Heavy tails: there are distributions which look qualitatively similar, such as Gaussian and Cauchy, however a simple plot does not make obvious the asymptotic behaviour of the tails. Gaussians and Exponentials have an asymptotic likelihood *exponential* in x, whereas Cauchy and other heavy-tail distributions have an asymptotic likelihood which is a *power* of x.

where the expected error in the sample mean is proportional to $\sigma/\sqrt{N}$. Therefore

- A modest number N of measured samples provides a meaningful sense of distribution. In other words, we can learn from history.
- It is not possible for a single outlier or bad data point to overwhelm or skew the rest of the data.

For example, adult human height has a distribution of modest variance, with an average adult height in North America of $5'7''$. Now suppose we are taking surveys of height, where we survey $N = 200$ people.

- Most of the time we end up with a sample mean near to the true mean $\mu_N \approx 5'7''$
- Now suppose we accidentally include the twenty members of a basketball team, height $6'5''$, in our survey:

$$\bar{\mu}_N = 0.9 \cdot 5'7'' + 0.1 \cdot 6'5'' = 5'8'' \tag{9.13}$$

Definitely an anomaly, but really only a relatively small bias in our survey result.

Because the error in the sample mean converges as $\sigma/\sqrt{N}$, as the measurement variance σ^2 increases the sample mean settles down more slowly. Therefore a greater number of data points N is required in order to reach a given level of confidence in the result, making it harder to learn from history, and making it easier for individual extreme events to skew the results.

In contrast to the previous example of human height, which is small-variance, personal financial wealth is a *large*-variance distribution, where the 2013 Canadian average wealth is $\mu = \$400,000$, however there are individuals with exceptionally large wealth, say $x = \$10^{10}$. Given a large survey of $N = 1000$ people,

- Without the high-worth individual, we might expect a sample mean $\mu_{1000} \approx \$400,000$
- With the high-worth individual in our sample,

$$\bar{\mu}_{1000} = \frac{999}{1000} \cdot \$400,000 + \frac{1}{1000} \cdot \$10^{10} \simeq \$10 \text{ million} \tag{9.14}$$

a huge anomaly from the mean. It would require a *much* larger sample size N to actually have a reliable estimate of the mean.

This example is perhaps contrived, however the conclusion is not: high-variance systems are in fact quite common in many systems. As seen in Figure 9.4, the high-variance distributions we most commonly encounter take a specific form, known as a power law:

A power law distribution is one in which the likelihood of some size of event is proportional to the size taken to some power,

$$p(x) \propto x^{-\gamma} \quad \textit{as opposed to} \quad p(x) \propto \exp(-x^{\gamma}). \tag{9.15}$$

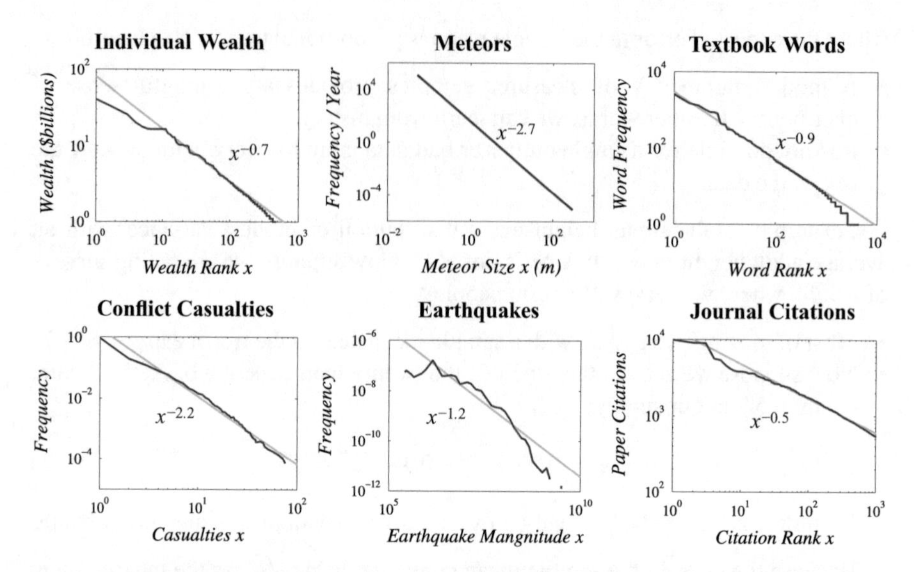

Fig. 9.4 There is an astonishing range of very different phenomena which exhibit power-law behaviour, including social systems (left), physical systems (middle), and language (right). All real systems are bounded, thus the associated power law can be present only over a finite range of scales. In ranked contexts the data (wealth, words, citations etc.) are first sorted, so that rank 1 refers to the wealthiest individual, rank 2 the second wealthiest etc.

Citation data from *CiteseerX*. Wealth data from the *Forbes World's Billionaires* list.
Conflict data from Friedman, "Using Power Laws to Estimate Conflict Size," *J. Conflict Resolution*, 2014.

So, what's the big deal?, you may wonder. As it turns out, there are major, important differences lurking here.

To begin, as illustrated in Figure 9.5, any exponential distribution is curved in the log-log domain,

$$p(x) = \lambda e^{-\lambda x} \quad \longrightarrow \quad \log p(x) = \log \lambda - \lambda e^{\log x} \tag{9.16}$$

so that for sufficiently large x the distribution drops very quickly. In contrast, a power-law distribution is straight in the log-log domain,

$$p(x) = \alpha x^{-\gamma} \quad \longrightarrow \quad \log p(x) = \log \alpha - \gamma \log x \tag{9.17}$$

so that the tails drop relatively slowly, making extreme events significantly more likely.

Since all valid probability densities must be normalized,

$$\int_{-\infty}^{\infty} p(x)\, dx = 1, \tag{9.18}$$

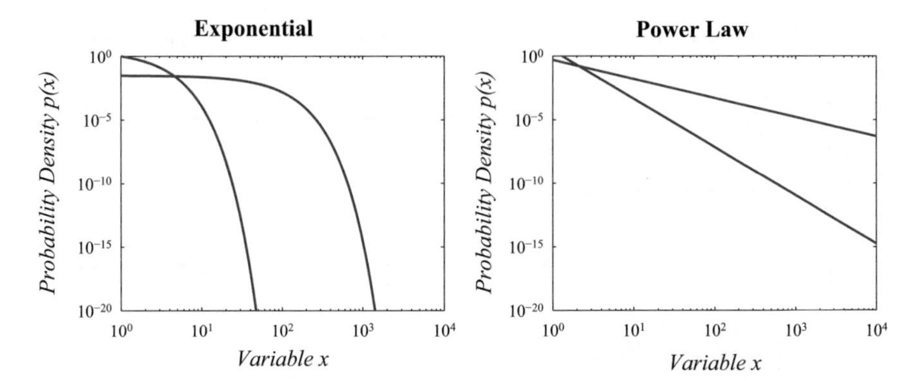

Fig. 9.5 The log-log domain much more clearly brings out the behaviour of rare events, where the probability density $p(x)$ is tiny. In the log-log domain, any exponential–like distribution (left) is curved, effectively truncating the distribution and limiting the size of x which can appear. In contrast, power-law–like distributions (right) are straight, leading to greater likelihoods at large values of x.

as a result it is not possible to have a power law over *all* sizes or scales, since the resulting distribution would be impossible to normalize:

$$\int_0^\infty \frac{\alpha}{x^\gamma}\,dx = \begin{cases} \alpha \ln x \Big|_0^\infty = \infty & \gamma = 1 \\ -\dfrac{\alpha(n-1)}{x^{\gamma-1}}\Big|_0^\infty = \infty & \gamma \neq 1 \end{cases} \tag{9.19}$$

Since $x \longrightarrow 0$ corresponds to more and more frequent events, in practice there must be some limit to frequency, thus it is reasonable to limit the power law as

$$p(x) \propto x^{-\gamma} u(x-1) \qquad \text{Step function } u(\tau) = \begin{cases} 1 & \tau \geq 0 \\ 0 & \tau < 0 \end{cases} \tag{9.20}$$

The truncated power law of (9.20) has the following properties:

$\gamma \leq 1$	Non-normalizable PDF		
$1 < \gamma \leq 2$	Normalizable PDF	Undefined mean	Infinite variance
$2 < \gamma \leq 3$	Normalizable PDF	Finite mean	Infinite variance
$3 < \gamma$	Normalizable PDF	Finite mean	Finite variance

Overwhelmingly nearly all power laws, including those in Figure 9.4, lie in the range $1 \leq \gamma \leq 3$. Therefore we are not just dealing with "large" variance processes rather, in principle, with *infinite* variance processes.

Figure 9.6 begins to illustrate the issues. The heavy-tail and exponential distributions appear qualitatively similar, and indeed frequently produce similar random points if only a few random samples are generated. However the heavy-tail distribution has a far greater variance and a far higher (although still small) probability of producing extreme events.

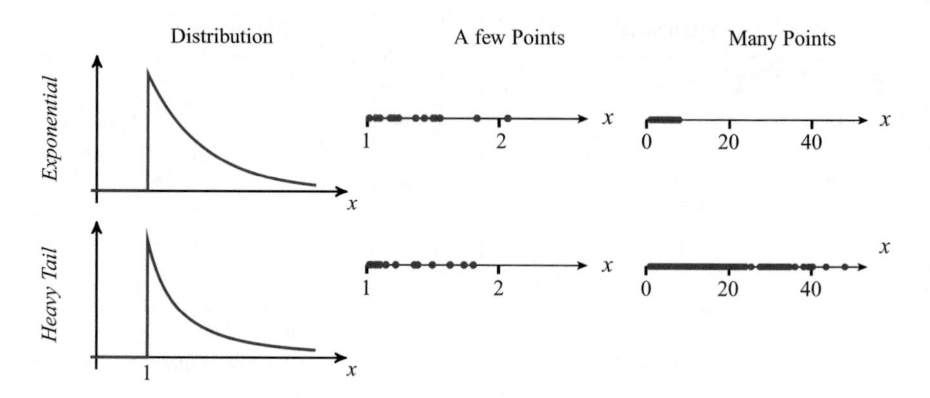

Fig. 9.6 Visually, the exponential and heavy-tailed distributions [now truncated per (9.20)] don't look particularly different, left, nor do the few random samples, middle. However given many samples, right, there *are* major differences in the rare events in the tails.

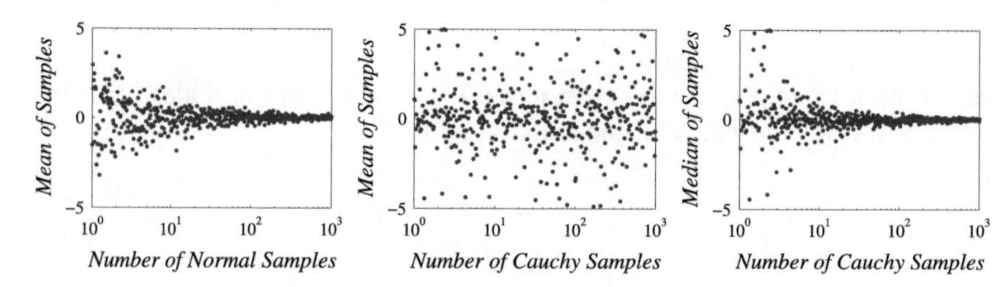

Fig. 9.7 As we get more and more samples, we expect the mean (average) to settle down, which is what happens with finite-variance distributions (left). However for heavy-tailed distributions the mean just does not settle down, no matter how many data points we have (middle). Of course we know the Cauchy distribution to be centred at zero, thus the median, which is insensitive to distant outliers, *does* in fact settle down (right).

With a finite-variance distribution, such as Normal, the mean converges as $\sigma/\sqrt{N}$ as illustrated in Figure 9.7; however the Cauchy[3] samples, which are infinite-variance, lead to a sample mean which simply does not settle down, even with more and more samples. So for the Cauchy distribution, as with infinite-variance power laws, you can *never* have enough data points to safely compute a mean, making it very challenging to learn from history.

Now, whether learning from history is challenging or not actually depends on what question is being asked. Given a power law x the following can be reliably estimated even from only modest numbers of samples:

[3]The Cauchy distribution is essentially a symmetric, two-sided heavy-tailed distribution. The Cauchy distribution was used in Figure 9.7, since its symmetric form leads us to expect a mean at zero. In practice the Cauchy distribution is encountered relatively rarely, and we will focus on regular power laws.

- the *median*, as illustrated in Figure 9.7,
- the *power law exponent*, as discussed in Section 9.5,

since both of these statistics are actually not a function of the extreme outlying samples.

In contrast, the following can*not* be reliably estimated:

- the *mean*, as illustrated in Figure 9.7,
- the *maximum*,

since both are a function of outlying samples; indeed, the maximum is, by definition, equal to the *most* extreme event allowed by the distribution.

Now for any perfect power law, of the form in (9.20), we know $\max(x)$ to be infinite. The issue arises in real problems which are power-law *like*, such as looking ahead to the rainfall data in Example 9.3, which have heavy tails but not indefinite ones. So if $p(x)$ describes the distribution of flooding events over a period of a year, then the "100 Year Flood" is that flood size, x_{100}, which has a 1 % chance of taking place in a year:

$$\int_{x_{100}}^{\infty} p(x)\, dx = 0.01, \tag{9.21}$$

which is not quite asking for the maximum of $p(x)$, but is certainly looking far into the tail. And it is precisely a concept like x_{100}, the size of an extreme event, which governments and policy makers find important to estimate, but which power laws make exceptionally difficult to learn from history, essentially since it is always possible for an even more extreme event to skew the results:

- If you wait long enough, there is always a larger meteor (dinosaurs ...);
- If you wait long enough, there is always a larger earthquake or tsunami (Fukushima ...);
- If you wait long enough, there is always a larger financial crash.

To be sure, a given distribution is not required to be either power-law or Gaussian-exponential. There are certainly *degrees* of distribution tail heaviness, a function of the power-law exponent γ. Furthermore many empirical distributions, such as the rainfall example just mentioned, will be power-law-like only over a limited range of scales, and not power-law outside of that range. Nevertheless, there is indeed an overwhelming prevalence of systems being one of memory-less/exponential, subject to the Central Limit Theorem/Gaussian, or obeying a power-law distribution, implying that the thin — heavy tail extremes of Figures 9.6 and 9.7 are indeed typical of a great many systems.

Finally, power laws possess scaling behaviour, as illustrated in Figure 9.8, such that the same power-law (Zipf's law) holds for word frequencies in the Bible, Wikipedia, and this textbook.

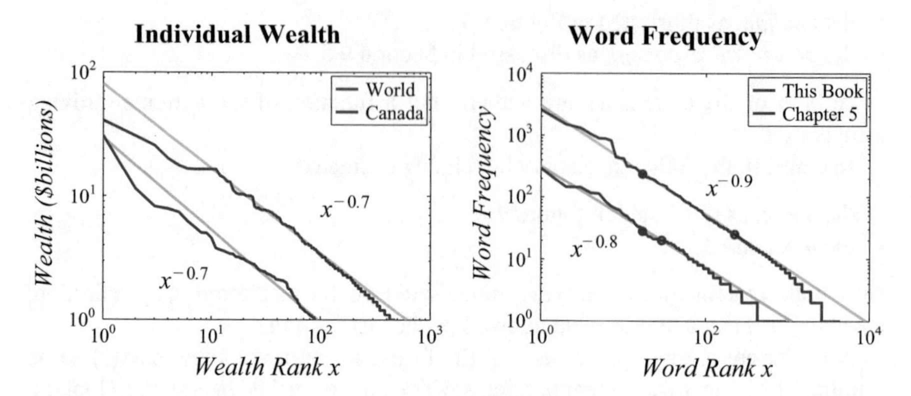

Fig. 9.8 Power law scaling: most power laws exhibit interesting self-similar or scaling relationships, such that the wealth in one part of the world (whether rich or poor) obeys a similar power to that of the whole world, or the word distribution in one chapter as opposed to this entire book. The actual individuals or words may be different, however the overall pattern is very nearly the same. So the word "which" (closed circle) happens to appear in exactly the same rank (17th) in Chapter 5 and in the whole book, whereas "eigendecomposition" (open circle) ranks very differently (29th and 224th).

Data from *Forbes* (world wealth data) and *Canadian Business* (Canada wealth data).

Indeed, the Pareto principle or 80-20 Rule stems from such an observation, of Pareto, that 80 % of the land in Italy was held by 20 % of the population, and that a scaled version applied to any subset of the population:

- In a very poor village no-one has much wealth, however *still* there will be 20 % of the people holding 80 % of what limited wealth there is;
- In an elite club every member is very wealthy, nevertheless there are a few who are that much *more* wealthy than the rest, such that 20 % of the wealthy hold 80 % of the assets.

Although much of the power law discussion may appear to focus on the negative (meteors, earthquakes, ...) the Pareto principle, when applied to social power laws, offers significant hope for effective policy by focusing attention on the small numbers of disproportionate events. For example, of the polluting cars on the road, 20 % of the cars produce 80 % of the pollution; by scaling, 4 % (20 % of 20 %) of the cars produce 64 % of the pollution. Therefore, rather than sweeping policies that try to test *all* cars, it may be far more effective to identify a strategy for focusing on the 4 %.

Pareto distributions similarly frequently appear in economics and business. For example, many businesses experience 80 % of their sales to 20 % of their customers and, again, 64 % of sales to only 4 % of customers, meaning that it makes exceptional economic sense to identify, and keep happy, those 4 %. The widespread

Example 9.2: Power Laws in Finance

Following up on the LTCM story of page 211, the presence of power laws in financial time series is trivial to demonstrate, making LTCM's use of short-term Gaussian risk models even more baffling. Consider the history of the Dow Jones Industrial Average $d(t)$ over time, one of the longest-established and best-known indices of American stocks:

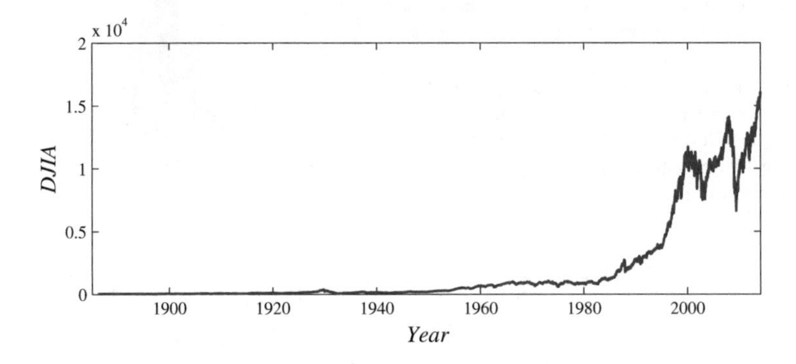

Clearly the index has grown exponentially over time, so that the raw index value from 2014 cannot be compared in any way to that of 1914. However if we examine fractional change f from day to day,

$$f(t) = \frac{d(t) - d(t-1)}{d(t)} \tag{9.22}$$

this normalizes the absolute size of the index,

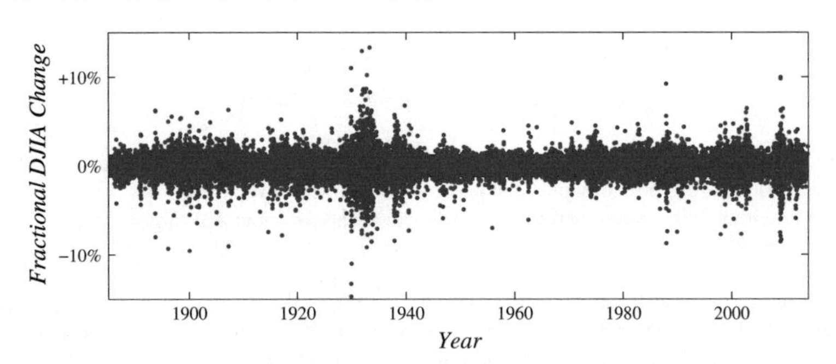

giving us a relatively stationary behaviour over time, where the financial shocks of 1929, 1987, and 2008 are all clearly visible.

Example continues ...

Example 9.2: Power Laws in Finance (continued)

We can fit a Gaussian distribution to f by learning the mean and variance of f (the DJIA data are real, so the sample mean and variance *are* finite). Two Gaussians are shown (dashed): the green one chosen to visually fit the plotted histogram, the red one having a larger variance equalling the sample variance of f, which should therefore take into account or at least be a function of the outliers in the tails. Based on a simple visual inspection both Gaussians appear to be a reasonable fit.

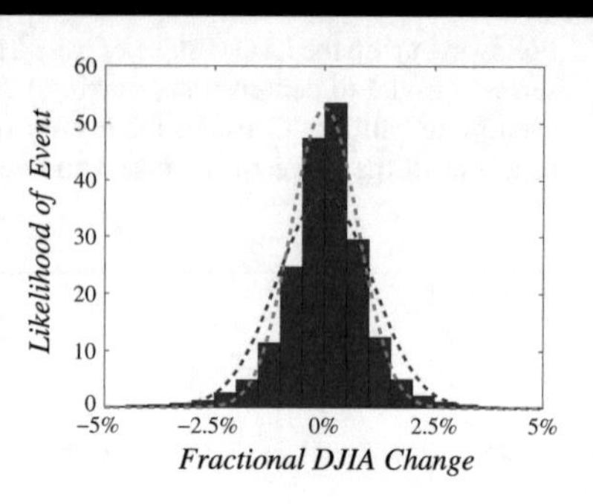

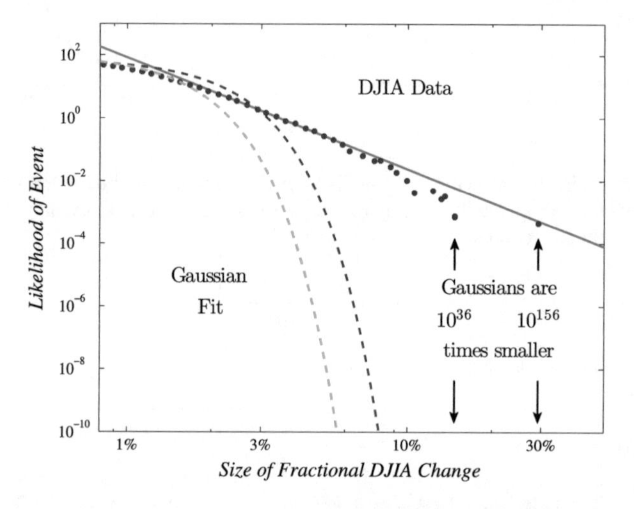

However appearances in a simple plot can be terribly deceiving. The tails of the distribution in the previous plot disappeared to zero, however those tails matter, a lot! The previous figure only went out to $\pm 5\,\%$, however we *know* that larger changes have taken place.

This figure looks at the tails in more detail, looking at the magnitude of the fractional change (ignoring the sign). We see the two Gaussians falling steeply, however the actual Dow Jones data is straighter, much more like a power law. (Oh oh . . .)

So for the most extreme actual one-day Dow-Jones changes of $15\,\%$ and $29\,\%$, the "conservative" Gaussian model suggested an event likelihood between 10^{36} and 10^{156} *times lower* than the actual data.

Now *that's* starting to look like the difference between 4 years and a billion times the lifetime of the universe!

Further Reading: Two recommended books on finance are by Taleb [11] and by Lowenstein [5].

use of customer loyalty programs (frequent flyer miles etc.) are, essentially, a reflection and recognition of such power laws.[4]

9.4 Sources of Power Laws

The range of systems subject to power law distributions is nothing short of remarkable:

Universe:	asteroid sizes, meteor impacts, solar flares
Physical Earth:	earthquakes, waves, avalanches, landslides, forest fires
Human Society:	city sizes, train derailments, wars and conflicts
Economics:	personal wealth, corporate profits, price variations
Language:	word use, journal citations

Given the amazing appearance of power laws in so many domains, surely there must be some common causes underlying these. Indeed, it turns out that there are a number of mechanisms which generate power laws:

Products and Powers of Random Variables:
We saw a brief demonstration of the Central Limit Theorem in Figure 9.1, such that the *sum* of independent random variables converges to a Gaussian.
Now, if we are given a *product* $y = \prod_i x_i$, then

$$\log(y) = \sum_i \log(x_i) \sim \text{Gaussian} \tag{9.23}$$

Now if $\log(y)$ is normal, then y is said to be *log-normal* which, although not precisely power-law, has relatively heavy tails and can exhibit power-law like behaviour over a range of scales.
Similarly if random variable $y = f(x) = x^\beta$, then the probability distribution for y can be computed to be

$$p_y(y) = \frac{p_x\big(f^{-1}(y)\big)}{f'\big(f^{-1}(y)\big)} = \frac{p_x\big(f^{-1}(y)\big)}{\beta y^{1-1/\beta}} \quad \longrightarrow \quad p_y(y) \approx \frac{\alpha}{y} \tag{9.24}$$

where the final approximation holds if p_x is sufficiently slowly varying and for a sufficiently high power β.
Related power laws: Over time, compound interest acts as an exponent on people's economic fortunes (investment) or misfortunes (debt), leading to power laws in personal wealth, as explored in Problems 8.12 and 9.9.

Scale Invariance:
Suppose that z is a random variable describing some measure, such as length or mass, over a vast range of scales. We say that z is *scale invariant* if rescaling z

[4] See Problem 9.12 for a related discussion on power laws, policy focus, and homelessness.

Example 9.3: Power Laws in Weather

Let us consider the statistics of a common phenomenon, such as rainfall. Here is a ten-year overview of daily rainfall for my home town of Waterloo, Canada:

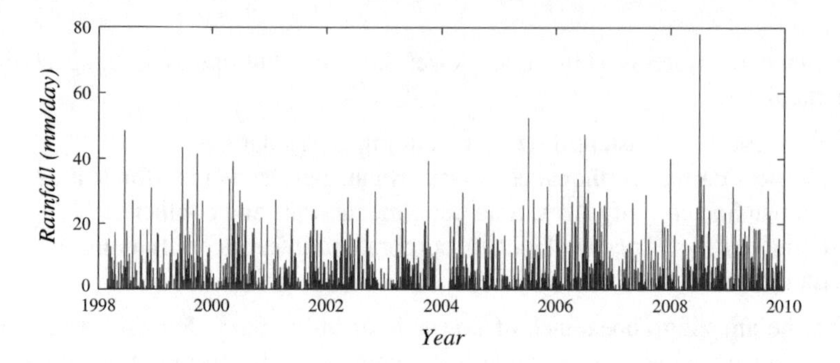

There are two basic questions that we would like to ask about rainfall:

1. *What are the statistics of amount of rain per day?*
   This question matters a lot to city planners, since the sewer/storm water management system needs to be able to accommodate most rain events, say up to the most extreme ten-year or fifty-year storm.
2. *What are the statistics regarding the length of time between rainy days?*
   This question matters a lot to farmers and gardeners, since irrigation requirements or what types of trees and plants you can grow may be dictated by the length of droughts.

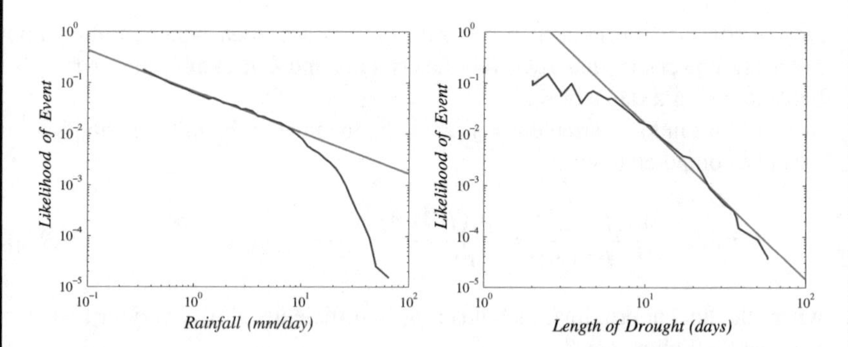

Neither is a perfect, indefinite power law. This isn't surprising: for the left plot, there are simply physical limits as to how much rain can fall in a day; for the right plot, it doesn't make much sense to talk about the likelihood of droughts less than a day in length.

Nevertheless, both plots clearly show domains of power-law behaviour. The drought plot, in particular, is somewhat unsettling, since the statistics of the tail suggest a continued power-law behaviour, meaning that rather longer droughts are very much possible. And this for an area like Southern Ontario, in Canada, which, compared to many other parts of the world, receives quite regular rain.

(multiplying z by a constant factor) does not change its probability:

$$\text{Prob}(a \le z < b) = \text{Prob}(a \le q \cdot z < b) \tag{9.25}$$

for any constant $q > 0$. If we shrink the width of the interval,

$$\text{Prob}(a \le z < a + \delta_a) = \text{Prob}(a/q \le z < (a + \delta_a)/q) \tag{9.26}$$

then in the limit as $\delta_a \longrightarrow 0$, and dividing through by δ_a, the probabilities in (9.26) become probability densities

$$p_z(a) = \frac{1}{q} p_z(a/q). \tag{9.27}$$

We can let $z = a/q$; since a is just a fixed value, therefore $ap_z(a)$ is also just a fixed value, not affecting the shape of $p_z()$, and can be absorbed as a constant of proportionality, in which case (9.26) becomes

$$p_z(z) = \frac{a}{z} p_z(a) \quad \longrightarrow \quad p_z(z) \propto \frac{1}{z}. \tag{9.28}$$

An interesting result! That is, essentially analogous to the exponential distribution for all memoryless processes in Example 9.1, *all* scale-invariant distributions must correspond to a $\gamma = 1$ power law.

Related power laws: Many financial/economic values, such as corporate revenues or profits, lie on a vast range (from near zero to billions). The scale invariance argument suggests the presence of power laws in such contexts, and indeed forms the basis of Benford's law (Example 9.4).

Evolving Fitness and Preferential Models:

Fitness and preferential models are dynamic systems whereby nodes/agents compete with each other for a resource, such that "fitter" nodes (by some definition of fitness) obtain more resources at the expense of others.

Related power laws: Video popularity on YouTube is an outstanding example of this phenomenon. The vast majority of videos are seen by very few people, however the occasional video "goes viral", with video links emailed between friends and posted on social media pages. Thus the more popular a video, the more popular it further becomes. There are many further, related examples in social systems, such as book popularity or research journal citations.

Criticality and Universality:

In very simple terms, stable systems stay unchanged over time, and unstable systems rapidly diverge and move to some other stable fixed point. But what happens right *at* the boundary between these two, referred to as a critical point, for systems that are just *barely* stable, but not yet unstable?

Such systems exhibit power-law behaviour, with relatively small responses most of the time (stable-like), with the occasional very large response (unstable-like). Although having a system carefully tuned between stable and unstable would appear to be degenerate, for reasons to be explained in Chapter 10, in fact a great many *Complex Systems* behave this way.

Related power laws: A wide variety of natural/physical systems exhibit power laws stemming from criticality, particularly systems consisting of a growth or forcing function and a limit or bound, as outlined in Case Study 10.

9.5 Synthesis and Analysis of Power Laws

Although infinite-mean and infinite-variance distributions sound exotic and complicated, power laws are actually quite straightforward to work with computationally.

We begin by understanding the process of random sampling in general. Suppose we are given a probability distribution function (PDF) $p_x(x)$ from which we would like to generate random samples. The *cumulative distribution* F_x is defined as

$$F_x(a) = \text{Prob}(x \le a) = \int_{-\infty}^{a} p_x(x)\, dx \tag{9.30}$$

Since F_x is a probability, clearly $0 \le F_x \le 1$, thus F_x is a non-decreasing function from random variable x onto $[0, 1]$, as shown in Figure 9.9. Therefore given a uniform random sample $u \in [0, 1]$, the desired random sample is just given by reversing the process, the inverse mapping $x = F^{-1}(u)$.

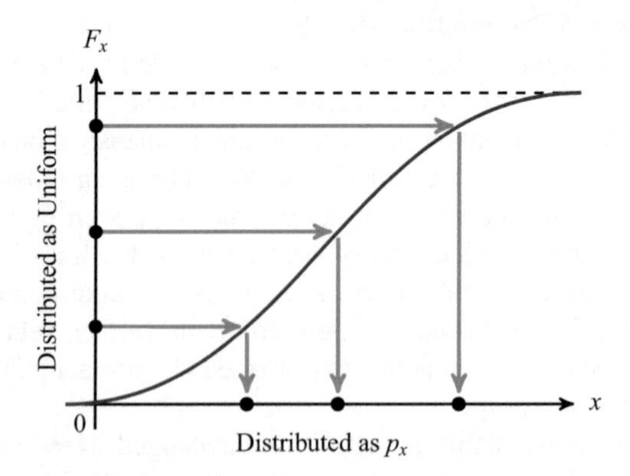

Fig. 9.9 Random sampling: we can construct random samples from *any* probability distribution p by constructing the cumulative distribution F, the integral of p, and then projecting random samples from a uniform distribution (left) onto F.

We can use the preceding strategy to synthesize power-law samples. In general, building on (9.20), we are interested in power-law distributions having a heavy tail of the form $p_x(x) \propto \alpha x^{-\gamma}$:

Case I: For $\gamma = 1$ the infinite-tail PDF is non-normalizable, therefore we can consider only the bounded power-law PDF

$$p_x(x) = \begin{cases} \alpha/x & 1 \leq x \leq \bar{x} \\ 0 & \text{Otherwise} \end{cases} \tag{9.31}$$

for some finite $\bar{x}$. We then have three steps to derive the synthesis scheme:

1. Solve for the normalization constant α in the PDF:

$$1 \equiv \int_1^{\bar{x}} \frac{\alpha}{x}\, dx = \alpha \ln(x)\Big|_1^{\bar{x}} = \alpha \ln(\bar{x}) \qquad \therefore \alpha = \frac{1}{\ln(\bar{x})}$$

2. Solve for the CDF, the cumulative distribution, by integrating:

$$F_x(a) = \int_1^a \frac{\alpha}{x}\, dx = \frac{\ln(x)}{\ln(\bar{x})}\Big|_1^a = \frac{\ln(a)}{\ln(\bar{x})}$$

3. Solve for the inverse function of the CDF:

$$u = F_x(x) = \frac{\ln(x)}{\ln(\bar{x})} \qquad \therefore x = F_x^{-1}(u) = \exp\big(u \ln(\bar{x})\big) \tag{9.32}$$

Therefore given uniform random u, $e^{u \ln(\bar{x})}$ will be $1/x$ distributed between 1 and $\bar{x}$.

Case II: For $\gamma > 1$ the unbounded power-law PDF is defined as

$$p_x(x) = \begin{cases} \alpha/x^{\gamma} & x \geq 1 \\ 0 & x < 1 \end{cases} \tag{9.33}$$

We again have the same three steps:

1. Normalize:

$$1 \equiv \int_1^{\infty} \frac{\alpha}{x^{\gamma}} dx = \frac{-\alpha}{\gamma - 1} x^{-\gamma - 1}\Big|_1^{\infty} = \frac{\alpha}{\gamma - 1} \qquad \therefore \alpha = (\gamma - 1)$$

2. Solve for the CDF:

$$F_x(a) = \int_1^a \frac{\alpha}{x^{\gamma}} dx = -x^{-(\gamma-1)}\Big|_1^a = \left(1 - a^{-(\gamma-1)}\right)$$

Example 9.4: Power Laws and Fraud Statistics

A great many economic indicators and financial transactions obey a power law distribution, thus it is possible to examine the statistics of *reported* financial values (e.g., on tax returns) to look for evidence of fraud, since people are subject to psychological biases (we like round numbers) and do not naturally think in terms of power laws.

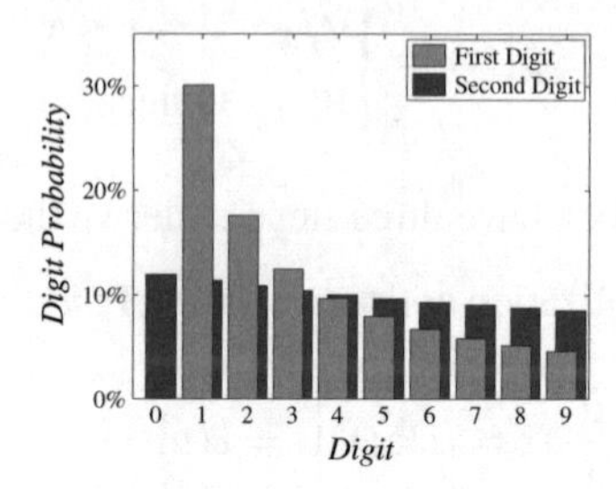

The best-known example is *Benford's Law*, which describes the distribution of digit statistics. Clearly not all numbers are subject to this law. Human adult height, measured in feet, will be dominated by first digits of 4 and 5, and in Canada the first digit of a person's age is relatively uniformly distributed between 1 and 6, and where 5 is actually the most likely first digit.

However $1/x$ power laws *do* follow Benford's Law. Why? Benford's law stems from scale-invariance (page 225), which applies to many economic data, meaning that the distribution of values is flat in the logarithmic domain:

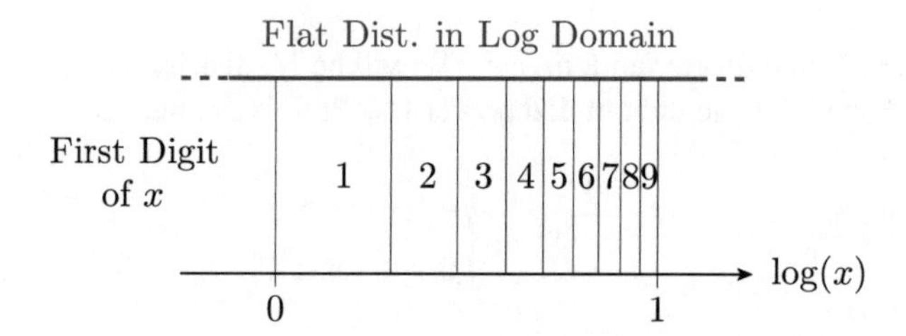

As a result,

$$\text{Probability}\big(\text{First Digit of } x \text{ is } i\big) = \log_{10}(i+1) - \log_{10}(i) \tag{9.29}$$

Example continues ...

Example 9.4: Power Laws and Fraud Statistics (continued)

Interestingly the DJIA index exhibits substantial departures from Benford's law, very likely because there is significant psychology present in stock trading, particularly in crossing certain thresholds, such as 1000 or 10,000:

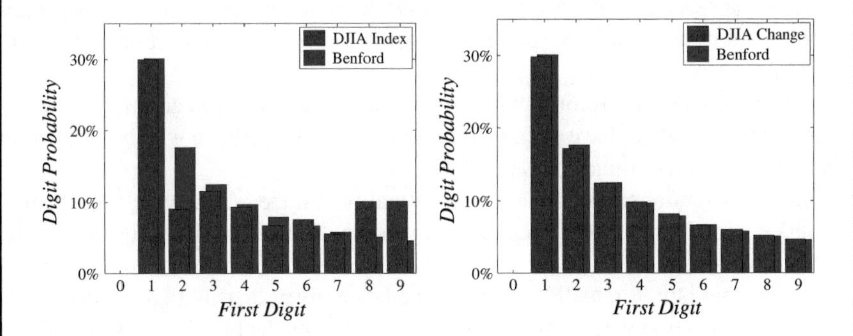

In contrast, the daily *change* in the DJIA, right, shows no evidence of bias and matches Benford's law amazingly well.

Further Reading: See Benford's Law in *Wikipedia*, Problems 9.5 and 9.10.

3. Solve for the inverse function, giving us the random sampler:

$$u = F_x(x) = 1 - x^{-(\gamma-1)} \qquad \therefore x = F_x^{-1}(u) = (1-u)^{-\frac{1}{\gamma-1}} \tag{9.34}$$

Therefore given uniform random u, $(1-u)^{-\frac{1}{\gamma-1}}$ will be $1/x^\gamma$ distributed.

Finally there are two further, related strategies:

A. The ratio of Gaussian random variables is Cauchy, thus

$$\text{Given } x_1, x_2 \sim \mathcal{N}, \quad \text{then} \quad \frac{x_1}{x_2} \sim \text{Cauchy} \tag{9.35}$$

which has $1/x^2$ tails and thus has infinite variance.

B. The exponent of Gaussian random variables is log-normal, thus

$$\text{Given } x \sim \mathcal{N}, \quad \text{then} \quad e^x \sim \text{log-normal} \tag{9.36}$$

which, although not a power law, can have very similar behaviour over a wide range of scales.

In contrast to *synthesis*, the *analysis* of power laws is somewhat more subtle. In principle we are just looking for a straight line in a log-log plot, however there are several issues to consider:

- Since power laws are associated with distributions having rare events, we would generally need a *lot* of data to actually observe these rare events.

  However our goal is not actually to wait for rare events. Rather our primary interest is to observe/validate that we have power-law behaviour, and to estimate the exponent γ from the prevalent (*non*-rare) data, in order to infer behaviour about the tails.
- We are observing *real* systems, which are bounded in space, time, and energy, which therefore can have power-law behaviour only over a certain range of scales.

  That is, there is a limit regarding how far into the tail we can infer; so we *do* need enough data to have events present in the tail to establish whether the power law behaviour continues or is truncated.
- Other distributions, particularly log-normal, can appear relatively straight in a log-log plot. How straight does a log-log distribution need to be for us to consider it a power-law?

  Our goal is not actually to *prove* whether a given distribution is necessarily exactly a power law. Our interest is in understanding the heaviness of the tails, to better understand the impact/consequences of the distribution.

The most straightforward are power-laws in rank, such as for word usage in Figure 9.8: the indicator (wealth, word frequency) is plotted against its sorted rank in a log-log plot, and the resulting plot is examined by eye.

A slightly more sophisticated approach involves what are known as quantile-quantile (Q–Q) plots which, in their simplest form, involve plotting samples from two distributions against each other. That is, given a set of N points $\{y_i\}$ and a hypothesized distribution p_x, we synthesize N random samples $\{x_i\}$ from p_x, sort both datasets and produce a scatter plot, plotting

$$\text{sort}\,\{y_i\} \quad \text{versus} \quad \text{sort}\,\{x_i\}$$

as illustrated in Figure 9.10:

1. If p_x is indeed the correct distribution, then the scatter plot should lie along the line $y = x$;
2. If p_x is from the correct distribution family (say with a different power) then the scatter plot should lie along a straight line;
3. If p_x is not a fitting distribution, the scatter plot will typically be curved.

The Q-Q plots are very easy to interpret qualitatively (straight or not straight), however the very limited number of data points in the tails can lead to irregular behaviour which is harder to assess (Figure 9.11).

To quantitatively assess a power-law distribution, we could use a simple statistical measure, such as the kurtosis parameter K, which measures tail heaviness. However our preference would be to explicitly estimate γ, to quantify the frequency of occurrence as a function of event size. Such a quantification may be undertaken by constructing a histogram but with bins logarithmically scaled, rather than linear,

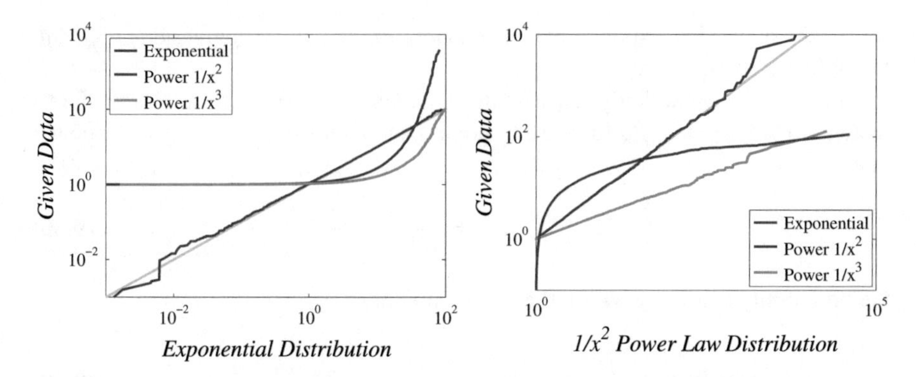

Fig. 9.10 Q–Q Plots: given sample data, here from an exponential and two different power laws, we can plot the data against data synthesized from a known distribution: exponential (left) and $1/x^2$ power-law (right). A straight-line relation suggests compatible distributions. In the left panel the exponential samples appear straight; in the right panel both power laws are straight, but it is the $1/x^2$ lying along the grey $y = x$ line, correctly identifying the sample distribution. The synthesis method of page 229 was used to generate these data.

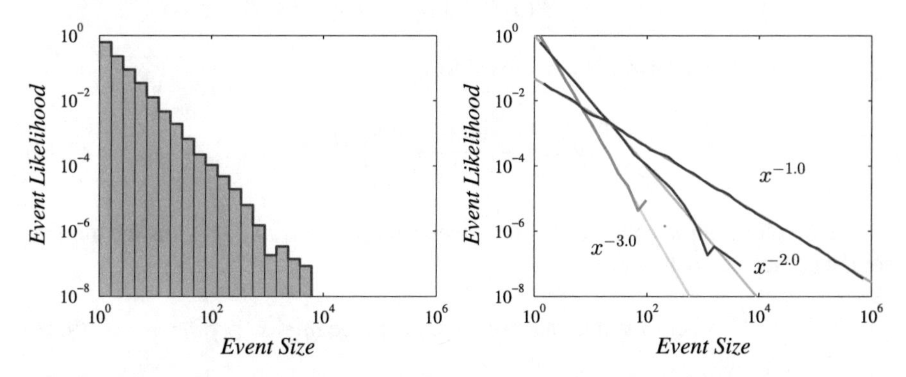

Fig. 9.11 Power law histograms: it is convenient to use histograms, via (9.37), to determine a power law distribution. Because of the wide range of values present in power laws a constant histogram bin width would not work, consequently logarithmically-spaced bins, left, should be used, shown here applied to $1/x^2$ data. The actual exponent is easily found via linear regression (lightly shaded lines) on the estimated distribution, right. As in Figure 9.10, the synthesis method of page 229 was used to generate the data.

so that a similar number of data points falls into each bin. The number n_i of samples in each bin i, relative to the bin size s_i, tells us the probability density of occurrence

$$\text{Probability Density}_i = p_i = \frac{n_i}{s_i \cdot N}, \tag{9.37}$$

the method which was used to generate the power-law plots throughout this chapter. If the centre of histogram bin i is located at x_i, then from (9.17) we know that

$$\log p_i = \log \alpha - \gamma \log x_i, \tag{9.38}$$

so that the power-law exponent γ can be determined by linear regression applied to $\log p_i$ versus $\log x_i$.

Finally, we can explicitly estimate the exponent γ in closed form. If we define a histogram bin $q \le x \le \beta q$ for starting point q and fixed ratio β, then given a power law

$$p(x) = \frac{\alpha}{x^\gamma} \tag{9.39}$$

the probability of x lying within the histogram bin is

$$P(q) = \int_q^{\beta q} p(x)\,dx = -\frac{\alpha}{\gamma - 1} q^{1-\gamma}\left(\beta^{1-\gamma} - 1\right). \tag{9.40}$$

Now for *two* such bins, the relative probability is a function of γ,

$$\frac{P(q_1)}{P(q_2)} = \left(\frac{q_1}{q_2}\right)^{1-\gamma} \tag{9.41}$$

from which we can solve for γ in closed form:

$$\gamma = 1 - \frac{\log\left[{}^{P(q_1)}/_{P(q_2)}\right]}{\log\left[{}^{q_1}/_{q_2}\right]}. \tag{9.42}$$

Since we will want to solve for γ from given data points $x_1, \ldots, x_N$, it is the number of data points in each bin,

$$M_i = \text{Number of given data points in the range } q_i \le x_j \le \beta q_i \tag{9.43}$$

which determines the probabilities, leading to an elegant estimate of the exponent from sample data:

$$\gamma = 1 - \frac{\log\left[{}^{M_1}/_{M_2}\right]}{\log\left[{}^{q_1}/_{q_2}\right]}. \tag{9.44}$$

In general, the further apart the bins and the greater the number of samples in each bin, the more accurate the estimation of γ.

Case Study 9: Power Laws in Social Systems

In Section 8.6 we briefly explored the question of social system simulation, however real humans are much more subtle and complex than agent models, so the types of social systems which we can meaningfully simulate is somewhat limited.

Example 9.5: Power Laws and Cities

Over half of the world's human population now lives in cities; as a social construct, cities have been a fantastically successful idea.

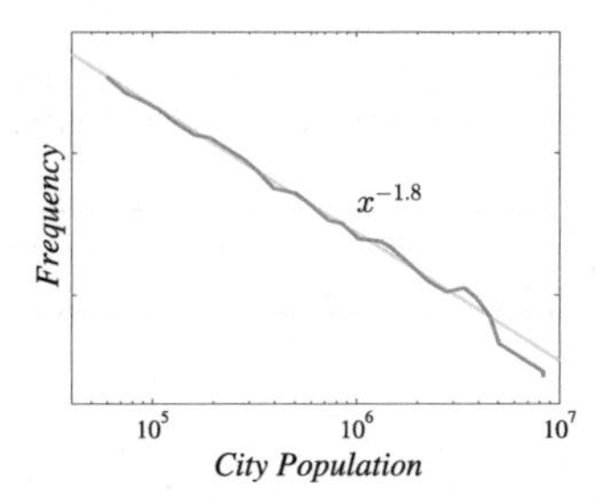

A plot of the distribution of city sizes, here shown for cities over 50,000 people in the United States, demonstrates a nearly perfect power-law fit. Almost certainly there is a fitness/preferential attachment phenomenon at work, in that larger cities have more to offer and will grow more quickly than smaller cities.

Per capita, larger cities presumably offer a more efficient use of infrastructure and offer more opportunities. A plot of road surface and GDP makes this clear:

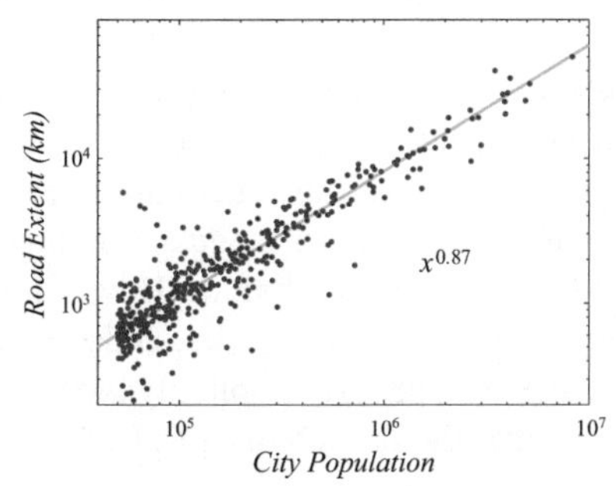

Data: Office of Highway Policy Information
Federal Highway Administration

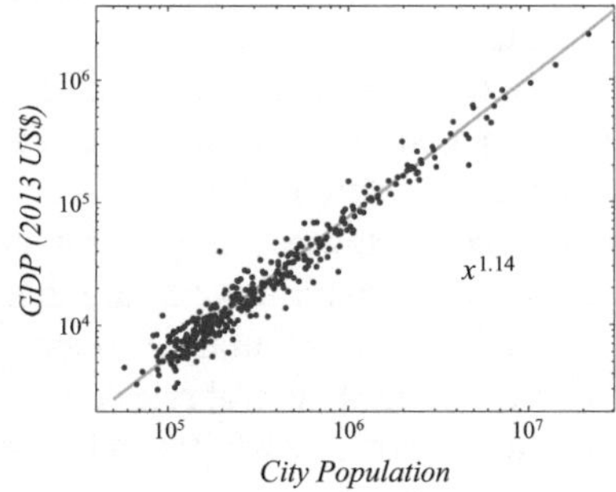

Data: U.S. Bureau of Economic Analysis

Amazingly, the dependence of road length, GDP, and many other urban indicators are *themselves* power laws. That is, nearly any indicator y can be modelled as a power law $y = y_o x^{\beta}$ as a function of city population x:

New Patents	$\beta \simeq 1.27$	GDP	$\beta \simeq 1.1$ to 1.3
Housing	$\beta = 1.00$	Electricity Use	$\beta \simeq 1.02$
Employment	$\beta \simeq 1.01$	Water Use	$\beta \simeq 1.01$
Road Surface	$\beta \simeq 0.83$	Gasoline Stations	$\beta \simeq 0.77$

Example continues ...

Example 9.5: Power Laws and Cities (continued)

In general, infrastructure costs (gas stations, roads) scale *sub*-linearly, slower than the growth rate of the city. Basic necessities (shelter, water, electricity) scale with the number of people, thus $\beta \simeq 1$. Whereas opportunities (patents, inventions, GDP) scale *super*-linearly, faster than the growth rate.

Why so many indicators behave as power laws is still only poorly understood, however recent compelling arguments and models from network theory (*Science* paper, below) begin to offer some clues.

Further Reading: Bettencourt, "The Origin of Scaling in Cities," *Science* (340), 2013
Bettencourt et al., "Growth, innovation, scaling, and the pace of life in cities," *PNAS* (104), 2007

The *analysis* of social systems is a completely different matter. There are many social networks (Internet, Facebook, Twitter, phone calls, physical neighbourhoods) for which data are available and which can be analyzed in detail.

In particular, for most social systems the connection statistics are power law, meaning that the probability $P(k)$ that a node (person, home page, computer, Facebook account) has k connections behaves as a power law, as

$$P(k) \propto k^{-\gamma} \tag{9.46}$$

also known as a Scale-free network, in most cases believed to be due to preferential attachment.

It is relatively easy to understand how preferential attachment leads to power law behaviour, since most of us can feel, within ourselves, a desire to hang out with popular ("preferential") people, or to watch the YouTube video that everyone is talking about. However the appearance of power laws in language, known as Zipf's law, as shown for this text in Figure 9.8, is much less obvious. How is it that language, an organic cultural development over thousands of years, ends up with word choices or grammatical rules that lead me, an author, to implicitly type words which follow a power law? Zipf's law states that the probability $P(w)$ of the wth most popular word obeys

$$P(w) \propto w^{-\gamma} \quad \gamma \approx 1 \tag{9.47}$$

In sharp contrast to the elegance and simplicity of (9.47) is the confusion and uncertainty as to why such a distribution should actually hold for language [6]. There are many ideas:

- *Stochastic Processes*: A purely random collection of letters and spaces leads to "words" which obey Zipf's law, so some people have argued that there is nothing special about language words obeying a power law in rank. However stochastic word models also imply that the number of unique words is exponential in word length, which is not at all the case for real languages, so language appears to be special after all.

Example 9.6: Power Laws and Discount Functions

One of the challenges of power laws is how they limit our ability to learn from the past. It turns out that there is a second factor, in parallel, thwarting our ability to plan for the future. Suppose you have a renewable resource (as in Problem 6.11), such as a forest, with a total present resource value of R. If managed, the forest could provide an indefinite sustainable annual utility or harvest of value u.

Because money earns interest i over time t, harvests further into the future are worth less than harvests now in the present:

$$\text{Net Present Value } P = \sum_{t=0}^{\infty} u(1+i)^{-t} \quad \text{Leading to} \ldots \quad P \underset{\text{Clear-cut}}{\overset{\text{Manage Sustainably}}{\gtrless}} R$$

So sustainability is challenged if u is low (old-growth forests, slowly reproducing animals such as elephants or whales) or i is high. That is, there are circumstances in which it is economically "rational" to clearcut a forest, to exterminate a species, or to cut that last standing tree on Easter Island [2].

One might complain that interest rates are artificial, manipulated by central banks, and furthermore that we don't know future interest rates. This is true; unfortunately we humans have a very similar effect wired into our brains, in what is known as the *discount function*, such that \$1 right now has a discounted value of $D(t)$ at a future time t.

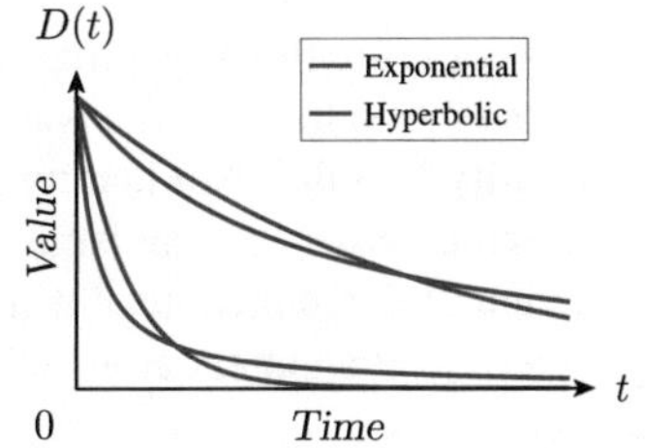

The figure shows two examples of the two most common choices of discount functions, plotted over time:

$$\text{Exponential } D(t) = 1/(1+k)^t \qquad \text{Hyperbolic } D(t) = 1/(1+kt) \tag{9.45}$$

In the exponential case the discount rate $\rho = -D'(t)/D(t)$ is constant over time, meaning that a passage of time does not affect preferences, essentially analogous to the memoryless property of Example 9.1. However psychological experiments have been devised to measure discounting, by assessing a person's indifference to a certain amount of food or money now and a larger amount in the future, showing that humans have hyperbolic discounting, a time-dependent discount rate which slows over time. Whereas pigeons discount within seconds, and rats within minutes, humans certainly have a longer time horizon, $k \simeq 0.2$ per year (but measurably much shorter for people with addictions, for example), which still means that even rather significant future damage, such as climate catastrophes decades into the future, can have a very limited effect on present decisions.

With regards to the spatial models of Chapter 8, certainly any agent-based model of humans vis-a-vis a resource extraction/depletion question would need to include discounting in its agent definition.

Further Reading: See Time preference, Time value of money, Hyperbolic discounting in *Wikipedia*, Brandt et al., "Tipping points and user-resource system ...," *Ecological Complexity* (13), 2013, Hardin, "The Tragedy of the Commons," *Science* (162), 1968

- *Preferential Attachment*: The frequency with which a word is used depends on context. If the probability of a word is based on the frequency with which that word has been used in the past, then a Zipf distribution appears.
- *Principle of Least Effort* or effort minimization, says that speakers and listeners do not want to have to work harder than necessary, and it is simpler to have relatively few commonly-used words.

The *Pareto Principle* or *80-20 Rule* is neither a "rule" nor an explanation, rather an observation that a great many social systems are subject to power laws. The scale-invariance of such distributions then further implies that the 80-20 behaviour can be expected to appear in subsets, thus in each of

Canada Province of Ontario Waterloo Region City of Waterloo

That is, in general, a small subset (of people, of companies, of special interest groups ...) will receive disproportionate attention, even without *any* biases, political lobbying etc. As a result,

- Most people are below average: Power laws are highly asymmetric distributions, and the sample mean will always be significantly larger than the median.
- Inequality is "fair": Not that inequality is *good*, rather that it arises even without systematic biases or cronyism.
- The rich and famous are not necessarily so unique: Random fluctuations and preferential attachment alone will cause "famous" people to appear in any social context.

The conclusion is not that abject poverty is inevitable, rather that caution needs to be exercised in addressing it (Example 7.8). Constructing a system which *enforces* equality, so to speak, directly opposes how complex social systems operate, and is more likely to lead to bad government than it is to an equitable society.

Further Reading

The references may be found at the end of each chapter. Also note that the textbook further reading page maintains updated references and links.

Wikipedia Links — Distributions: Normal Distribution, Central Limit Theorem, Exponential Distribution, Memorylessness, Poisson Distribution, Power law, Black swan, Heavy-Tailed Distributions

Wikipedia Links — Power Laws: Zipf's Law, Benford's Law, Pareto Principle, Pareto Distribution, Gutenberg-Richter Law

Wikipedia Links — Principles: Scale invariance, Scale-free network, Preferential attachment, Fitness model

One of the most popular books regarding heavy-tailed distributions, certainly in the area of finance, is *The Black Swan* by Taleb [11]. Those readers interested in following up on the LTCM disaster of page 211 are referred to the book by Lowenstein [5].

A highly accessible text on power laws is that by Bak [1], with a somewhat more advanced/technical treatment by Sornette [9]. Although focusing on Zipf's law, the paper by Manin [6] offers an engaging and fairly broad treatment of the subject.

There are many mathematicians and scientists who claim that power laws have been over-promoted. Most of these people do not deny that heavy-tailed phenomena exist and need to be accounted for, however they do suggest some caution in claiming that an observed behaviour is power law [10].

In any event, the goal of this chapter has *not* been to suggest that nearly all social phenomena are power law, rather that there are a great many social, financial, and natural phenomena which are heavy tailed (whether power law, log-normal, or some other distribution), and that significant caution and prudence is required in encountering systems where it is challenging to learn from history and where significant, unanticipated outliers can take place.

Sample Problems

Problem 9.1: Short Questions

(a) Define "power law"
(b) Give a brief description of Zipf's law
(c) Give a brief description of Pareto's law
(d) Suppose that NASA claims to have developed a system to deflect/destroy asteroids which might endanger the earth. What do power laws have to say about this?
(e) Suppose that power law behaviour is observed in a *natural/ecological* system. What sorts of mechanisms might commonly lead to such power law behaviour?
(f) Suppose that power law behaviour is observed in a *human social* system. What sorts of mechanisms might commonly lead to such power law behaviour?
(g) Any real, physical system may be approximately power law, but never *exactly*. What is it that prevents exact power law behaviour in real systems?

Problem 9.2: Conceptual — Distributions

Name the most likely distribution associated with each of the following, with one or two sentences explaining why:

(a) The time between successive lightning strikes in a thunderstorm.
(b) The height of students in a class.
(c) The human intelligence (IQ) in a randomly selected group of people.
(d) The frequency and severity of asteroid impacts.
(e) The number of online social–media connections that a person has.

Problem 9.3: Analytical — Linear Functions and Gaussians

On page 213 it was stated that

Any linear function of Gaussian random variables is still Gaussian

At first glance this may seem obvious. What else should it be?

It turns out, however, that for most distributions there are very simple linear functions which produce *different* distributions:

(a) Given two *uniform*, independent random variables u_1, u_2, find a simple linear function

$$q = f(u_1, u_2)$$

such that q is *not* uniform.

(b) Repeat part (a) for two *exponential* random variables e_1, e_2.

Problem 9.4: Analytical — Power Laws

On page 219 there was a discussion of power law attributes, with the following table:

$\gamma \le 1$	Non-normalizable PDF		
$1 < \gamma \le 2$	Normalizable PDF	Undefined mean	Infinite variance
$2 < \gamma \le 3$	Normalizable PDF	Finite mean	Infinite variance
$3 < \gamma$	Normalizable PDF	Finite mean	Finite variance

Show that the claims in the table are correct by deriving the normalization, mean, and variance behaviour for each of $\gamma = 0.5, 1.5, 2.5, 3.5$. You may wish to refer to Appendix B, particularly (B.5),(B.6).

Problem 9.5: Analytical — Scaling Laws

Example 9.4 argued that a scale-invariant distribution should obey Benford's law. We wish to explore this a bit further.

Let z be a random variable obeying a distribution invariant to scaling, and let p_i be the probability that z begins with digit i. Because of scale-invariance, multiplying z by some factor must not cause a change in the digit statistics.

It is obvious that

$$1 \le z < 2 \quad \rightarrow \quad 2 \le 2 \cdot z < 4$$

Therefore if the value of z begins with the digit "1", then the doubled value of z *must* begin with a digit "2" or "3", from which it follows that

$$p_1 = p_2 + p_3 \tag{9.48}$$

(a) Validate that (9.48) satisfies Benford's law from (9.29).

(b) Using an argument similar to the above, list seven further assertions on digit probabilities for scale-invariant distributions.

Problem 9.6: Analytical — Power Law Synthesis

In Section 9.5 we derived the synthesis scheme for power law distributions. Since the $\gamma = 1$ power law is non-normalizable, in (9.31) we considered a bounded power law $1 \le x \le \bar{x}$, whereas for $\gamma > 1$ in (9.33) we derived the synthesis for the *unbounded* power law $1 \le x$.

Using the same strategy as in Section 9.5, derive the synthesis method for *bounded* power laws

$$p_x(x) = \begin{cases} \alpha / x^{\gamma} & 1 \le x \le \bar{x} \\ 0 & \text{Otherwise} \end{cases}$$

for $\gamma > 1$.

As an optional validation, simulate your result numerically and then use (9.44) to deduce γ.

Problem 9.7: Analytical — Power Laws

The three panels below show three different societal power-law behaviours for income:

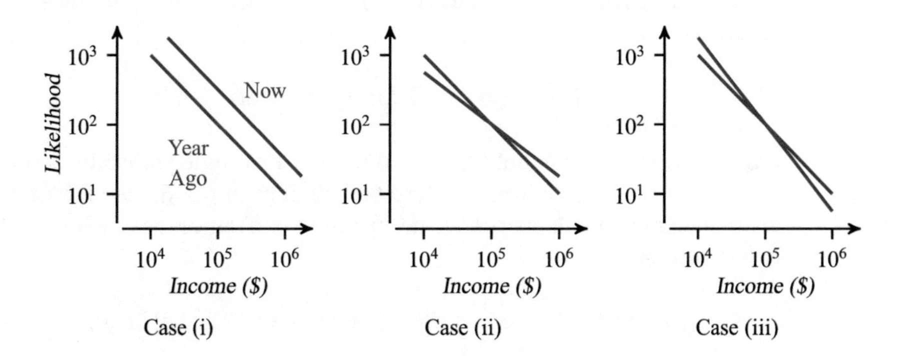

In each case we have a society starting at the red line and a year later described by the blue line. For *each* of the three cases, give a brief description of the change that took place in the society over the last year.

Problem 9.8: Analytical — Power Laws

Given the following distribution

$$p(x) = \begin{cases} 0 & x < 1 \\ 1/x^2 & x \ge 1 \end{cases}$$

(a) What do we call this sort of distribution?
(b) Derive the mean of the distribution.
(c) Derive the median of the distribution.

(d) For any Gaussian distribution, the mean and the median are the same. Explain clearly how they can be so different in this case.

Problem 9.9: Numerical/Computational — Poverty and Power Laws

We would like to build on the non-spatial agent model of Problem 8.12 and [3] to understand some of the reasons for persistent wealth inequality.

Implement the algorithm of Problem 8.12 and comment on the observed distribution of wealth.

In particular,

(a) What is the slope of the power law?
(b) What is the minimum number of agents (people) required to have a recognizable power law distribution?
(c) What is the minimum number of iterations, as a function of the number of agents, in order to have a converged distribution?

Problem 9.10: Numerical/Computational — Benford's Law

Example 9.4 discussed Benford's law, the statistical behaviour of digits in many numerical data.

(a) Develop a function which measures digit statistics; that is, we need a function

```
[pd, pb] = Benford( num )
```

for `num` a vector of numbers, for which the first-digit probabilities are returned in `pd` and the ideal Benford probabilities in `pb`. Since uniformly distributed random numbers have no first-digit preference, use such values to test your code:

```
[puniform, pb] = Benford( rand(1,1e5) );
```

(b) Values which obey Benford's Law should be *scale invariant*: if the prices of a group of objects obeys Benford's law, then doubling all prices should not affect the digit statistics.

Test the scale-invariance of four distributions: uniform, Gaussian, exponential, and heavy-tail. Given a random vector $\underline{r}^d$ from distribution d, let $p^d_{i,\alpha}$ be the probability of digit i in $\alpha \cdot \underline{r}^d$, such that the random numbers in $\underline{r}$ are multiplied by scale factor α.

Report the scale sensitivity of each distribution d, as a function of i, by plotting the minimum and maximum values of p over varying α. Discuss your observations.

(c) Now we will examine real data. Online, or via other sources, select two data sets,

(i) One which you believe should obey Benford's law,
(ii) One which you believe should *not* obey Benford's law.

For each data set, offer a brief rationale as to why the first digit should/should not follow Benford's law, show the actual digit statistics, and interpret your results.

Problem 9.11: Reading and Policy — Power Laws I

There are two basic aspects of power laws of which we need to be aware in formulating policy:

- How to be *cautiously aware* about power laws, extreme events, and our limited ability to learn from history.
- How to *take advantage* of power laws, in that Pareto's law allows us to focus on the relatively few cases that have a disproportionately great effect.

In this problem we wish to consider the former issue.

For any type of planning, say for floods or droughts, in a true power law there is always a bigger flood and always a longer drought. Even if floods and droughts are not exactly power laws, that they are *like* power laws means that historically unprecedented, extreme events are certainly more likely than what Gaussian models would suggest.

Pick a specific power-law behaviour, whether economic, social, or natural, that has some connection to or impact on humans. Briefly overview or provide a bit of context regarding the behaviour and then discuss a few political or policy implications, focusing on the significance of the power-law behaviour, and how that power-law leads to problems/challenges/specific policy outcomes.

Problem 9.12: Reading and Policy — Power Laws II

There are two basic aspects of power laws of which we need to be aware in formulating policy:

- How to be *cautiously aware* about power laws, extreme events, and our limited ability to learn from history.
- How to *take advantage* of power laws, in that Pareto's law allows us to focus on the relatively few cases that have a disproportionately great effect.

Following on Problem 9.11, we now wish to consider the latter issue. Read the article "Million Dollar Murray" by Malcolm Gladwell[5] .

What are your perspectives on this essay?

Identify a few other social/environmental issues that might fall into the same category, whereby disproportionate gains could be had by thoroughly addressing a small subset of cases.

[5]The article can be found online; a link is available from the textbook reading questions page.

References

1. P. Bak. *How Nature Works*. Copernicus, 1996.
2. J. Diamond. *Collapse: How Societies Choose to Fail or Succeed*. Penguin, 2011.
3. B. Hayes. Follow the money. *American Scientist*, 90(5), 2002.
4. G. Lawler. *Introduction to Stochastic Processes*. Chapman & Hall, 2006.
5. R. Lowenstein. *When Genius Failed: The Rise and Fall of Long-Term Capital Management*. Random House, 2001.
6. D. Manin. Zipf's law and avoidance of excessive synonymy. *Cognitive Science*, 32, 2008.
7. S. Resnick. *Adventures in Stochastic Processes*. Birkhäuser, 2002.
8. M. Schroeder. *Fractals, Chaos, Power Laws: Minutes from an Infinite Paradise*. Dover, 2009.
9. D. Sornette. *Critical Phenomena in Natural Sciences*. Springer, 2006.
10. M. Stumpf and M. Porter. Critical truths about power laws. *Science*, 335, 2012.
11. N. Taleb. *The Black Swan: The Impact of the Highly Improbable*. Random House, 2010.

Chapter 10
Complex Systems

2003 coincidentally saw two major electrical system failures: the Northeast Blackout in Ontario, Canada and the northeastern United States, affecting 55 million people, and the Italy Blackout in most of Italy and parts of Switzerland, affecting 56 million people.

In both cases a tree falling on power lines disabled only a comparatively small portion of the electrical system. In principle the blackout would not need to have been larger than the amount of power lost on the failed transmission lines. In actual fact, the small perturbation led to cascading failures and huge disruptions.

To be fair, electrical system control is quite challenging: electricity is not stored, so production and consumption need to be matched in real-time. Inadequate or excessive loads lead to increasing or decreasing system frequency, respectively; turbines can be damaged by even modest frequency disruptions, so to protect infrastructure generating stations are disconnected from the grid and shut down relatively rapidly in the event of a loss of frequency control.

Certainly the North American system was stressed, the power outage taking place in a weekday afternoon on a hot summer day. On a day with low demand, losing one transmission line is easily accommodated by transferring the power to another line. As transmission flows increase, the flexibility to move power from one line to another certainly decreases, nevertheless a localized blackout would have resolved the stress associated with the failed line.

However the electrical system is actually a complex system consisting of many interacting parts, and in cases where complex systems are pushed towards instability they tend to fail in quite unanticipated ways. Although the actual fault in the North American case was traced to a software error in the alarm system for an energy corporation in Ohio, in some ways this doesn't matter: there will always be bugs and glitches, what matters is the resilience of the complex system as a whole.

The challenges of a system near capacity has parallels throughout western society. The push for efficiency in many sectors of the economy has led to ever more tightly coupled systems. Examples include the just-in-time delivery of auto parts at

P. Fieguth, *An Introduction to Complex Systems*,
DOI 10.1007/978-3-319-44606-6_10

manufacturers, the constant stream of deliveries to grocery stores, and particularly air travel, where flight delays in one part of the world are regularly blamed on storms somewhere far away.

Complex systems are essentially a manifestation of power laws due to the behaviour of coupled, distributed, nonlinear systems: the culmination of the five preceding chapters.

10.1 Spatial Nonlinear Models

Suppose we have a continuous-state spatial nonlinear system. If we spatially discretize the system onto a $n \times n$ domain, we have a very high-dimensional state $\underline{z} \in \mathbb{R}^{n^2}$. It should be readily apparent that we cannot possibly compute and plot the fixed points, bifurcations, and basin boundaries in n^2 dimensions.

We have two basic options:

1. We can examine the behaviour of a low-dimensional sub-system, or
2. We can examine the statistical behaviour of the large system as a whole.

We have actually *already* seen a number of simplified, reduced-dimension or lumped-parameter systems:

- Global climate fixed points and bifurcations in Figure 6.19, representing climate by the single state variable of ice-extent.
- The Thermo-Haline system, in Figure 6.23 and the lumped-parameter two-box model of Example 8.3, with a few state variables of temperature and salinity.
- The *Panarchy* model in Example 7.8, illustrating the nonlinear dynamics of a large, complex society in terms of three state variables of capital, connectedness, and resilience.
- The power-law behaviour of Example 9.5 did not have a dynamic model, however the principle is similar: taking a large complex system (city) and characterizing it in terms of the variables of population, road length, and GDP.
- Even the rabbits and goats problem of Example 7.1 and here in Figure 10.1 is, in principle, a spatially distributed high-dimensional agent model with a state for the age, health, gender, and geographic location of each animal, but which was represented in (7.11) via a two-dimensional lumped state of populations.

A low-dimensional representation is attractive, as concepts of fixed points, bifurcations, and basins are once again easy to interpret, even for large complex systems. However we need to be cognizant of the limitations: first and foremost, the corresponding low-dimensional phase diagram is at best a cartoon, an interpretation or approximation of the underlying system.

Secondly, most large, physical systems will be stochastic, as illustrated in Figure 10.1, so the system will not actually settle indefinitely at fixed points, thus we should not expect fixed points and basins to have the sharp, explicit interpretations which they had for deterministic systems in Chapter 7.

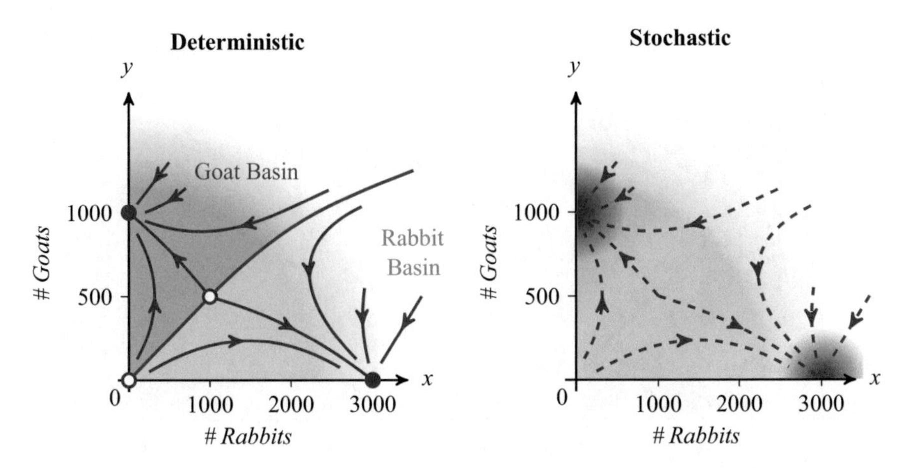

Fig. 10.1 Stochastic models: in Example 7.1 we had a nice, clean, deterministic model, left, of goat and rabbit populations. However most systems, including the reproduction of goats and rabbits, are stochastic, right, in which case basin boundaries and fixed points become blurred.

Our alternative, then, is to examine the statistical behaviour of a large, complex, interacting system as a whole.

10.1.1 Phase Transitions

In the same way that a bifurcation represents a spontaneous change in fixed-point behaviour in a nonlinear system, similarly a phase change represents a change in the aggregate statistical behaviour of a complex system.

Phase transitions are not that unusual, of course, the most familiar being the transition of water between frozen and liquid forms, or between liquid and vapour. The phase transitions, as a function of system parameters such as pressure or temperature, can be summarized in a transition diagram, such as in Figure 10.2.

Characterizing a vast array of molecules by their macroscopic state is, admittedly, in some ways not so different from a lumped model; indeed people frequently interchangeably refer to the phase or state of a system. There should be a distinction, however: the phase is an *observation*, a *deduction* from a complex, highly interacting underlying state. That is, we have a high-dimensional underlying complex system, which gives rise to the macroscopic phase, however we do not have a dynamic model for the phase on its own.

The most common phase transitions are known as *first order* or *discontinuous*, phase transitions which involve a transfer of energy, such as the energy to melt ice (the heat of fusion) or to evaporate water (the heat of vaporization). Because of the need for an energy transfer the first-order transitions are associated with fairly obvious changes, even at the microscopic level, so that individual water molecules

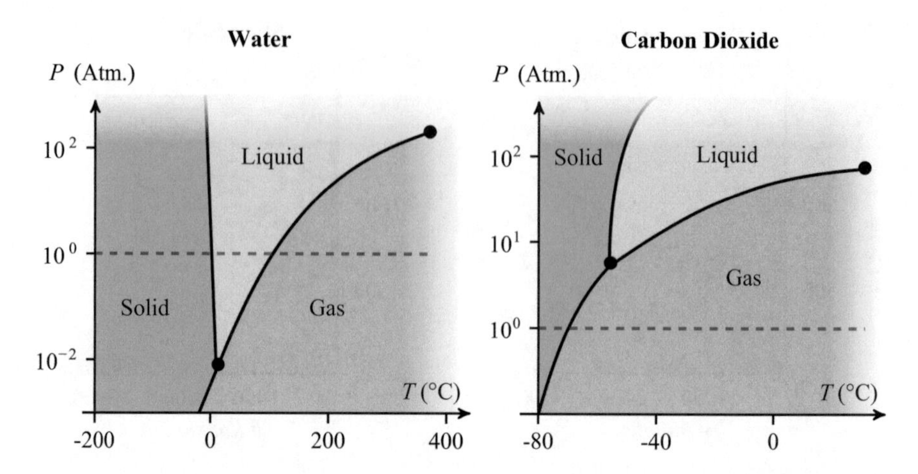

Fig. 10.2 Phase transitions: phase transition diagrams are shown for both water and CO_2. The grey dashed line shows the atmospheric pressure at the surface of the earth, for which water melts at 0 °C and boils at 100 °C. In contrast, at atmospheric pressure there is *no* temperature at which CO_2 is liquid, for which a pressure of at least 5.2 atmospheres is required. Note that these phase diagrams illustrate *steady-state* behaviour: a glass of water in which the water slowly evaporates is a *dynamic* system and is not in steady state. The diagrams omit many details, such as the dependence on volume, or the phase transitions *within* the solid domain (at least ten different solid-water phases have been identified).

(or, at most, very small groups of molecules) know whether they are part of solid ice, involving fixed hydrogen bonds, or have experienced an energy transfer to break those bonds, becoming liquid water with time-varying hydrogen bonds.

10.1.2 Criticality

The first-order transitions are certainly very important in thermodynamics and in understanding the physical properties of materials, however these transitions are relatively unsurprising: we have distinctly different behaviours in the two phases. There is, however, a second type of phase transition, known as a *second-order* or *continuous* transition, which is much more subtle than the first-order transitions just discussed, and much more significant in the context of complex systems.

Look again at Figure 10.2; there are two solid circles plotted in each:

- One is the triple-point, that unique point where all three of the solid, liquid, and gas phases simultaneously coexist;
- The other is the critical point, the end of the liquid-gas phase transition, located at the intersection of the critical temperature T_c and pressure P_c.

What is this critical point? Is it “illegal” or impossible for water or CO_2 to be at a temperature higher than the critical temperature T_c? Not at all; however above the critical temperature there is no longer a discrete transition between liquid and gas, rather a *continuous* one, whereby a liquid gradually becomes more of a froth and then a gas. The critical point is thus *itself* a meta-transition:

$$\begin{matrix} \text{Distinct liquid,} \\ \text{gas phases} \end{matrix} \quad T < T_c \qquad \overset{\text{Criticality}}{T_c} \qquad T_c < T \quad \begin{matrix} \text{Continuous froth,} \\ \text{compressible liquid} \end{matrix} \tag{10.1}$$

The behaviour at the critical point is best explained using the Ising model of Figures 8.16 and 8.17, as discussed in Example 10.1.

The Ising model is characterized in terms of a single parameter β, which measures the strength of coupling between spatially adjacent states. At $\beta = 0$ we have uncoupled states, which are therefore random with no spatial correlation. As β *increases* the coupling grows, creating larger magnetized clumps, but globally the overall net magnetization remains zero.

Similarly at $\beta = \infty$ the state is forced to be -1 everywhere or $+1$ everywhere, out to infinity, meaning that there is a non-zero global magnetization. As β *decreases* the coupling shrinks, permitting tiny clusters of exceptions which grow larger as β decreases, however there remains a global magnetization.

There is a critical point between these two β regimes, the transition between globally magnetized or not:

$$\begin{matrix} \text{Zero global} \\ \text{magnetization} \end{matrix} \quad \beta < \beta_c \qquad \overset{\text{Criticality}}{\beta_c = \frac{\ln(1+\sqrt{2})}{2}} \qquad \beta_c < \beta \quad \begin{matrix} \text{Globally} \\ \text{magnetized} \end{matrix} \tag{10.2}$$

At the pixel level there is nothing special taking place at β_c; this is, after all, a *continuous* state transition. In contrast to frozen and liquid water, simulated samples of an Ising model just above and just below the critical point would look essentially identical. Rather, what characterizes the critical phenomenon here is its *non*-local behaviour, whereby criticality is characterized by structures on *all* length scales, out to infinity. Thus water, which up to the critical temperature is characterized by transparent liquid and gas phases, becomes opaque at the critical temperature since there are bubbles/fluctuations at *all* scales, from nanometre to macroscopic, which scatter light of all wavelengths. Similarly as the Ising model approaches criticality, clumps (correlations) appear of *all* sizes out to infinity.

Spatial dynamic processes having structures or correlations on *all* scales, such as at criticality, are said to be *scale-free*, in that there is no characteristic scale where activity takes place, rather that there is behaviour over *all* scales, from which two consequences emerge:

- Wildly different fine-scale models can give rise to the same large-scale behaviour, and
- Given that there are fluctuations taking place, from tiny to large scales, it should not be surprising that power laws appear.

Example 10.1: Ferromagnetic Phase Change

The Ising model, shown in Figure 8.16, is attractive because it is one of the simplest systems to possess a phase change: when strongly coupled ("frozen") the state is crystallized, with occasional exceptions, to infinity; when weakly coupled ("gaseous") the state is essentially random.

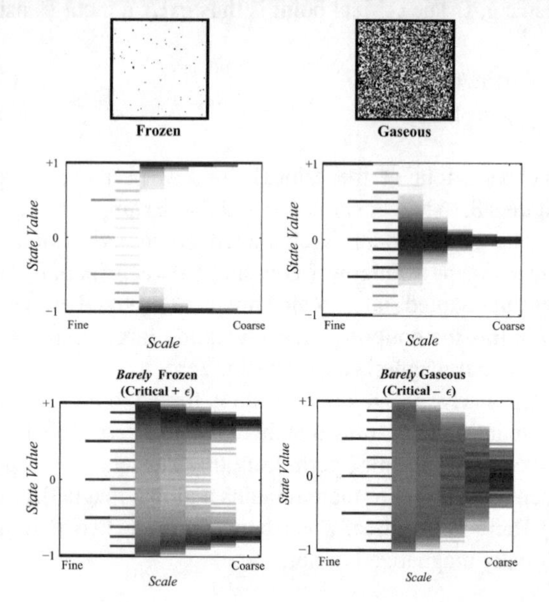

The histograms show the state distribution as a function of scale (by averaging over larger and larger regions). In a frozen state at coarser scales the local exceptions average out, and the average state is either $+1$ (spin up) or -1 (spin down). In contrast, in a gaseous state at coarser scales the essentially random state averages to 0 (zero net spin). The *critical point* occurs at a coupling of 0.4407, the phase transition between globally magnetized and unmagnetized. Near criticality ($\pm\epsilon$) the phase is ambiguous at fine scales, but becomes progressively clearer at coarser scales.

We can understand the critical point as that moment where the correlation or structure size goes to infinity (in principle). A frozen domain consists of a single global magnetization (that is, "white" or "black" out to infinity), with occasional exceptions (small structures); a gaseous domain similarly consists of random magnetizations forming small structures. In between the structure size is increased, with larger and larger clumps forming as the critical point is approached.

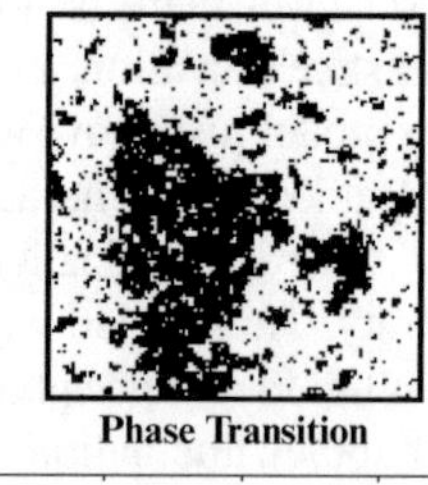

The phase transition figures were computed numerically, on a 2D domain of limited size and limited number of iterations, therefore the *empirical* structure size is finite, as seen here.

If you are interested, the Ising domain size in the figure, given 2D state matrix S, was computed as `regionprops(bwlabel(S>0,4), 'area')`

Further Reading: Wikipedia Ising model; Hill, *Introduction to Statistical Thermodynamics*, Dover, 1987

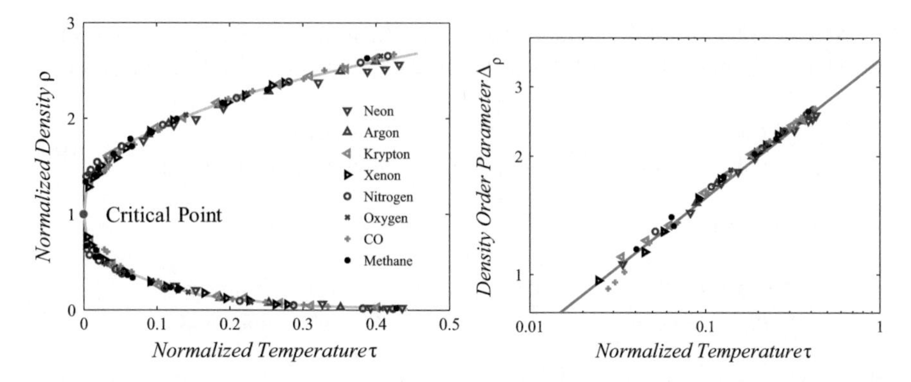

Fig. 10.3 Universality: What do neon (inert), nitrogen (nonreactive) and methane (inflammable) have in common? Normally not a lot! However around the critical point the fluctuations on all length scales cause the molecular-scale structure to become nearly irrelevant at macroscopic scales, such that the eight molecules (left) have the same normalized density–temperature relationship, which is in fact a power law (right). The density order parameter is the *difference* in density between the liquid and vapour phases, which goes to zero at the critical point.

Figure adapted from Zemansky and Dittman, *Heat & Thermodynamics*, McGraw-Hill, 1996.

Since there is structure on *all* scales, for a sufficiently large domain we can't really "see" the finest scale any more. That is, whereas the macroscopic hexagonal pattern of a snowflake *does* reveal the nano-scale crystal structure of ice, at the critical point the macroscopic behaviour is essentially disconnected from the model/molecules at the finest scale, meaning that very different models/molecular structures can end up having very similar behaviour at criticality.

In thermodynamic systems, if we normalize temperature to the critical point,

$$\tau = \frac{T_c - T}{T_c} \tag{10.3}$$

then near criticality even very different molecules will have nearly identical macroscopic power-law behaviour as a function of τ, as shown in Figure 10.3. This convergence is one aspect of *Universality*, whereby the power-law exponent is primarily a function of problem dimensionality, meaning that there are a great many different two-dimensional models which share a common behaviour at criticality.

The relevance to complex systems is that criticality and universality *also* appear in graphs, networks, and Agent-Based models. In particular, the edges in a graph can represent the connectivity between internet servers, which directors and CEOs sit on each other's corporate board of directors, transmission lines in an electrical grid, and whether two people know each other on Facebook.

We can create a random graph, made up of n nodes, by randomly connecting each pair of nodes with probability p. In this case, there is one sudden transition

$$\begin{array}{cccc} \text{Almost surely} & p < \frac{1}{n} & \frac{1}{n} < p & \text{Almost surely} \\ \text{many small fragments} & & & \text{one giant component} \end{array} \tag{10.4}$$

and a second one:

$$\begin{array}{cccc} \text{Almost surely} & p < \frac{\ln n}{n} & \frac{\ln n}{n} < p & \text{Almost surely} \\ \text{not fully connected} & & & \text{fully connected} \end{array} \tag{10.5}$$

The idea here is that we can use random graph theory to test the sensitivity of a given network to failure (the removal of edges or nodes). The points of criticality are of particular interest, since right around criticality even very few failures (meaning a slight reduction in p) could cause significant network fragmentation.

The problem with random graphs is that they are poor models of nearly all human-created networks. Specifically, let d_k be the *degree distribution*, the fraction of nodes having k connections. Creating edges at random, as in a random graph, causes d_k to be *Poisson* distributed, as in (9.5), whereas complex systems and human networks are characterized by a scale-free behaviour, in which the degree distribution is *power law*. Scale-free/power-law graphs still have a critical transition; if we define the degree feature

$$Q = \sum_{k=1}^{\infty} \frac{d_k}{n} k(k-2) \tag{10.6}$$

then the transition in (10.4) becomes

$$\begin{array}{cccc} \text{Almost surely} & Q < 0 & 0 < Q & \text{Almost surely} \\ \text{many small fragments} & & & \text{one giant component} \end{array} \tag{10.7}$$

For example, a particularly well studied example of critical transitions in human networks comes from epidemiology/immunization, whereby there is, for each disease, some critical fraction q_c of the population that must be immunized for the disease to die out:

$$\begin{array}{cccc} \text{Almost surely} & q < q_c & q_c < q & \text{Almost surely} \\ \text{disease epidemic} & & & \text{disease dies out} \end{array} \tag{10.8}$$

where q is the actual immunized fraction of the population.

In conclusion, we have established close connections between complex interacting systems, critical phase transitions, universality, and power laws. The remaining question, then, is why criticality actually appears so commonly in observed complex systems.

10.2 Self-Organized Criticality

As was discussed in the previous section, power law behaviour appears *at* criticality, a razor-thin phase transition/bifurcation between stability and chaos. In the Ising model criticality had to be finely tuned, setting the coupling β to an *exact* value; how is it therefore that critical systems and associated power-law behaviours show up robustly, all over nature?

Hopefully the reader has had opportunities to sit on a beach and play in the sand.[1] Wet sand will clump, however dry sand flows quite freely, forming a neat cone if you pour it from your hands. Given a continuous flow of sand the pile will slowly build up, with avalanches forming from time to time, some smaller, some larger. Even if the flow is reduced to individual grains of sand, one at a time, the same variable avalanche behaviour is seen, such that a single grain can trigger an avalanche the extent of the entire pile.

The avalanche distribution does in fact follow a power law: for most grains of sand almost nothing happens, there are frequent tiny avalanches, infrequent larger ones, and quite rare global cascades. For the sand pile there does appear to be critical behaviour at the transition between equilibrium and chaos:

Equilibrium:	Stable, fixed, robust to disturbance	← Criticality
Chaos:	Unstable, random, divergent	

Right *at* criticality the system is *nearly* unstable, with tiny disturbances possibly leading to global effects. However this notion of criticality would appear to be degenerate, in the sense of lying in an infinitesimal domain between stability and instability. The robustness of criticality in nature is due to what is known as *self-organized criticality*, for which there are two key ingredients:

1. A forcing function, pushing the system;
2. A bound, limiting the system, beyond which the system is unstable.

Since the system is always pushed towards instability, but cannot actually rest in the unstable domain, by definition it settles right *at* criticality, as illustrated in Figure 10.4. Systems subject to self organized criticality are characterized by a constant input energy (the forcing function), but where the energy dissipation is power law, subject to extreme outliers.

---

[1] If not, perhaps it is time for a "complex systems" field trip to broaden your education!

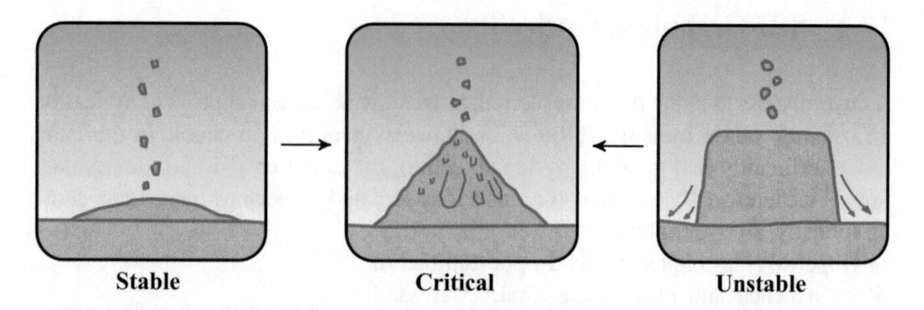

Fig. 10.4 Self-organized criticality: suppose we allow dry sand to fall on a neat pile. The angle formed by the resulting pile will be right *at* the critical angle: any flatter and the system is stable, left, allowing additional sand to build up; any steeper and the system is unstable, right, causing the pile to spontaneously slump. As long as we continue to add sand, the sand pile will remain quite robustly *at* criticality.

The macro/large–scale dynamics are reflected in the angle at which the pile settles, its angle of repose:

Wheat kernels 27 ° Dry sand 34 ° Snow 38 ° Wheat flour 45 °

This angle is a (terribly complicated) function of the micro/fine–scale dynamics of the individual particles in the system, with rounder particles piling less steeply, and more irregular, sticky particles piling more steeply.

For our sand example, playing with sand at the beach, we have

1. Forcing function: continuous stream of sand dropped onto pile;
2. Bound: critical angle of repose;
3. Interaction: friction between contacting grains of sand.

The physics of an actual sand pile would be terribly difficult to simulate well, however nearly any algorithm possessing a forcing function, a bound, and interacting elements exhibits self-organized criticality. One of the simplest is the abstracted sand-pile model, discussed in Problem 10.6, whereby units of sand are dropped onto a grid (*forcing*), and no grid element may have more than three grains of sand on it (*limit*), at which point the grains are distributed to neighbours (*interaction*). Further examples are discussed in the case study to this chapter on page 262.

The challenge in studying self-organized criticality is that the response of the system at criticality to some disturbance depends exceptionally subtly on the detailed state $\underline{z}$ of the entire system, implicitly representing the past history of inputs and responses. That is, for a given sand pile, the addition of a single grain of sand could lead to

- No response,
- A small, local response, or
- A global, system–wide response,

clearly a highly nonlinear system. Deducing *which* of these takes place is terribly difficult, since the response emerges from the coupled interaction of a great many sand grains. The typical strategy for *non*critical large-scale systems involves either a local model or a low-resolution one, however neither approach works at criticality:

- A *local* model is ill-suited to revealing power-law effects, since a large-scale sand-pile avalanche requires the particular alignment/interaction of *many* sand grains, a global alignment which cannot be deduced locally.
- A *low-resolution* or big-picture overview of the system is similarly ill-suited, since the response of the system depends on the details of the sand-grain interactions, which a low-resolution view cannot reveal.

Essentially the large-scale structure emerges through the finer scales; all scales are coupled, again essentially the scale-free concept of graphs in Section 10.1.

Back to the electrical power grid at the start of this chapter, and related to the resilience question in Example 10.4, is whether complex societies are themselves examples of self-organized criticality:

- Forcing Function: world-wide population growth with respect to fixed limits, or even a fixed population with respect to shrinking limits (depleting oil and mineral reserves, depleting groundwater aquifers);
- Bound: resources, energy, land, fresh water, pollution sinks, and capital.

There are many societal responses to approaching limits: theft from other societies (war), theft from rich to poor (revolution), limit deferral by substitution and technological development, and limit deferral through efficiency. Clearly energy and water efficiency are a good thing, however a networked system optimized further and further towards efficiency possesses a decreasing margin for error, making such systems particularly poorly suited to respond to power-law influences (floods, storms), which will almost certainly possess outliers exceeding the error margin.

Building on earlier discussions in Sections 5.5 and 7.4, we are motivated to consider the issue of controlling complex systems subject to self–organized criticality. Given that SOC systems are characterized by a power law in energy dissipation, the concept of *control* is understood to mean not the explicit control of the whole system itself, rather as the exercising of some influence on the energy dissipation pattern by reducing or preventing the most extreme outliers. To be sure, we know that *some* SOC systems can be controlled, in principle, as illustrated by the widespread practice of proactively reducing the risk of avalanches, themselves SOC phenomena, by detonations or blasting.

Since SOC appears in contexts of interconnected systems subject to some limit and driven by a forcing function resulting in power-law dissipation, it is these attributes which proposed control strategies seek to influence:

Change the Interconnection: How can the interactions between system elements be changed to move the critical point, eliminate it, or reduce extreme events?

Change the Deposition: Are there ways to be deliberate or strategic about the forcing function, adjusting the pattern of energy deposition to avoid large dissipation events?

Proactive Energy Release: Can we deliberately create dissipation events to avoid the future possibility of even larger ones? The dissipation strategy may take one of two forms:

- by observing the past pattern of dissipation events,
- by actually observing and analyzing the current state of the system.

In general the latter approach would be expected to be superior, however the inference of the system state may not be possible or practical, depending on what sort and how many measurements can be taken of the system.

To be sure, not all strategies are possible or reasonable in all situations. The laws of physics underlying the interconnection of snowflakes cannot really be changed, whereas setting off small explosives leads to a relatively straightforward control of snow avalanches via proactive energy release. In contrast, the interconnectivity pattern of the electricity network is very much under human control, and there is some evidence [12] that random upgrades of transmission lines, essentially introducing a degree of heterogeneity into the system, prevents the self-organizing near criticality of large subsets of the electrical grid.

10.3 Emergence

In principle, Newton's laws plus a friction model should describe a sand pile: each sand particle has mass, position, velocity, and forces and points of contact with other sand particles.

In contrast, the sand pile as a whole has its own *meta* dynamics, dynamics such as a critical angle of repose and power-law distributions of sand slides, which would appear to be scarcely related to the underlying Newtonian dynamics. The meta-behaviour of the system as a whole, induced by the nonlinear coupling of many elements, is referred to as an *emergent* phenomenon.

An emergent behaviour is the classic "whole greater than the sum of the parts" idea, a large-scale or complex phenomenon which arises from simple small-scale rules, for which several examples are listed in Table 10.1. There are two key challenges in working with emergence:

1. The emergent behaviour is terribly difficult to anticipate, since the large-scale effects are the consequence of an enormous number of interactions, which are essentially impossible to model analytically.
2. The emergent behaviour can certainly be simulated, and many nonlinear distributed-state models (Section 8.6) give rise to emergent/power-law phenomena. However the *predictive* power of such simulations is a very different matter; certain aspects of the emergent behaviour may or may not be subtly or

Table 10.1 Emergence: complex systems composed of large numbers of interacting elements tend to possess a large-scale *emergent* behaviour that is only weakly related to or unanticipated from the fine-scale behaviour of the individual.

Individual units	Emergent behaviour from group
Sand particles, water molecules	Sand dunes, water waves
Amoeba slime mould cells	Mobile slug and fruiting body
Ants, bees	Colony "intelligence", optimization
Termites	Mound structures
Birds	Complex flocking dynamics
Neuron	Consciousness, self-awareness
Human	Societies, civilizations

overtly dependent on details of the fine-scale model[2], so actually predicting the critical angles of repose for wheat kernels, sand, and flour is exceptionally difficult.

Because it relies on the interaction of many elements, emergence appears most strikingly in colonies or clusters, such as flocking dynamics in birds and schooling in fish, and the near-intelligent behaviour seen in ant, bee, and termite colonies.

10.4 Complex Systems of Systems

We have come a long way since the preliminary discussion of Systems of Systems in Section 3.3, from which the overview table is repeated in Table 10.2.

The concept of *Systems of Systems* is thus more of a mental construct or picture, in that all systems of system examples are essentially complex systems, possessing attributes of nonlinearity, interconnectedness, and emergence, as discussed in this chapter.

The issue of interconnectedness, in particular, is now understood to be very subtle. The degree distribution of a network, discussed in Section 10.1, is clearly of great significance in determining the robustness of a network to disturbances. However true systems of systems include dependencies between *different* systems [5, 6, 17]; these interdependencies may be physical (direct connection), geographical (local influence), or informational (critical reliance on transmitted data), such that pumps in the water distribution system clearly depend on the electrical network, however furthermore the stability of the electrical network depends in part on

[2]This is not a contradiction of universality: universality says that, normalized to the critical point, many systems exhibit a common power-law behaviour independent of fine-scale details. However the actual location of the critical point itself, i.e., the value of the critical temperature, *does* depend on fine-scale details.

Table 10.2 Systems of systems: having studied linear, nonlinear, and complex systems, we are now far better able to interpret Table 3.1, repeated here. Interacting, heterogeneous systems are characterized by complex, emergent behaviour, and would most commonly be studied via computer simulation.

	System	System of systems
Similarity	Homogeneous	Heterogeneous
Operation	Autonomous	Interacting
Behaviour	Straightforward	Emergent
Geography	Local	Distributed
Concepts	Simple	Interdisciplinary
	↓	↓
Methods	Linear systems	Complex systems
	Nonlinear systems	Agent-based models
	Partial diff. equations	Cellular automata
Strategy	Analysis	Simulation

communications via the internet, etc. Understanding the likelihood or patterns of cascading failure requires some understanding of the statistics or topology of network interdependence.

Finally, if we are going to talk about *complex* systems, in principle we would like to have some sort of measure of complexity, to talk about some systems as being *more* complex than others. Any measure of complexity probably needs to refer to a particular system *model*, since the underlying physical system is, in principle, essentially infinite dimensional: an *exact* representation of the ocean or of the human brain would require arbitrarily fine steps in space and time.

Human consciousness is perhaps the ultimate emergent phenomenon. Researchers have made significant progress in understanding individual neurons and in simulating groups of neurons corresponding to simple life forms. An individual neuron possesses essentially no intelligence, and undertakes a relatively simple nonlinear signalling operation. However somehow 100 billion such nonlinear dynamic elements, coupled together, amazingly produce consciousness.

A great deal has been written on the subject of complexity [1, 20, 23], and most measures of complexity focus on one of the following:

Size Complexity: How compactly can an algorithm representing the system be written; that is, how many bytes are required to describe the computer code and data variables in simulating the system? Clearly a homogeneous system can be represented much more compactly than a heterogeneous one.

Time Complexity: How much time is required to simulate or compute the behaviour of the system? In particular, it is the time complexity of the most efficient strategy in which we are interested; thus for a *non*-complex system, such as a large, linear, homogeneous one, the optimally efficient strategy should make use of eigendecompositions or other transformations to allow the system to be solved relatively simply.

Example 10.2: Spatial SOC — Earthquakes

We encountered the Stick-Slip model in Case Study 7, where we saw that the nonlinearity of friction allowed constant forcing to lead to limit-cycling behaviour. The stick-slip system possessed two of the key ingredients for self-organized criticality:

1. Forcing function: the constant movement of the end of the spring
2. System bound: the coefficient of static friction limits the maximum spring extent, beyond which the system is forced to slip

However in Case Study 7 we had only a single mass, whereas criticality and power-law behaviour require a third criterion:

3. Interaction and coupling.

That is, two tectonic plates, moving past one another, will be in contact at *many* points. We can imagine the main bulk of the plates moving past each other at constant velocity, however the various points of contact will, at any given time, be subject to varying degrees of tension, leading to a spatial ensemble of coupled stick-slip models:

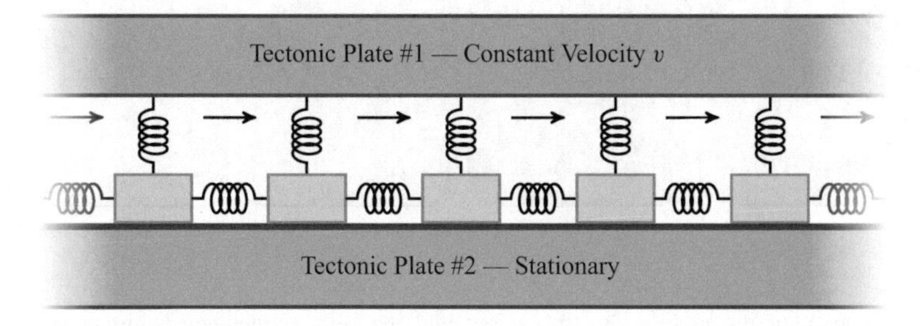

The figure shows a one-dimensional line of blocks, however a more realistic model would couple blocks on a two-dimensional grid.

The resulting model combines stick-slip (limit cycles) and self-organized criticality, such that there are very frequent small shifts of individual blocks (small tremor), infrequent shifts of small groups (minor earthquake) and, on rare occasions, shifts of a large cascade of blocks (large earthquake). That is, the introduction of coupling in the stick-slip model produces the power-law behaviour widely expected with earthquakes.

Further Reading: The book by Bak [2] or see Self-Organized Criticality in *Wikipedia*.

Example 10.3: Slime Mould Emergence

For most people, slime moulds aren't very high on their list of conversation topics. However the single-celled slime mould amoeba *Dictyostelium Discoideum* is a truly astonishing example of emergence.

When food supplies are plentiful, the amoeba remain as individual cells and consume bacteria. However presented with inadequate food supplies, a very large number of undifferentiated amoebae can spontaneously aggregate to form a macroscopic "animal" (slug), clearly with no brain or nervous system, which is however capable of large-scale motion and can sense light, temperature, humidity, and pH.

Now clearly a regular garden slug is *also* composed of many cells, however normal slugs, like human beings, start as a *single* fertilized cell which then develops a brain, heart, and nervous system. However the *Dictyostelium* slug seen under a microscope was, just hours earlier, roughly 100,000 independent cells wandering around a Petri dish, which came together and spontaneously differentiated to form a structured animal. The slug is 2–4 mm long, so definitely macroscopic in size and visible to the human eye.

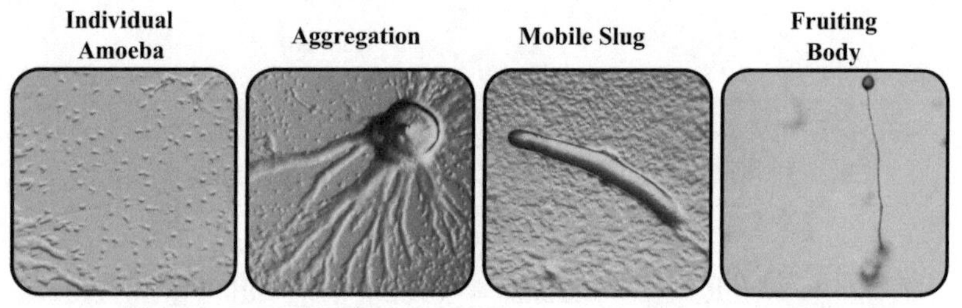

Dictyostelium images by Bruno in Columbus, Public Domain, Wikimedia Commons

At some point the mobile slug takes root and the cells spontaneously organize and differentiate. A fruiting body is formed, whereby 20 % of the cells from the slug self-sacrifice, so to speak, to form a tall stalk, at the top of which are hardy spores. These spores can then be picked up and transported elsewhere, possibly to more plentiful food supplies, by larger invertebrates such as worms, snails, or flies.

Thus Dictyostelium is able to reproduce asexually, either by mitosis if food supplies are plentiful, or via a fruiting stalk as described above. As if this behaviour was not remarkable enough, Dictyostelium is also able to reproduce sexually, in which two cells of opposite types combine, engulf surrounding cells, and form a thick protective wall.

Further Reading: See Dictyostelium in *Wikipedia*. A text description of the life cycle of this fascinating organism really does not do justice to its remarkable behaviour, and the reader is strongly encouraged to watch videos online.

Example 10.4: Resilience III — Complex Systems

How does our understanding of complex systems influence our perspective on system *resilience*?, a topic we examined earlier in Examples 5.3 and 7.3.

Probably one of the most striking aspects of advanced human societies is the number of complex networks:

- Distribution of water, gas, electricity
- Collection of garbage, recycling, sewage
- Shipments of food, auto parts, raw materials, energy
- Movement of people by car, public transit, airplanes
- Movement of information through telephone, internet

…to name only a few. As was discussed in Section 10.1, the *degree distribution* d_k (in parentheses) characterizes network topology:

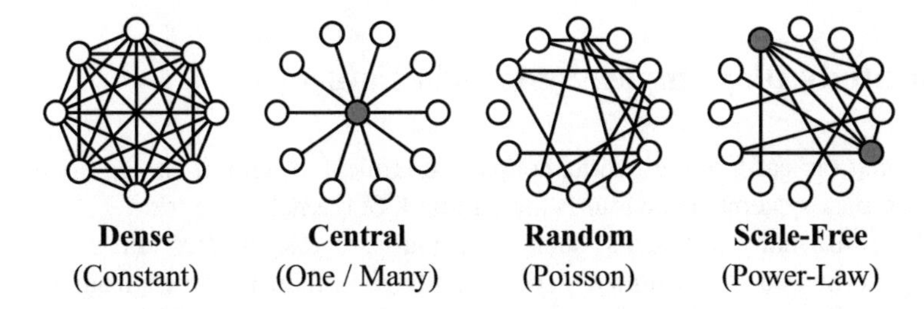

where the distinction between random and scale-free admittedly becomes striking only for larger networks.

Although water and gas have complex piping systems, the actual organization of the network is central, with one or few water treatment plants and many homes. Clearly the resilience of the network depends crucially on the central node, however that has always been obvious, and appropriate security/resilience for central nodes is standard practice.

Of particular interest to us are scale-free topologies, such as the internet, where the presence of critical/power-law behaviour is *not* obvious, and where the steps needed to ensure adequate resilience for the network as a whole may similarly not be obvious.

Consequently one measure of resilience is the distance to the critical phase transition of the associated complex network, analogous to measuring resilience as the distance to the basin boundary in Example 7.3.

In practice the question of resilience is at least one step more subtle. Passing a phase transition in a complex network implies a loss of connectivity: frustrating, but not necessarily a disaster. The disaster of a *cascading* failure, such as in widespread electrical blackouts, is based on the system *response* to broken connectivity and is frequently induced by the interconnection of multiple complex systems, so resilience in such contexts needs to measure cascade likelihood via simulation.

Further Reading: J. Greer, "Salvaging Resilience," Online link
P. Smith et al., "Network Resilience: A Systematic Approach," *IEEE Communications Magazine*, 2011
See also [6, 7, 18, 19, 21, 22].

Energy Complexity: How much *energy* is required to simulate or compute the behaviour of the system? Closely related to time complexity, Landauer's Principle is a thermodynamic measure of the minimum energy required to carry out a computation.

Time and energy complexity are effective measures when there is a particular problem to solve, such as the optimization of a travelling salesman problem. However in seeking to characterize the complexity of a system there isn't actually anything to *solve*, so to speak; thus in comparing the complexities of two systems, such as a pond ecosystem and a manufacturing distribution system, over what period of time and with which time discretization would a simulation be run in order to measure its execution time?

In contrast, a measure of code size is more concerned with representation than with solution, and is therefore more closely aligned with notions of system complexity.

Case Study 10: Complex Systems in Nature

Given that the earth is a gigantic, coupled, nonlinear system, it can be no surprise that complex systems behaviour is the hallmark of the natural world.

There are many mammals, such as leopards, tigers, giraffes and zebras who have patterned coats, presumably for camouflage in their native environment. The patterning is semi-random, like human fingerprints. The embryonic form of these animals is at some point covered in melanocytes (pigment cells), particular cells which can be in one of two states, either differentiated or undifferentiated, where the ultimate pigmentation of the skin is a function of cell type. For some period of time during embryonic development, the melanocytes are able to switch their state, based on the presence of a morphogen (a chemical signal), produced by only the differentiated cells:

Morphogen signal too strong: differentiated switches to undifferentiated
Morphogen signal too weak: undifferentiated switches to differentiated

This model starts to closely resemble a spatial cellular automaton, and the model will be further developed and explored in Problem 10.4. This class of spatial behaviour goes well beyond the signalling between embryonic cells; similar patterns have been found in the spatial growth habits of shrubs and grasses, and there is evidence that the phase transition between spatial patterns of growth (stripes, spots) may form one early indicator of approaching bifurcations [19].

Particularly intriguing is the case of the sea snail *Conus Textile* and its curious resemblance to Wolfram Rule 30, from page 193, as illustrated in Figure 10.5.

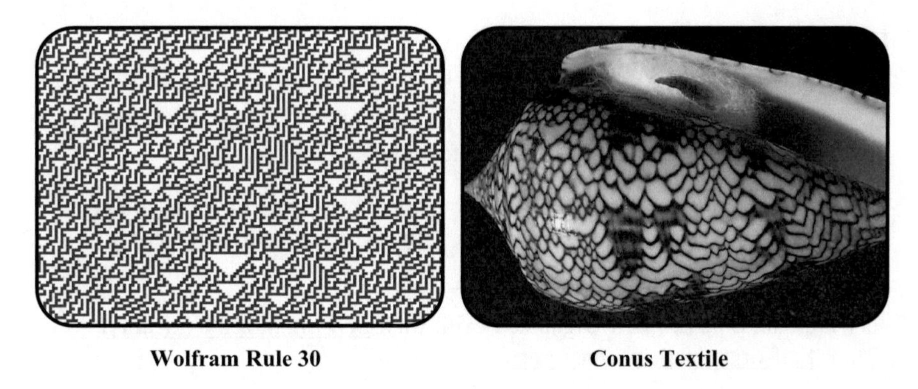

Fig. 10.5 Conus textile: the sea snail *Conus Textile* has an unusually patterned shell. The patterning is almost certainly driven by stochastic processes more complex than the deterministic 1D Rule 30 cellular automaton, however the process is most likely some type of local discrete-state spatial model.

Conus Textile image by Jan Delsing, Public domain, Wikimedia Commons.

The natural world is replete with forces and limits, so there is no surprise that elements of self-organized criticality appear in many places:

Earthquakes:

- Forcing Function: Continental drift
- Limiting Function: Limits to crust deformation

Snow Avalanches:

- Forcing Function: Continued snowfall on steep terrain
- Limiting Function: Snow depth is limited by a function of terrain angle, snow depth, and temperature

Forest Fires:

- Forcing Function: Growth of forest, accumulation of dead wood, drying over time
- Limiting Function: How much wood and how dry, before high probability of ignition by lightning

Essentially all of nature is characterized by emergent phenomena, whether in the origins of DNA, of single- and multi-cellular organisms, or higher life forms. Certainly there are some common examples, such as ants and birds, that many of us will see on a regular basis even in cities, far away from nature.

Clearly single ants and bees are very much limited in their intelligence, such that they are incapable of understanding the colony dynamics which emerge from their collective behaviour. It is interesting to contemplate whether such a rule holds in general, that regardless of the intelligence of the individual unit, the emergent behaviour associated with millions to billions of such individuals contains subtleties and complexities that is inherently beyond the grasp or perception of the individual. Food for thought.

Further Reading

The references may be found at the end of each chapter. Also note that the textbook further reading page maintains updated references and links.

Wikipedia Links — Complex Systems: Complex system, Phase transition, Critical phenomena, Critical point, Universality, Percolation theory, Network theory, Chaos theory

Wikipedia Links — Self Organized Criticality: Self-organized Criticality, Self-organization, Emergence

Wikipedia Links — Complex Behaviours: Swarm behaviour, Flocking

For next steps in complex systems, two very accessible books are those of Holland [13] and Mitchell [15] and the outstanding overview papers by Strogatz [21] and Dorogovtsev et al. [8].

The classic text on spatial simulation and self–organized criticality is the highly readable work by Bak [2] or the companion article [3] in *Scientific American*. A useful juxtaposition is the article by Frigg [10], which offers a somewhat critical perspective on the extent of self-organized criticality in nature.

The control of complex systems subject to self–organized criticality is certainly a somewhat more specialized topic, however for readers of this chapter the paper by Hoffman and Payton [12] and the following two journal papers should be fairly accessible: D. Cajueiro, R. Andrade, "Controlling self-organized criticality ...," *Eur. Physical B* (77), 2010;P. Noël et al., "Controlling Self-Organizing Dynamics on Networks ...," *Phys. Rev. Lett.* (111), 2013.

Since complex systems and emergence are fairly hot topics, there are a great many popular books written for mainstream audiences, including [9, 11, 14, 15]. Books for a more technical audience include the texts by Bar-Yam [4], Cohen and Havlin [6], and Scheffer [18]. The related areas of fractals and chaotic systems have a similarly large literature [16, 18, 21].

In terms of the interaction and cascading failure of complex systems, two papers [5, 17], both motivated by challenges such as the blackout example at the beginning of this chapter, offer an accessible and engaging treatment of the subject; the more substantial text by Cohen and Havlin [6] develops the material much further and at a more advanced level. The question of understanding critical transitions in scale-free/power-law graphs, and in particular the degree feature of (10.6), can be found in W. Aiello, F. Chung, L. Lu, "A random graph model for power law graphs," *Experiment. Math.* (10), 2001.

With regards to universality, an excellent reference is the classic paper by Nobel prize winner Kenneth Wilson: K. Wilson, "Problems in Physics with Many Scales of Length," *Scientific American*, 1979.

To follow up on the specific connection between electrical blackouts and complex systems, the reader may be wish look at one of Carreras et al., "Evidence for Self-Organized Criticality ...," *IEEE Circuits and Sytems I* (51), 2004; Dobson et al., "Complex Systems Analysis of Series of Blackouts ...," *Chaos* (17), 2007.

Sample Problems

Problem 10.1: Short Questions

Provide a short answer for each of the following:

(a) What is the relationship between critical phenomena and power laws?
(b) Give two examples of scale-free networks.
(c) The two key elements for SOC are a forcing function and a limit. Why are both required? What happens if one of these is absent?
(d) Why is SOC so commonly studied via simulation, instead of analytically (mathematically)?
(e) Explain why it is essential that the grid be *large* for SOC simulation.
(f) Define the distinction between *continuous* and *discontinuous* phase transitions. How do such phase transitions relate to complex systems?
(g) Briefly explain the principle of universality.

Problem 10.2: Analytical — SOC

Figure 10.6 shows results from a sand-pile simulation (Problem 10.6). Two-dimensional grids of sizes 10×10, 20×20, 50×50, and 200×200 were each simulated to one million iterations, with the first 200,000 iterations discarded to give the system time to settle down.

Please explain, in fairly substantial detail, all of the phenomena which you observe in this figure.

Problem 10.3: Numerical/Computational — Patterning

In Case Study 10 we encountered the patterning of *Conus Textile* and its curious resemblance to Wolfram Rule 30, from page 193.

Develop a program to implement Rule 30, and show results based on initializing the first row with

(a) A single black pixel in the middle, the remainder white.
(b) Black and white pixels at random.

Problem 10.4: Numerical/Computational — Patterning

Case Study 10 discussed the randomly striped and patterned coats of many animals, such as leopards, tigers, giraffes and zebras. In their embryonic form, the animals are at some point covered in pigment cells which can be in one of two states:

State $Z_{\underline{x}} = 1$ Cell at location $\underline{x}$ is *Differentiated*
$Z_{\underline{x}} = 0$ Cell at location $\underline{x}$ is *Undifferentiated*

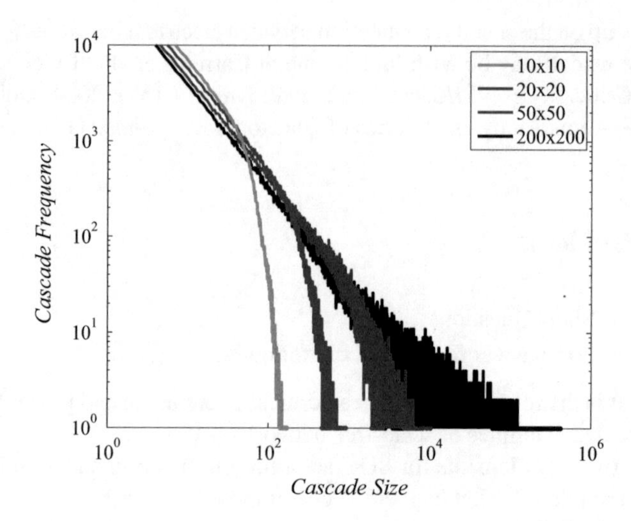

Fig. 10.6 Frequency versus size of cascade in a sand-pile simulation for Problem 10.2.

The resulting pigmentation of the skin is reflected by the spatial pattern of this state. For some period of time during embryonic development, the melanocytes are able to switch their state, based on the presence of chemical signals produced by only the differentiated cells:

Activation signal:	Causes cells to become differentiated
Inhibition signal:	Causes cells to become undifferentiated

The key is that the spatial behaviour of the two signals is different: the activation signal is stronger, but diffuses more slowly (acts more locally) than the inhibition signal.

The cellular automata model proposed by Young is parameterized as follows:

r_a	Radius of activation	w_a	Strength of activation
r_i	Radius of inhibition	w_i	Strength of inhibition

The net signal at some point $\underline{x}$ is then

$$S_{\underline{x}} = \sum_{|\bar{\underline{x}}-\underline{x}|<r_a} w_a Z_{\bar{\underline{x}}} - \sum_{|\bar{\underline{x}}-\underline{x}|<r_i} w_i Z_{\bar{\underline{x}}}$$

where $|\cdot|$ represents the standard Euclidean distance. The signal S determines the state evolution:

Current state $Z_{\underline{x}}^{n-1}$	Signal $S_{\underline{x}}$	Next state $Z_{\underline{x}}^{n-1}$	
$\times$	> 0	1	(Differentiated)
z	0	z	(Unchanged)
$\times$	< 0	0	(Undifferentiated)

(a) Let $r_a = 2.5, r_i = 5.0, w_a = 1.0, w_i = 0.3$. Initialize a 100×100 state z to values of zero and one at random. Implement the cellular automaton and iterate to produce a simulated texture.
(b) Observe the simulated texture as w_i is increased from 0.25 to 0.35. Explain, in qualitative terms, how changing the inhibition weight w_i leads to the observed behaviour.
(c) There is something of a phase transition, between patterned and unpatterned (solid). Identify the value of w_i corresponding to the transition.
(d) This model is isotropic (looks the some horizontally and vertically). However striped patterns (tigers) are *an*isotropic. Propose a modification to the model to make its behaviour anisotropic.

Skin model from *D. Young, "A local activator-inhibitor model…,"* Mathematical Biosciences *(72), 1984*

Problem 10.5: Numerical/Computational — Forest fire revisited

A spatial forest-fire model was described in Problem 8.10, where we limited our attention to looking at simulation results.

However the forest-fire model is an example of self-organized criticality and power laws. We would like to examine the power-law behaviour by measuring forest fire cluster sizes; to do so will require some familiarity with image processing, so the question is recommended only for readers with such experience.

There are some subtleties in how to define a single forest fire:

- Any burning cell at time n is considered part of the same forest fire as any spatially adjacent cell that was burning at time $n-1$.
- Two separated forest fires could burn together, in which case the total size would be the sum of the two sizes.
- You may wish to look up the `bwlabel` and `regionprops` commands in Matlab/Octave.

Using the model from Problem 8.10, plot the distribution of forest fire size versus frequency. Discuss your results.

Problem 10.6: Numerical/Computational — Self Organized Criticality

For self-organized criticality we need some sort of forcing function and limit. We will implement a simplified sand-pile model:

- Start with a 100 by 100 (or larger) of state elements $z(i, j)$, all set to zero
- Then iterate …
  - Select a random grid location and add one to it (*the input*).
  - If any location reaches four (*the limit*) or more, subtract four and add one to each of its four neighbours (or tossed away if on the boundary). Keep repeating this step until no location has a value of four or more.

Thus at every iteration i precisely one sand grain is dropped, and zero or more piles topple, where we will record in $q(i)$ the number of different pixels that reach a value of four or more.

Plot q as a function of i. Identify the ranges of iterations corresponding to the initial build-up (where the sand pile is growing) and equilibrium (where the process has reached SOC); you may find the total state sum

$$\sum_i \sum_j z(i,j)$$

to be a good indicator for this purpose. Because of the very different system behaviour in the initial build-up, as opposed to the later equilibrium, signal $q(i)$ is nonstationary with iteration, bringing us back to the baseline problem of Chapter 4.

Produce a log-log plot of the frequency of events in q as a function of the event size. For the plot, use only that range in q corresponding to equilibrium.

Comment on your observations.

Finally, do something creative with the simulation: change the rules, or change the limit of four, or vary the size of the domain, perhaps something else. Discuss your observations.

Problem 10.7: Reading — Self Organized Criticality

The book[3] by Per Bak [2] was quite influential in making people think of the role that critical phenomena played in familiar contexts.

Select any one chapter beyond Chapter 3 in [2] and offer a summary of the systems being studied and the manifestation of criticality. Offer your own critique or perspective of the chapter.

Problem 10.8: Reading — Emergent Behaviour

There are many fascinating social examples where nonlinear dynamics at a fine scale leads to unusual or complex behaviour at a coarse scale (ant colonies, bee hives, human mob psychology, group-think). You may also want to search keywords such as "swarm" or "emergence".

Select a specific emergent behaviour, whether economic, social, or natural, and offer a summary in no more than one page. Be quite specific regarding the definition of the fine-scale individual unit, the nature of interactions, and the degree to which the emerged behaviour can or cannot be anticipated from the fine scale.

Problem 10.9: Policy

An Internet search on *policy and complex systems* reveals an enormous amount of interest, including research journals and government think-tanks.

Find an article, book chapter, or policy description of interest to you and prepare a brief overview:

- What do the authors actually mean when they refer to *complex systems*?
- What policy is being discussed or impacted?

[3]Excerpts of most books can be found online; links are available from the textbook reading questions page.

- In what way are the complex systems being studied — models, simulations, mathematical analysis, by analogy, philosophically?

References

1. S. Arora and B. Barak. *Computational Complexity: A Modern Approach.* Cambridge, 2009.
2. P. Bak. *How Nature Works.* Copernicus, 1996.
3. P. Bak and K. Chen. *Self-Organized Criticality.* Scientific American, 1991.
4. Y. Bar-Yam. *Dynamics of Complex Systems.* Addison-Wesley, 1997.
5. S. Buldyrev et al. Catastrophic cascade of failures in interdependent networks. *Nature Physics*, 464, 2010.
6. R. Cohen and S. Havlin. *Complex Networks: Structure, Robustness and Function.* Cambridge, 2010.
7. G. Deffuant and N. Gilbert (ed.s). *Viability and Resilience of Complex Systems: Concepts, Methods and Case Studies from Ecology and Society.* Springer, 2011.
8. S. Dorogovtsev, A. Goltsev, and J. Mendes. Critical phenomena in complex networks. *Reviews of Modern Physics*, 80, 2008.
9. L. Fisher. *The Perfect Swarm – The Science of Complexity in Everyday Life.* Basic Books, 2009.
10. R. Frigg. Self-organised criticality — what it is and what it isn't. *Studies in History and Philosophy of Science*, 34, 2003.
11. J. Gribbin. *Deep Simplicity: Bringing Order to Chaos and Complexity.* Random House, 2005.
12. H. Hoffmann and D. Payton. Suppressing cascades in a self-organized model with non-contiguous spread of failures. *Chaos, Solitons, & Fractals*, 67, 2014.
13. J. Holland. *Complexity: A Very Short Introduction.* Oxford, 2014.
14. S. Johnson. *Emergence.* Scribner, 2001.
15. M. Mitchell. *Complexity - A Guided Tour.* Oxford, 2009.
16. H. Peitgen, H. Jürgens, and D. Saupe. *Chaos and Fractals: New Frontiers of Science.* Springer, 2004.
17. S. Rinaldi et al. Identifying, understanding, and analyzing critical infrastructure interdependencies. *IEEE Control Systems Magazine*, 21, 2001.
18. M. Scheffer. *Critical transitions in nature and society.* Princeton University Press, 2009.
19. M. Scheffer, S. Carpenter, V. Dakos, and E. van Nes. *Generic Indicators of Ecological Resilience: Inferring the Chance of a Critical Transition.* Annual Reviews, 2015.
20. M. Sipser. *Introduction to the Theory of Computation.* Cengage Learning, 2012.
21. S. Strogatz. Exploring complex networks. *Nature*, 410, 2001.
22. B. Walker and D. Salt. *Resilience Thinking: Sustaining Ecosystems and People in a Changing World.* Island Press, 2006.
23. S. Wolfram. *A New Kind of Science.* Wolfram Media, 2002.

Chapter 11
Observation and Inference

The modern era in medical imaging began in 1895 when Roentgen, a German physicist, first produced and detected X-rays. Since that time the technology of medical imaging has made truly amazing advances, to the point that we take it for granted that high–resolution three-dimensional MRI, CT, or PET scans can be acquired to offer unprecedented visualization of the interior of the body. But X–rays pass *through* the body, fewer of them through bone, more easily through flesh, to produce a projected 2D image on the other side of one's body; how *do* we produce visual slices through the interior of the body?

Very similar questions can be raised with regards to sampling the planet. Consider, for example, the town of Staufen, in Germany, an attractive medieval settlement at the southern end of the Black Forest.

The town has recently acquired some infamy due to substantial changes in ground height. In 2007, deep underground drilling was undertaken to provide geothermal heating to the city hall. It is believed that the drilling cut through an anhydrite layer, causing it to come into contact with groundwater, leading to swelling and raising parts of the town by an astonishing 30 cm. As a consequence there has been significant damage to buildings, including the city hall.

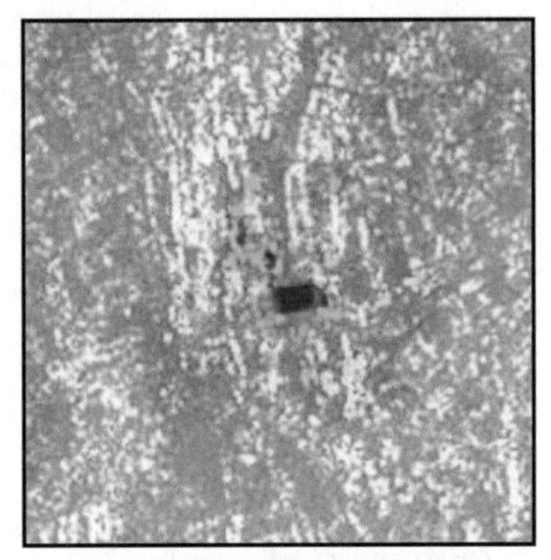

Data from TerraSAR-X / DLR

Now a height change of 30 cm might be measured by careful land-based surveying from a distant reference point, but how it is possible for the TerraSAR-X

P. Fieguth, *An Introduction to Complex Systems*,
DOI 10.1007/978-3-319-44606-6_11

satellite, orbiting at a distance of over 500 km, to produce such a map, as shown, with colours implying changes in height accurate to less than one cm?!

Truly a wide range of phenomena can be sensed at a distance. Given such measurements, the *reversal* of the measurement process, deducing the properties of an object from distant measurements, is known as an *inverse problem.*

11.1 Forward Models

There is a huge variety of spatial phenomena that we might want to study:

- A 3D map of ocean salinity (saltiness);
- The temperature of the atmosphere (as in Case Study 11B);
- The location and flow of groundwater;
- A mapping of human land use (e.g., agricultural, suburban, city);
- Winds, precipitation, ocean currents, vegetation density

and many more. Each such system could be observed or studied by taking measurements directly, in-situ, or remotely, at a distance. A great many kinds of measurements can be taken in-situ, however such measurements are both time consuming and expensive, such that in very many problems it is just not possible to collect sufficient in-situ data to adequately study the system.

In contrast, there are actually rather *few* things that we can measure remotely, at a distance. Essentially we are limited to measuring the *strength* or *time* of propagating waves, mostly electromagnetic. As a result, you are not just handed a map of the quantities you wish to study (salinity, temperature, flow, land use), rather you have to *infer* such a map from sensed measurements which will be somehow related, *indirectly*, to your quantities of interest, as illustrated in Example 11.1.

The process by which measurements $\underline{m}$ are generated from an underlying spatial process $\underline{z}$ is normally relatively straightforward, and is referred to as a *forward*

Table 11.1 Examples of forward problems: generally, knowing some regular, well-structured, fundamental set of quantities (left) allows derived quantities (right) to be inferred relatively easily.

Knowing ...		*It is easy to determine ...*
The arrangement and sizes of all organs, bones, blood vessels etc.	$\longrightarrow$	The appearance of a measured X-ray, MRI, CAT scan etc.
The location and shapes of below-ground clay, sand, gravel, and rock layers	$\longrightarrow$	The flow of groundwater and other pollutants
The temperature and current flow at all points in the ocean	$\longrightarrow$	The acoustic (sound) delay between two points in the ocean
The temperature of the atmosphere as a function of altitude	$\longrightarrow$	The strength of emitted radiation as a function of wavelength (see Case Study 11B)
A sharp image, in focus	$\longrightarrow$	A blurry image, out of focus

Example 11.1: Indirect Inference

The concept of indirect inference is as common and everyday as reading a thermometer. When you read an old-fashioned thermometer, do you actually *observe* the temperature? No, you do not. What you observe is the height of alcohol in a glass tube.

The forward model of this system comes from physics, that the density of alcohol is a decreasing function of temperature, such that the alcohol expands as it warms, pushing the coloured alcohol up the glass tube.

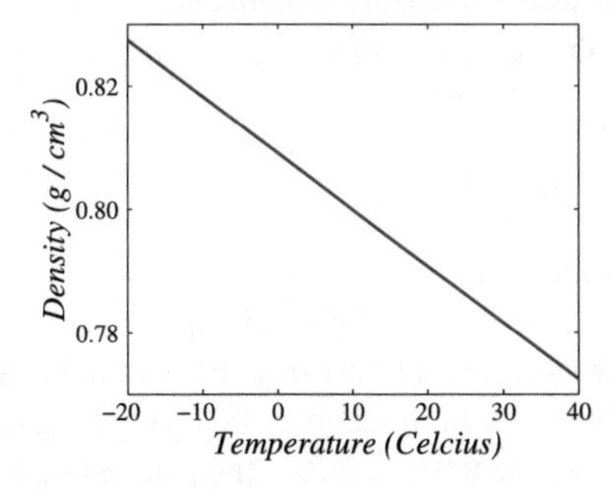

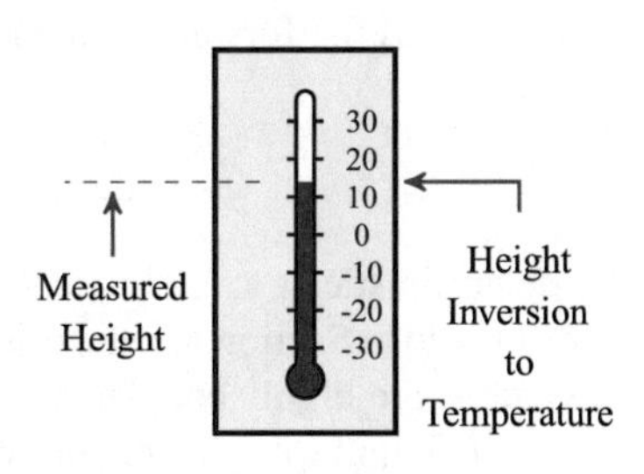

Because the density is very nearly a linear function of temperature, the inverse problem is comparatively straightforward. Indeed, the inverse function is printed right on the thermometer, allowing you to infer temperature, indirectly, from an observation of alcohol column height.

On this basis it is straightforward to generalize indirect inference to significantly more subtle problems, such as identifying forest type via satellite, based on the difference in how deciduous leaves reflect the full spectrum of sunlight compared to coniferous needles.

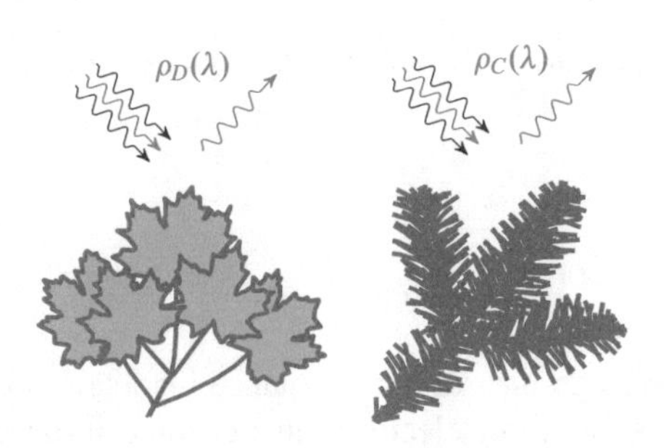

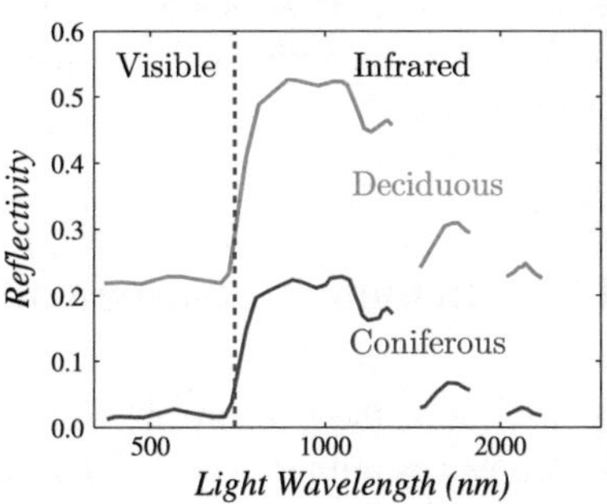

In general, deciduous leaves are shinier (reflect more light), however there are strong reflectance patterns in the infrared, which our eyes cannot see, but which satellite instruments can detect.

The key in indirect inference is to establish the forward model, which determines the relationship between the quantity of interest (temperature, forest type) and that which you can conveniently measure (height, reflectance).

Further Reading: Cipar et al., "Distinguishing Between Coniferous and Deciduous Forests ...," *IGARSS*, 2004

problem, as in Table 11.1. Far more difficult is the *inverse problem*, to be discussed in Section 11.4, the inference of the process of interest from the measurements.

We will assume our measurements $\underline{m}$ to be some mathematical function

$$\underline{m} = C(\underline{z}) \tag{11.1}$$

of some more basic, underlying quantity $\underline{z}$, where forward function C could be deterministic or stochastic. Note that (11.1) is not a dynamic model, rather a transformation from an unknown state $\underline{z}$ to observed measurements $\underline{m}$.

In some cases the forward problem might be linear, in which case (11.1) could be expressed as

$$\underline{m} = C\underline{z} \qquad \text{or} \qquad \underline{m} = C\underline{z} + \underline{w} \tag{11.2}$$

for the deterministic or stochastic case, respectively.

As discussed in Section 4.2, the role of a stochastic term $\underline{w}$ can be quite important. Earth systems are highly coupled and interacting, so almost *any* measurement will be a function of multiple phenomena, only one of which you may be trying to assess. A classic example comes from using satellite altimetry data to study ocean surface height, where the height is influenced by a variety of factors:

$$\begin{array}{lcccccc} \text{Height} = & \text{Sphere} & + \text{Rotation} & + \text{Gravity} & + \text{ Tides} & + \text{Pressure} & + \text{Currents} \\ & \text{1000's km} & \text{100's km} & \text{100's m} & \text{10's m} & \text{m} & \text{m} \end{array} \tag{11.3}$$

Whatever phenomenon you are studying (say, ocean currents), the other effects (tides and atmospheric pressure) are nuisance terms, which you either need to model and estimate, or treat as noise in $\underline{w}$.

So what quantities can we measure, and to which forward models do they correspond?

11.2 Remote Measurement

Nearly all of the phenomena which we can measure at a distance involve propagating waves, either acoustic (sonar, seismic) or electromagnetic (light, infrared, microwave, radar). In most cases we have two categories of instruments:

1. **Passive**: The scene is not illuminated, rather the emitted radiation is observed passively, like the human auditory and visual systems. Clearly the scene needs to possess sufficient signal (to be sufficiently loud or well lit, for example) to allow meaningful measurements to be taken.
2. **Active**: The scene is illuminated by the measuring instrument, such as measuring speed with a radar pulse or walking around at night with a flashlight. The scene needs to be as dark as possible at the signal wavelength to allow the illumination signal to be detected.

For an instrument operating at a particular electromagnetic wavelength λ, whether the medium is essentially clear, or whether the medium itself participates in the measurement process, distinguishes between two fundamental approaches to measurement:

1. **Tomographic**: Measurements are taken *through* a medium, such as a human body, a planet, or the atmosphere. The medium should be neither too opaque, in which case too much signal is lost, or too transparent, in which case there is too little influence on the signal.
2. **Surface**: Measurements are taken based on the reflecting surface of an object. The medium *between* the measuring instrument and the measured object should be as transparent as possible, otherwise additional tomographic nuisance signals appear, as in (11.3).

All four permutations of active/passive and tomographic/surface measurements are possible, with examples shown in Table 11.2. In principle, all four of these permutations fall into the forward-model framework of (11.1); the challenge is how to select an appropriate measuring instrument for a particular problem, and how to characterize its corresponding forward model.

The forward model for the propagation of electromagnetic radiation through a medium is described by what is known as *radiative transfer*, mostly characterized by four effects:

1. The **absorption** of radiation by matter, where each atom or molecule has a characteristic absorption spectrum as a function of wavelength.
2. The **emission** of radiation by matter, in many cases a blackbody radiation spectrum which is a strong function of temperature.
3. The **scattering** of radiation by matter, essentially a reflection off a surface.
4. The **refraction** of radiation by matter, essentially a bending of the radiation in passing from one medium to another.

Aside from medical imaging and certain problems in atmospheric inference (Case Study 11B), most problems of interest are surface-based, so it is scattering which

Table 11.2 We have four categories of remotely sensed measurements, depending on the type of instrument (active/passive) and whether we observe an object's surface or its interior.

	Passive	Active
Surface	Ocean surface radiometry Satellite imaging Human visual system	Altimetry Radar Imaging Scatterometry Sonar Bat echolocation
Tomographic	Atmospheric radiometry Planetary seismology	Ground penetrating radar Medical imaging Subsurface acoustic imaging Reflection seismology

is of greatest importance to us. The behaviour of scattering depends strongly on the relative sizes of the wavelength λ and that of the scattering object θ:

- $\lambda > \theta$: *Rayleigh* scattering by tiny objects (atoms, molecules) has a scattering intensity proportional to $1/\lambda^4$, therefore scattering more strongly at shorter wavelengths, which is why the sky is blue.
- $\lambda \approx \theta$: *Mie* scattering by small objects (dust, droplets) has a scattering intensity insensitive to λ, which is why clouds are white.
- $\lambda < \theta$: *Geometric* scattering by large objects will have a behaviour dependent on surface shape, roughness, and material.

In most cases we are looking at relatively large objects, so we will focus on geometric scattering. The scattering from a surface, illustrated in Figure 11.1, is characterized in terms of two principles:

Angular Dependence $\rho(\psi_i, \psi_r)$: The reflectivity as a function of incident angle ψ_i and reflected angle ψ_r.

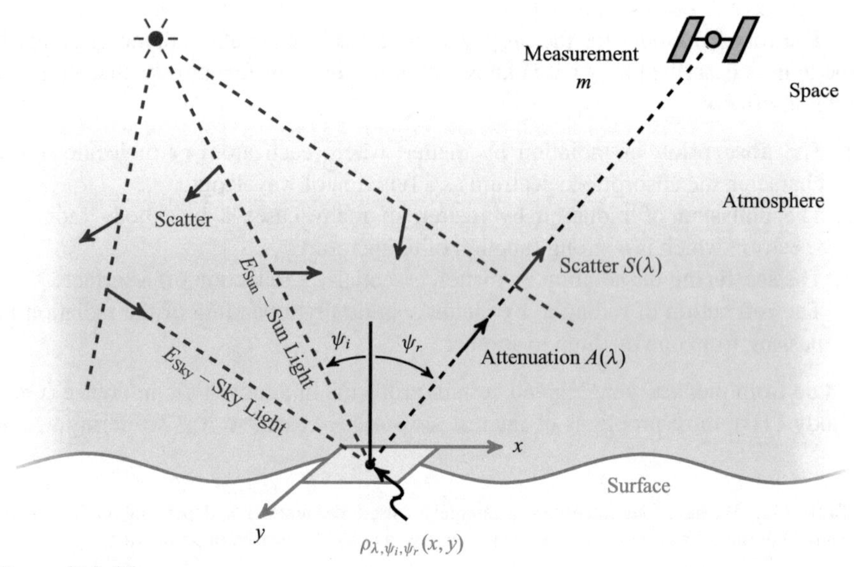

Forward Model:

$$m(x, y, \lambda) = \underbrace{S(\lambda)}_{\text{Non-Surface}} + \frac{1}{A(\lambda)} \cdot \left\{ \underbrace{E_{\text{Sun}}(\lambda)\rho_{\lambda,\psi_i,\psi_r}(x, y)}_{\text{Surface – Sun Light}} + \underbrace{E_{\text{Sky}}(\lambda) \int \rho_{\lambda,\bar{\psi}_i,\psi_r}(x, y)\, d\bar{\psi}_i}_{\text{Surface – Sky Light}} \right\}$$

Fig. 11.1 What does a satellite measure: although it is really only the surface characteristic $\rho(x, y)$ which we would like to measure, the forward model to the satellite involves tomographic (non-surface) scatter, signal attenuation, and the illumination of the surface directly by the sun and indirectly via scatter from the sky.

At the atomic level, a given atom has no "idea" whether it is part of a mirror or an eggshell; indeed, each atom scatters light at random in all directions. However given a sequence of atoms aligned in a straight row, light scattered at random from all of the atoms leads to destructive interference in all directions *except* at $\psi_r = \psi_i$. That is, a mirror-like surface is just one in which the atoms are sufficiently well aligned, where *"sufficiently well"* is defined by the Rayleigh surface criterion:

$$\text{Root Mean Square Surface Height Variation} \underset{\text{Smooth}}{\overset{\text{Rough}}{\gtrless}} \frac{\lambda}{8} \cdot \frac{1}{\cos\psi} \tag{11.4}$$

The two extreme and most common models in angular dependence are that of a mirrored surface, where the reflected and incident angles are equal, and a diffuse surface, scattering light such that the surface appears equally bright[1] at all angles:

$$\underbrace{\rho(\psi_i, \psi_r) = \delta(\psi_i, \psi_r)}_{\text{Mirror/specular reflectance}} \qquad \underbrace{\rho(\psi_i, \psi_r) = \cos(\psi_r)}_{\text{Diffuse/Lambertian reflectance}} \tag{11.5}$$

where δ is the Dirac delta or impulse function, zero for all angles except when $\psi_i = \psi_r$.

Spectral Dependence ρ_λ: The reflectivity as a function of wavelength λ.

The atoms on or near the surface of an object may or may not have spectral preferences. The variation in ρ as a function of wavelength

$$\rho_\lambda(\psi_i, \psi_r) = \frac{\text{Energy Reflected at angle } \psi_r \text{ at wavelength } \lambda}{\text{Energy Incident at angle } \psi_i \text{ at wavelength } \lambda} \tag{11.6}$$

is what we mean by "colour." A few simple cases should make the relationship clear:

$$\underbrace{\rho_\lambda = 1}_{\text{White}} \qquad \underbrace{\rho_\lambda = \tfrac{1}{2}}_{\text{Grey}} \qquad \underbrace{\rho_\lambda = \begin{cases} 1 & 600\,\text{nm} < \lambda < 800\,\text{nm} \\ 0 & \text{otherwise} \end{cases}}_{\text{Red}} \tag{11.7}$$

The combination of angular and spectral effects is what characterizes a surface, thus eggshell and a mirror have the same spectral dependence (white) but with different surface roughness, whereas polished gold and silver rings are both mirrored, but with different spectral responses.

In practice, many materials will possess an overall reflectivity which is actually a combination of specular and diffuse components: think of an apple, which is

[1] See Problem 11.7 to look at this effect more closely.

shiny, but certainly nothing like a mirror. Furthermore, these specular and diffuse components frequently have differing wavelength dependence ρ_λ:

- The specular component will typically reflect all wavelengths equally, leading to a white glint or glare on polished objects;
- The diffuse component is much more strongly controlled by the internal body colour of the reflecting object itself.

Therefore a polished apple will appear red (diffuse component) except for the glint of the reflected sun, which will appear yellow/white (specular component).

The dependence of reflectivity ρ on λ is precisely what makes multi-spectral instruments so effective, such that two surfaces may look similar at one wavelength, but appear very different at another. For example, smooth ice and water look very different at visual wavelengths, but are difficult to discriminate in radar maps; on the other hand, smooth and rough ice both appear similar in visual images, but are very distinct at radar wavelengths.

Figure 11.1 offers a relatively comprehensive picture of the definition of a forward model corresponding to a remotely sensed signal.

11.3 Resolution

In solving an inverse problem we use measurements $\underline{m}$ to infer an unknown state $\underline{z}$. What determines the resolution of a measurement, and how does this affect or constrain the resolution of the state?

11.3.1 Measurement Resolution

Given measurements, whether over space, time, wavelength or something else, there are four fundamental factors which determine the achievable resolution r, as illustrated in Figure 11.2:

- The underlying signal strength z,
- The strength of the signal noise σ,
- The degree of blur b,
- The period of time integration T,
- The sampling interval δ.

The sampling interval δ is an engineering decision, and normally we would design $\delta \ll b$, such that δ is not a limiting factor in resolution.

We would like to consider the ability to resolve two signals, separated by an offset r in dimension x, as sketched in Figure 11.2. The forward model, describing the physics of the measurement process, normally introduces some degree of blur,

$$s(x) = z(x) * \phi(x, b) \tag{11.8}$$

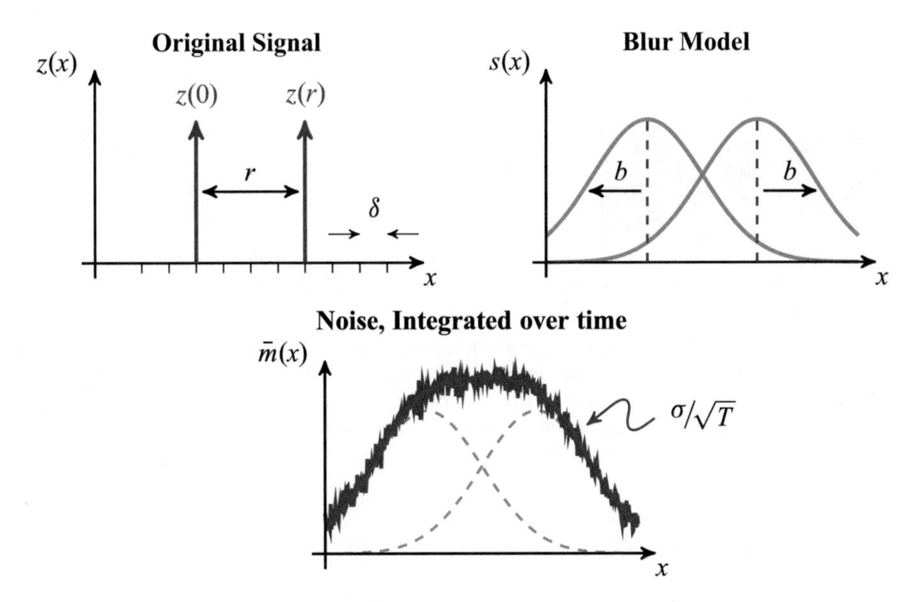

Fig. 11.2 Resolution parameters: the effective resolution r along some dimension x is determined by the underlying signal strength z, the sampling resolution δ, the degree of blur b, the period of time integration T, and the amount of noise σ in the measurements.

where ϕ controls the shape of the blur, with the spatial extent of the blur controlled by parameter b. In most cases we would expect the resolution to be limited based on blur, $r \geq b$, however in practice the question of resolution is more subtle.

We will acquire noisy measurements $m(t, x)$ of $s(x)$ over time,

$$m(x,t) = s(x) + w(x,t) \tag{11.9}$$

where w is Gaussian white noise[2] with constant power spectral density σ^2. We are assuming the signal $s(x)$ to be constant, at least over sufficiently short instants of time. In order for the deterministic signal to reveal itself as distinct from the random noise, we can integrate the measurements over some period of time T, as illustrated in Figure 11.3. The strength of the integrated signal grows linearly with time, but

[2]White noise is a very tricky concept in continuous time, and the details are well outside the scope of this text. In principle, perfect white noise has *infinite* variance, however *any* real process of physical measurement must have an upper limit on the signal frequencies which can be captured, necessarily similarly limiting the frequency range of the captured noise, making the measured result subject to a *finite* variance noise process. Therefore in Figure 11.3 the measurements $m(t)$ are actually in discrete time, with a sampling period of δ_t, in which case the noise standard deviation would actually be $\sigma/\sqrt{\delta_t}$, which can be seen to become infinite as $\delta_t \longrightarrow 0$. It is only the representation in the left panel of Figure 11.3 which is problematic; the representation in the right panel, with a standard deviation of $\sigma\sqrt{T}$, and the corresponding results in (11.10),(11.11) are correct.

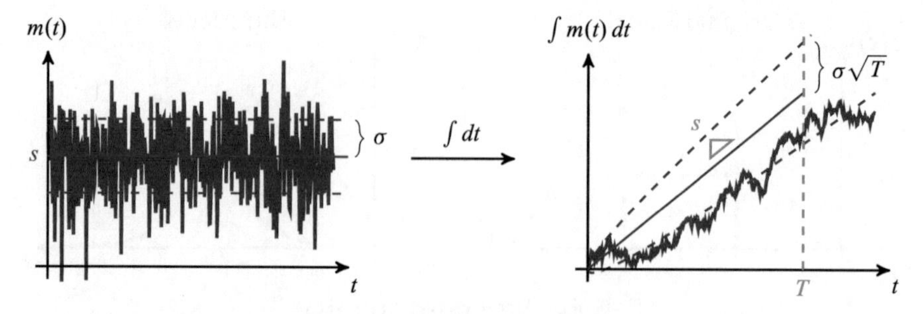

Fig. 11.3 Time integration of noise: suppose we are given noisy measurements $m(t)$, left, of some underlying constant signal s. We can integrate this signal over some period of time T, right, in which case the white noise in $m(t)$ turns into a so-called random walk. Since the integrated signal grows linearly with time, but the random walk only as the square-root of time, as T increases the discernibility of the signal relative to the noise, the signal-to-noise ratio, grows as $\sqrt{T}$.

that of the random noise only as the square-root[3] of time,

$$\tilde{m}(x) = \int_T m(x,t)\,dt \quad \longrightarrow \quad \tilde{m}(x) \sim \mathcal{N}\big(Ts(x), \sigma^2 \cdot T\big). \tag{11.10}$$

Therefore averaging over time

$$\bar{m}(x) = \frac{1}{T}\int_T m(x,t)\,dt \quad \longrightarrow \quad \bar{m}(x) \sim \mathcal{N}\big(s(x), \sigma^2/T\big), \tag{11.11}$$

reveals the signal, with the noise decreasing as T increases.

Now suppose that we have an original signal consisting of two spikes (e.g., distant stars) separated by an offset r, as illustrated in Figure 11.2; to what extent are we able to resolve these on the basis of measurements $\bar{m}(0), \bar{m}(r)$? The steps of blur, noise, and integration lead to the following forward model:

$$\begin{aligned} \bar{m}(0) &\sim \big(\phi(0,b)z(0) + \phi(r,b)z(r),\ \sigma^2/T\big) \\ \bar{m}(r) &\sim \big(\phi(r,b)z(0) + \phi(0,b)z(r),\ \sigma^2/T\big) \end{aligned} \tag{11.12}$$

Assuming a normalized blur function $\phi(0,b) = 1$, we can simplify (11.12) as

$$\begin{bmatrix} \bar{m}(0) \\ \bar{m}(r) \end{bmatrix} = \begin{bmatrix} 1 & \beta \\ \beta & 1 \end{bmatrix} \begin{bmatrix} z(0) \\ z(r) \end{bmatrix} + \begin{bmatrix} w(0) \\ w(r) \end{bmatrix} \tag{11.13}$$

where $\beta = \phi(r,b)$ and w represents the noise. Section 11.4 will discuss the principle of an inverse problem more formally, however for now we will estimate the original

[3]The *variance* of the integrated noise grows linearly with T, however it is the *standard-deviation* of the noise, having the same units as the signal, which grows as the square root of T.

signal by matrix inversion:

$$\begin{bmatrix} \hat{z}(0) \\ \hat{z}(r) \end{bmatrix} = \begin{bmatrix} 1 & \beta \\ \beta & 1 \end{bmatrix}^{-1} \begin{bmatrix} \bar{m}(0) \\ \bar{m}(r) \end{bmatrix} = \frac{1}{1-\beta^2} \begin{bmatrix} 1 & -\beta \\ -\beta & 1 \end{bmatrix} \begin{bmatrix} \bar{m}(0) \\ \bar{m}(r) \end{bmatrix} \tag{11.14}$$

Substituting (11.12) into (11.14), then the estimate of the first signal component becomes

$$\hat{z}(0) = \frac{\bar{m}(0) - \beta\bar{m}(r)}{1-\beta^2} \tag{11.15}$$

$$= \frac{(z(0) + \beta z(r) + w(0)) - \beta(\beta z(0) + z(r) + w(r)}{1-\beta^2} \tag{11.16}$$

$$= \underbrace{z(0)}_{\text{Original Signal}} + \underbrace{\frac{w(0) - \beta w(r)}{1-\beta^2}}_{\text{Noise}} \tag{11.17}$$

This is an attractive result: in our computed estimate $\hat{z}(0)$ we have successfully recovered the original signal $z(0)$, *unmixed* from $z(r)$, although clearly subject to uncertainty due to noise. From the variance manipulation rules of (B.8), we find the noise variance in $\hat{z}(0)$ to be

$$\text{Variance}\left(\frac{w(0) - \beta w(r)}{1-\beta^2}\right) = \frac{(\sigma^2/T + \beta^2\sigma^2/T)}{(1-\beta^2)^2} = \frac{\sigma^2}{T}\frac{1+\beta^2}{(1-\beta^2)^2} \tag{11.18}$$

The ability to discern the signal is determined by the signal to noise *ratio* (SNR), the size of the signal relative to the noise standard deviation, the square-root of the variance in (11.18):

$$\text{SNR} = \frac{\text{Signal}}{\text{Noise}} = \frac{z(0)\sqrt{T}}{\sigma}\frac{(1-\beta^2)}{\sqrt{1+\beta^2}}, \tag{11.19}$$

from which it is clear that signal discernibility is improved by signal strength $z(0)$ and integration time T, and is compromised by a larger noise σ and blur β.

To connect (11.19) back to the concept of resolution, recall that $\beta = \phi(r, b)$ is a function of the signal separation r:

$$\text{SNR}(r) = \frac{z(0)\sqrt{T}}{\sigma}\frac{(1-\phi(r,b)^2)}{\sqrt{1+\phi(r,b)^2}}, \tag{11.20}$$

therefore the resolution limit is the point where we can no longer discern the signal relative to the noise:

$$\text{Resolution Limit} = \text{minimum } r \text{ such that } \Big\{\text{SNR}(r) > \Gamma\Big\} \tag{11.21}$$

for some minimum signal-to-noise ratio $\Gamma \approx 1$.

So achieving an *arbitrarily* fine resolution r appears to be relatively straightforward, since from (11.20) we can always make the SNR large by allowing for a sufficiently long integration time T, or by controlling signal and noise. In practice, there are limitations:

- The integration time does indeed allow for noise to be reduced by averaging the measurements over a longer period of time, however it may not be possible to integrate indefinitely. The factor which fundamentally limits integration time[4] is that of stationarity: over what period of time are both the behaviour under study, and the forward measurement model, actually constant? Factors which limit time-stationarity could be day/night cycles, changes in weather, satellite movement etc.
- The noise will have some component under our control, in the measuring instrument itself, and a component not under our control, stemming from the physics of the measurement process [see, for example, the discussion regarding clutter/noise terms around (11.3)]. Even the noise under our control cannot be made arbitrarily small, since noise is reduced by cooling, and there will be practical limits on the degree of cooling which can be achieved.
- The strength of the measured signal will be dominated by four factors:
  1. Growing, relative to the noise, as the square root of integration time $\sqrt{T}$, as already discussed.
  2. Growing with instrument size, since a larger instrument (telescope, antenna etc.) allows for more signal to be collected. However there are substantial engineering challenges in making larger instruments, particularly the cost in launching larger and heavier satellites into space.
  3. Growing with bandwidth B, meaning the extent in the frequency domain which is preserved in the measured signal. B can only be made so large and is normally dictated by the physics of the problem: a colour digital camera has relatively broad bandwidth (large B), like the human eye, whereas the study of atmospheric temperature (Case Study 11B) requires B to be tiny, only a small fraction of a spectral line. The more wavelength-specific the phenomenon we are measuring, the smaller B needs to be.
  4. Finally, in the context of *active* measurements, the strength of the received signal will be proportional to the transmitted signal strength. However, as with instrument size, there are significant engineering limitations on the transmit power available, particularly for spaced-based satellite platforms.

How resolution is limited, in practice, depends significantly on the context, so we will briefly examine each of spatial resolution, temporal resolution, and signal resolution in turn:

[4]Readers familiar with photography will most definitely have encountered integration time in the context of shutter speed. A nonstationary scene (fast motion) will require a short integration time (fast shutter), whereas a stationary scene *and* stationary measurement model (camera on tripod) will allow for very long integration times (slow shutter).

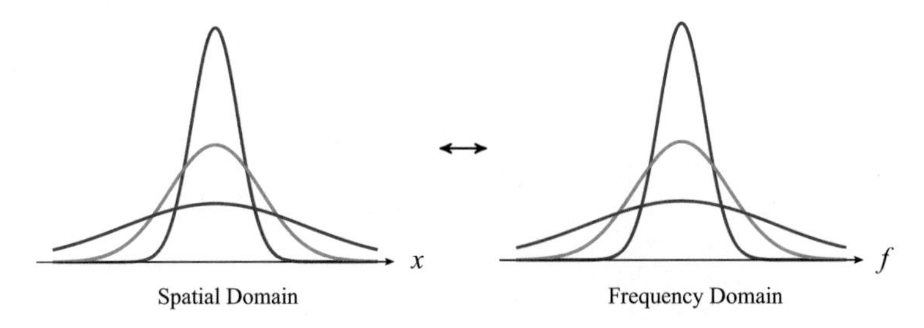

Fig. 11.4 Space vs. frequency: there is an interesting tradeoff between resolution in the spatial and frequency domains: the more localized a signal is in space, the more it is spread out in frequency; the more narrowly a signal is defined in frequency, the broader the extent of the signal in space.

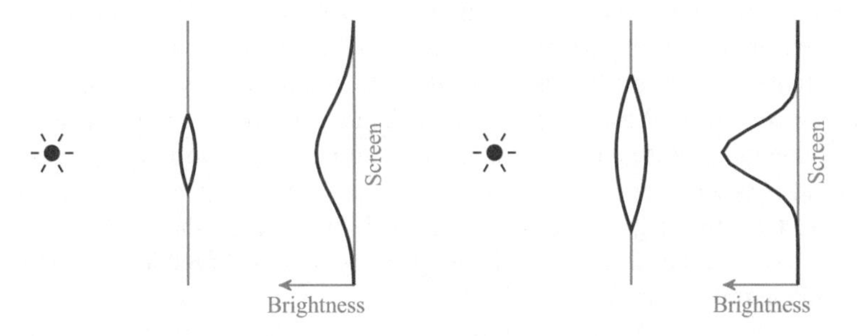

Fig. 11.5 Consider a point–source of light, whose image is projected onto a screen through a lens. All lenses are subject to diffraction, leading to blurring. Very much analogous to Figure 11.4, a smaller lens corresponds to a greater blur (left), and a larger lens corresponds to a finer focus (right).

Spatial Resolution: Figures 11.4 and 11.5 illustrate the fundamental duality between an instrument's size and its spatial resolving power: a larger telescope can visually distinguish more closely spaced objects. The limit is caused by the blur associated with diffraction effects, quantified by the *Rayleigh criterion*

$$\text{Angular Resolution} \quad \theta = 1.22\,\frac{\lambda}{d}, \tag{11.22}$$

such that an instrument of diameter[5] d operating at wavelength λ can distinguish two objects if their angular separation exceeds θ, as illustrated in Figure 11.6.

[5]Technically d measures the diametric *extent* of the instrument, which could be one large telescope of mirror diameter d, or several smaller interferometrically–coupled telescopes spaced apart by separation d. That is, rather than a single gigantic telescope, a few small widely spaced telescopes can have the same angular resolving power, however the interferometric coupling requires the location/position of the telescopes to be known to a small fraction of a wavelength, a significant technical challenge at shorter (visible light) wavelengths. The interferometric principle is widely applied in arrays of radio telescopes.

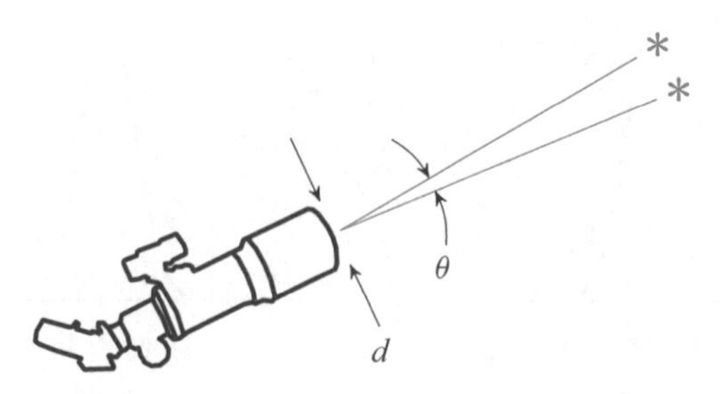

Fig. 11.6 What can we see: from Rayleigh's criterion (11.22), two stars, viewed through a telescope, can be resolved if their angular separation θ exceeds $1.22\lambda/d$.

Therefore for an instrument at a range R from a surface or medium under study, the measurement footprint or minimum spatial resolution is $R \cdot \theta$.

Note that (11.22) describes the *visual* distinguishability of two points, as opposed to the *mathematical*/inverse-problem distinguishability of (11.20), which is why components of noise and time integration do not appear in (11.22). Essentially (11.22) is describing the nature of the blur function ϕ, the degree of separation required for two adjacent blur functions to be distinguishable. For a circular mirror or lens, the diffraction blur is given by a Bessel function, which can be roughly approximated as

$$\phi(r) \approx \text{sinc}^2\Big(2.4\frac{dr}{R\lambda}\Big) \quad \text{where} \quad \text{sinc}(x) = \frac{\sin(x)}{x}. \tag{11.23}$$

It is important to recognize that (11.22) applies equally well to signal *reception*, such as through a telescope, as it does to signal *projection*, for example describing the beam width θ as a function of the size d of a radar antenna. Indeed, as shown in Figure 11.7, for a two-dimensional radar antenna, (11.22) is applied separately in the horizontal and vertical directions, such that a rectangular antenna can quite deliberately possess different horizontal and vertical beam widths θ_H and θ_V, respectively.

Time and Range Resolution: How well can we measure *when* an event takes place? Time and range are essentially analogues, in that range is measured via a signal propagation time, like a radar antenna: a signal is sent, and the time delay until the signal returns measures the distance to the reflecting object.

We like to think of a radar pulse as a short infinitesimal spike, however it is very hard to push enough power into an exceptionally short signal, and we lack the integration time T to build up a robust response to noise. Instead, a signal spread over time allows for greater power/time integration, but would appear to lead to a blurring over time. A clever choice of transmitted signals (what are known as *chirp*-like, having a time-varying frequency) allows for fine time resolution.

In practice, range can be measured very finely, to a fraction of a wavelength λ, and may more likely be limited in accuracy by the geometry of the reflecting object, for example the ambiguity in defining the distance from a satellite

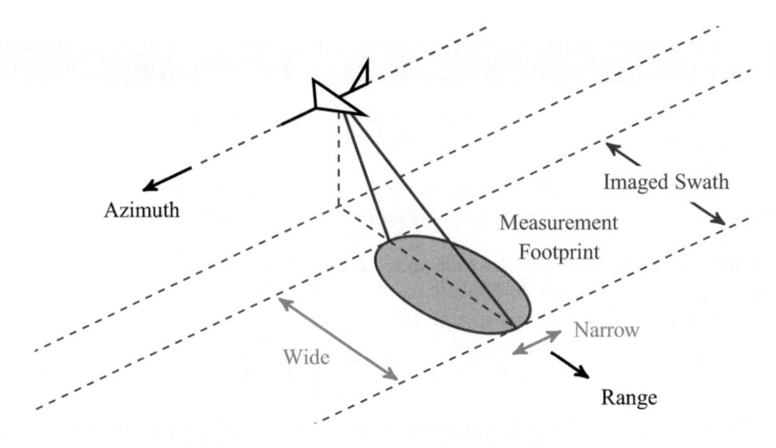

Fig. 11.7 The one-dimensional illustration of resolution in Figure 11.5 can be generalized to two dimensions. Suppose we wish to acquire an image from a radar antenna mounted on an airplane. For the radar beam to illuminate both near and far range requires a relatively *large* angle vertically, corresponding to a *small* antenna height. In contrast, to produce resolution in azimuth requires a relatively *small* angle horizontally, corresponding to a *large* antenna width.

microwave altimeter to the ocean surface, where the ocean surface isn't flat because of waves.

Brightness Resolution: The fineness with which we can measure a radiation strength, such as the brightness of a pixel in a digital camera or in a microwave sensor, depends on the signal-to-noise ratio

$$\mathrm{SNR} = \sqrt{T}\,\frac{\mathrm{Signal}}{\mathrm{Noise}}. \tag{11.24}$$

The measured signal grows with underlying signal intensity I, instrument diameter d, and frequency bandwidth B:

$$\mathrm{SNR} = \sqrt{T}\,\frac{Id^2B}{\mathrm{Noise}}, \tag{11.25}$$

that is, similar to before, but without a blur. Therefore the resolution along the intensity axis I is given by

$$r_I \approx \frac{\sigma}{\sqrt{T}d^2B}, \tag{11.26}$$

where σ describes the measurement noise standard deviation. The instrument noise is a complex subject depending on many factors, however all electronic circuits do experience temperature dependent noise, which is why many digital cameras attached to large telescopes are cooled.

Example 11.2: Can a Space Telescope Read a Newspaper?

It is popular in certain TV shows and movies to show space telescopes resolving fantastic details on the earth's surface, such as identifying cars or people.

The Hubble Space Telescope, one of the largest and most expensive space-based telescopes, has a primary mirror 2.4 m in diameter and orbits at an altitude of 560 km.

We are imaging in the visible domain, so let us assume that we are observing yellow–green light at a wavelength of 550 nm, the wavelength where the sun is brightest. The diffraction limit (11.22) gives us a resolvable angle

$$\alpha = 1.22\frac{\lambda}{d} = 1.22\frac{550\,\text{nm}}{2.4\,\text{m}} = 2.8 \cdot 10^{-7} \tag{11.27}$$

For the orbital altitude of 560 km, this corresponds to a resolvable point separation of

$$\Delta = \alpha \cdot D = 2.8 \cdot 10^{-7} \cdot 560 \cdot 10^{3}\,\text{m} = 0.16\,\text{m} \tag{11.28}$$

That is, under ideal circumstances, the Hubble telescope can resolve details 16 cm apart, definitely enough to see a car, but nowhere near the resolution to recognize faces, license plates, or newspaper headlines.

Furthermore the above calculation ignores atmospheric distortions which, if anything, are *significantly* worse than the diffraction limit, and is the reason for putting telescopes into orbit, above the atmosphere, in the first place.

Further Reading: Space Telescope in Wikipedia or HubbleSite.

11.3.2 State Resolution

In contrast to measurements, where the geometry of the sensing device places a limit on resolution, in principle we can choose *any* desired spatial grid resolution for the unknown state. However there are a few key constraints in choosing a state resolution:

Computational Complexity: A finer resolution implies a greater number of state elements to estimate, so the computational resources available may limit the feasible resolution. For a linear inverse problem having n unknowns[6], the computational complexity grows somewhere between $\mathcal{O}(n^{1.5})$ and $\mathcal{O}(n^3)$.

Information Content: It is normally not useful or meaningful to have a spatial grid significantly finer than the information actually present in the problem. For an out-of-focus image from a digital camera, the image doesn't look any better at 10 mega-pixels than it did at 5.

[6]Keep in mind that n represents the *total* number of unknowns. Thus in a 3D problem, *doubling* the resolution actually multiplies n by *eight*, corresponding to an increase in complexity between 20 and 500 times.

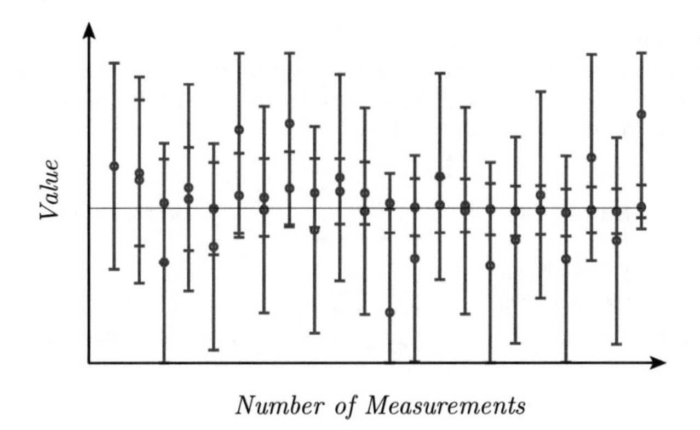

Fig. 11.8 Super-resolution: sufficiently many poor measurements can lead to very accurate estimation. Here poor measurements (red) of an unknown value (solid line) are combined (by averaging) to produce estimates (blue) which become progressively more accurate.

The total information present in a given problem is given by the information in the prior model (what you knew about the problem ahead of time) plus that of all of the measurements.

Prior/Dynamic Model: The underlying model describing the problem may require a certain resolution. Particularly in the context of nonlinear coupled spatial models, there may be a certain fineness required of the spatial grid in order for phenomena of interest to emerge, as we saw with the eddy-resolving challenge of Chapter 8.

Although it seems counterintuitive at first, it *is* possible for the state to be estimated at a resolution significantly finer than that of the measurements, known as *super-resolution*, one example of which appears in Case Study 11A. Noisy (low resolution) measurements clearly have less information than accurate (high resolution) ones, however with a sufficient number of measurements we can, in theory, recover a high resolution state, as illustrated in Figure 11.8. Actually recovering this super-resolved state, *in practice*, requires us to solve an inverse problem.

11.4 Inverse Problems

Our underlying state $\underline{z}$ is the ideal, complete representation of the system: detailed, noise-free, and regularly structured (e.g., spatially gridded), whereas the measurements $\underline{m}$ are incomplete and approximate: possibly noise-corrupted, irregularly structured, or limited in number, as summarized in Table 11.3.

So what makes an inverse problem hard? The key problem is that the process of producing an estimate $\hat{\underline{z}}$ of the unknown state, going from the forward model to the inverse function

Table 11.3 Forward vs. inverse: an inverse problem is, by definition, the difficult inversion of a comparatively-straightforward forward problem. Inverse problems are common because the available measurements in any given problem normally have the attributes on the right, whereas what we *want* are the attributes on the left (from [6]).

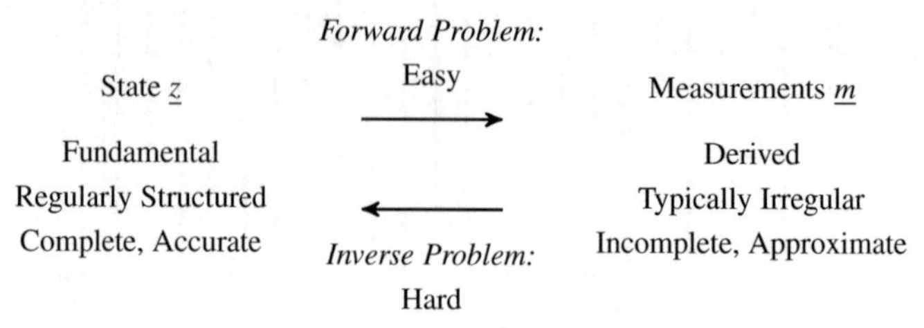

$$\underline{m} = C(\underline{z}) \qquad \longrightarrow \qquad \hat{\underline{z}} = C^{-1}(\underline{m}) \tag{11.29}$$

is not known. In most cases, $C()$ isn't a mathematical function which you can write down, rather a big, complex algorithm. How do you invert an algorithm?

If the forward problem is, indeed, invertible, then we could try to characterize the inverse by guessing, as

$$\hat{\underline{z}} = C^{-1}(\underline{m}) \triangleq \{\underline{z} \mid C(\underline{z}) = \underline{m}\}, \tag{11.30}$$

that is, to try all possible values of $\underline{z}$ to find the one which produced the observed measurements $\underline{m}$.

However for a 1000×1000 eight-bit-depth colour image, the number of possible images, the number of possible $\underline{z}$ to try, is enormous:

$$|\{\underline{z}\}| = \left(2^{3 \cdot 8}\right)^{(1000 \cdot 1000)} \qquad \longrightarrow \quad \text{Huge!} \tag{11.31}$$

In practice we would use an iterative strategy,

$$\text{Find estimates } \hat{\underline{z}}_1, \hat{\underline{z}}_2, \ldots \text{ such that } \lim_{i \to \infty} C(\hat{\underline{z}}_i) = \underline{m}. \tag{11.32}$$

Such an approach is used in finding the roots of a function $f()$, for example, where bisection or Newton's method are applied to find an iterative sequence of states such that $f(\hat{\underline{z}}_i) \rightarrow 0$.

An even greater problem is that, in most cases, the quantity or quality of the measurements is inadequate to allow a reconstruction of $\underline{z}$, implying that the inverse function C^{-1} does not actually exist:

$$\begin{array}{c} \text{3D Ocean} \\ \text{Temperature} \end{array} \xrightarrow[\checkmark]{C()} \begin{array}{c} \text{2D Measured} \\ \text{Image} \end{array} \xrightarrow[\times]{C^{-1}()} \begin{array}{c} \text{3D Ocean} \\ \text{Temperature} \end{array} \tag{11.33}$$

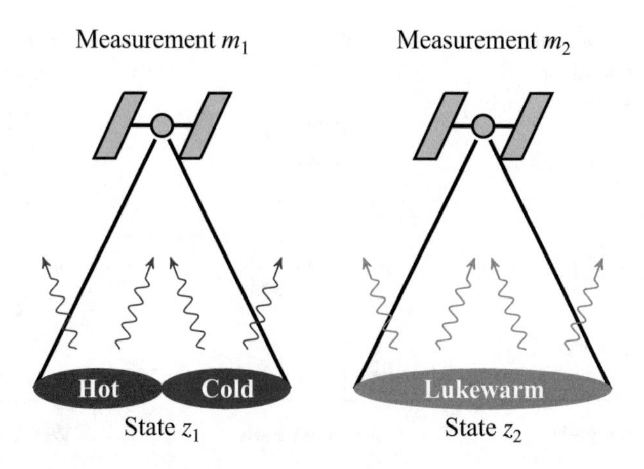

Fig. 11.9 The ambiguity of an inverse problem: the sensor on a satellite measures thermal radiation over a region (essentially a surface integral). Scalar measurements m_1, m_2 are observed for the two ocean patches, characterized by underlying state z_1, z_2. Although the two patches are clearly very different, they result in identical measurements.

We simply cannot estimate a 3D temperature field from a single 2D image. As a simpler example, a single thermal measurement, as sketched in Figure 11.9, cannot distinguish between two different distributions of ocean temperature — both scenarios result in the same measurement.

For us to be able to find a solution, an inverse problem must be *well-posed*:

Existence: For each meas. $\underline{m}$, there must be at least one state $\underline{z}$
Uniqueness: For each meas. $\underline{m}$, there may not be more than one state $\underline{z}$

In the somewhat restrictive case that our forward problem is linear and deterministic,

$$\underline{m} = C\underline{z} \qquad \text{for some matrix } C \tag{11.34}$$

then we just have a linear system of equations, for which the issues of existence and uniqueness are fairly simple to interpret, as sketched in Figure 11.10:

System is Overdetermined or Overconstrained: There are *more* equations than unknowns, so there is no $\underline{z}$ satisfying all of the equations, thus existence fails.
System is Underdetermined or Underconstrained: There are *fewer* equations than unknowns, so there are infinitely many possible solutions for $\underline{z}$, thus uniqueness fails.
System Is Invertible: We can solve the system of equations and get a unique answer, thus the problem is well posed.

In what sorts of contexts do we encounter each of these scenarios? Let's briefly examine each in turn.

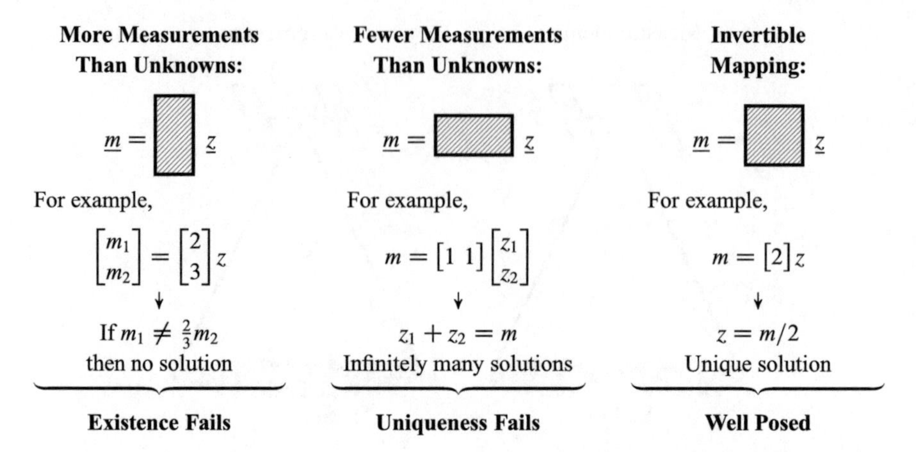

Fig. 11.10 A linear system falls into one of the three categories, as shown. Having a square matrix (same number of measurements and unknowns) does not automatically guarantee problem well-posedness; the matrix representing the linear system must be invertible to ensure posedness.

Invertible Problems

For the problem to be invertible, the number of measurements must equal the number of unknowns. This is commonly encountered in image processing, where our measurements are an $n \times n$ input image, and the unknown state is some enhanced/processed/analyzed $n \times n$ image.

In practice, even simple problems such as image denoising suffer from *uniqueness failure*, since many original images could have led to the same observed, noisy one. Outside of image processing invertible problems are uncommon, therefore our focus is on the cases of existence and uniqueness failure, below.

Existence Fails

A failure of existence means that the problem is overconstrained, that there are more measurements than unknowns, a circumstance which most commonly occurs in model fitting problems, for example

$$\text{Linear Regression:} \quad \underbrace{\text{Fitting a line}}_{\text{2 unknowns}} \quad \text{to} \quad \underbrace{\text{a given data set}}_{\text{many measurements}} \tag{11.35}$$

All model learning problems should fall into this scenario, since the number of model parameters should always be much less than the number of observations, otherwise we are likely to be overfitting the data, like fitting a ninth-order polynomial to ten data points.

You may wonder whether any "real" inverse problems fail existence. Certainly, however only those cases in which the unknown state consists of very few values, such as

What is the rate of global sea-level rise in mm per century?

Example 11.3: Measurements and Constraints

Suppose we have a stochastic (random) model that allows straight lines in a 2D plane. Any given information, such as a measurement, constrains the model. If the measurements are exact (left), then the mathematics assert a hard constraint on the permitted lines; in contrast, measurements subject to some uncertainty (right) permit a wider range of lines.

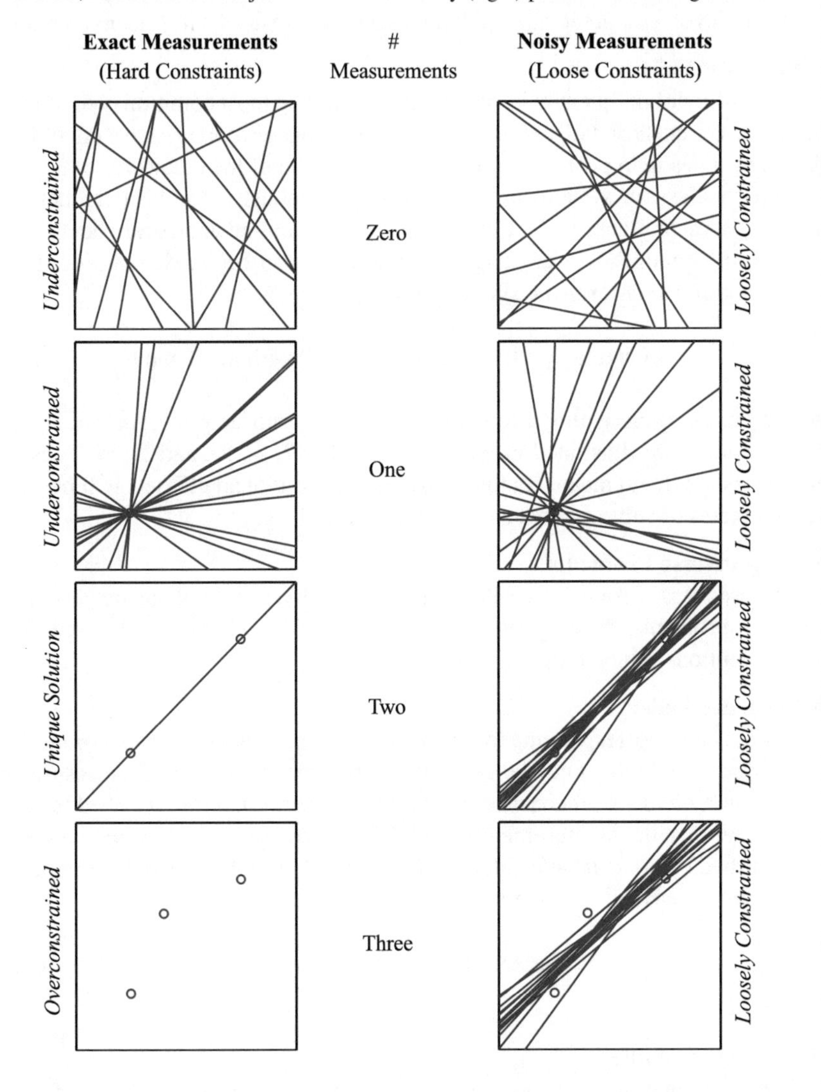

Two points determine a line, therefore for exact measurements the problem becomes overconstrained, and unsolvable, with three or more points. If the measurements are not exact, then having three or more measurements in no way contradicts the model.

How many hectares of tropical forest have been lost since 1990?
What fraction of the world has public internet access within 1 km?

All of these are important questions, where there is only a single unknown to be inferred from large data sets.

However certainly any problem where we wish to see a gridded map or an image, the result involves a great *many* unknowns and will instead suffer from uniqueness.

Uniqueness Fails

The failure of uniqueness is, by a wide margin, the most common type of inverse problem. Uniqueness fails, almost by definition, since we *never* have enough data, driven by a desire to push the limits on the resolution of the unknown state $\underline{z}$.

In particular, our systems under study are enormously complex, whether a city, an ecosystem, or planet-wide. Clearly there will always be unresolved details which are not measured, thus every large-scale system must be *undersampled*. Indeed, every acquired measurement can be characterized by its

$$\text{Location } (x, y), \quad \text{Height } h, \quad \text{Wavelength } \lambda, \quad \text{Time } t$$

and any or all of these dimensions can be subject to subsampling, for which typical circumstances are illustrated in Figures 11.11, 11.12, 11.13, and 11.14. These four categories are by no means mutually exclusive, and typical datasets will be subject several forms of subsampling, as shown in Figure 11.15.

Solving Inverse Problems

Ill-posedness is not necessarily a terrible difficulty. Indeed, there are very simple, common problems, such as linear regression (11.35), which are ill-posed but for which solutions can be proposed. In general:

If Existence Fails:

There is no $\underline{z}$ corresponding to the given $\underline{m}$: the problem is *overdetermined*.
We don't need more knowledge about $\underline{z}$, rather the issue is that the measurements are contradictory or inconsistent with one another. Therefore we need to change or generalize the measurement model. We propose to understand $\underline{m}$ to have been perturbed by noise from its ideal value, so we choose that $\underline{z}$ which comes *closest* to being consistent with the measurements:

$$\text{Select } \hat{\underline{z}} \text{ by minimizing } \|\underline{m} - C\hat{\underline{z}}\| \tag{11.36}$$

If Uniqueness Fails:

For the given observation $\underline{m}$ there are infinitely many possible solutions for $\underline{z}$: the problem is *underdetermined*.
The measurements and the measurement model are fine, however we somehow need additional information on $\underline{z}$, above and beyond what we know from the measurements. Such information is called a *prior model*, an assumption or knowledge derived from the physics of the problem, possibly expressed in the

Systematic Subsampling
Measured locations determined by satellite orbit
ex. Satellite altimeter sampling in the Pacific

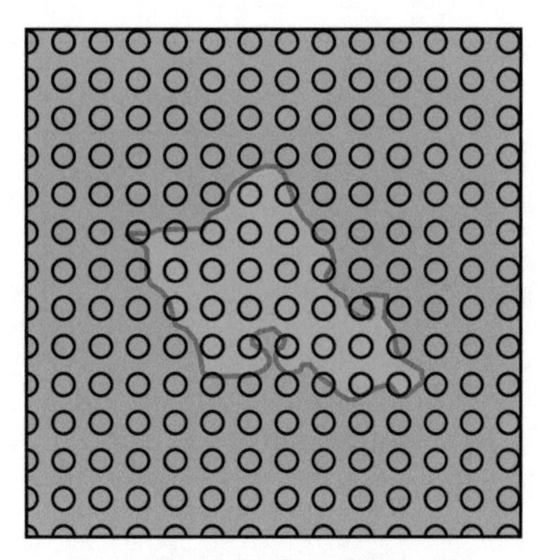

Low Resolution Sampling
Large-footprint measurements
ex. Weather satellite pixels over Hawaii

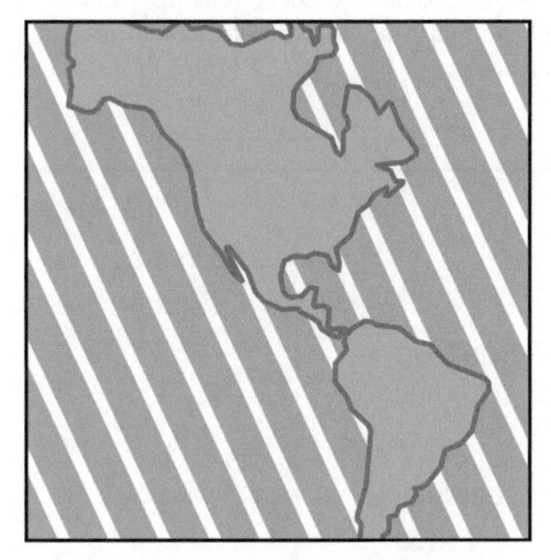

Irregular Subsampling
Measurements only over water or only over land
ex. Sea Surface Temperature swaths blocked by land

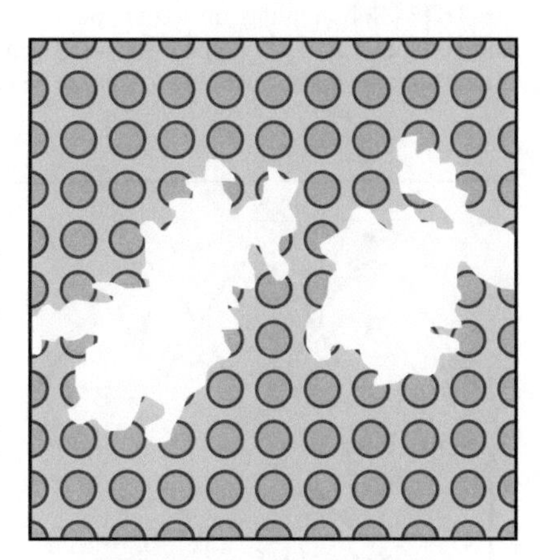

Random Subsampling
Measurement absence due to clouds, dust etc.
ex. Visual land mapping data blocked by clouds

Fig. 11.11 Subsampling in space: there are many factors which limit how densely in space an instrument can sample, including the satellite orbit, instrument design, and objects such as land or clouds which get in the way of the signal.

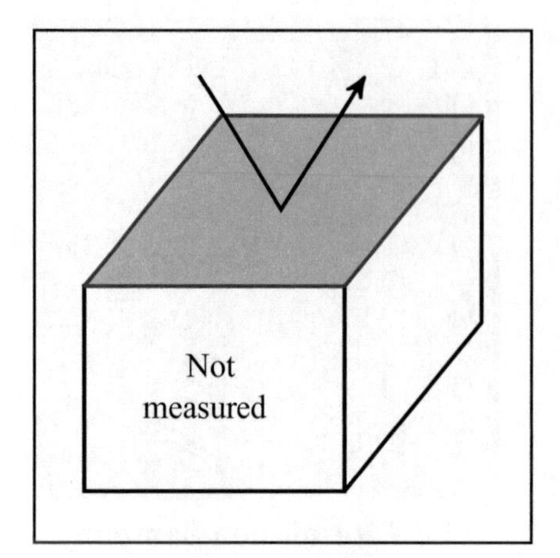

Fig. 11.12 Subsampling in height: most remote-sensing satellites observe only the ocean or land surface, left. Tomographic measurements, right, normally in contexts of medical imaging or atmospheric studies, are able to sample height in layers.

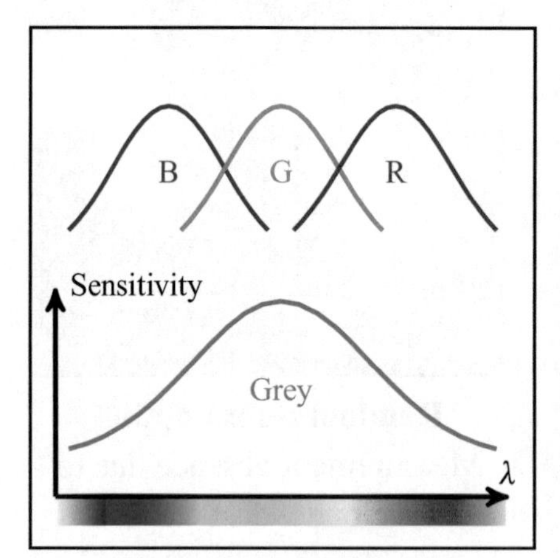

Fig. 11.13 Subsampling in wavelength: in most applications spatial resolution is more important than spectral, so that relatively few wideband (left) or narrowband (right) sensors suffice.

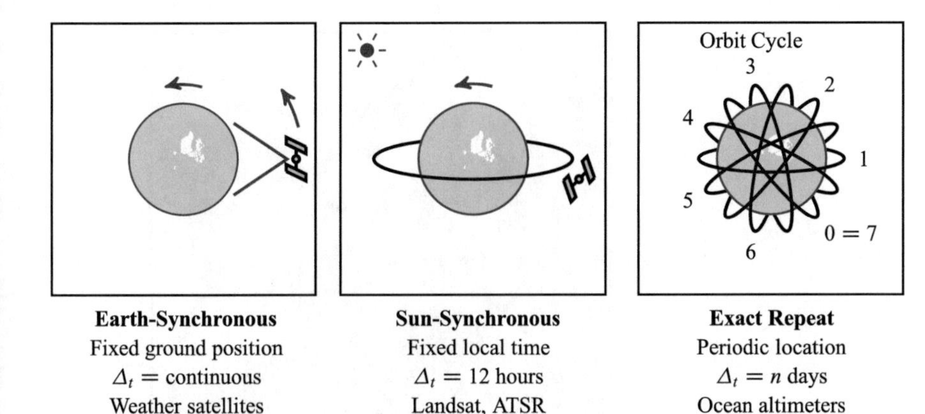

Fig. 11.14 Subsampling in time: for satellite remote sensing, the time-dependence of samples depends greatly on the orbital characteristics of the satellite. In general, the maximum time sampling rate in any context is dictated by the Nyquist limit, such that sampling needs to occur at least twice as often as the highest temporal frequency of interest.

form of a probability density function $p(\underline{z})$, a covariance $\Sigma_{\underline{z}}$, or a constraint $L\underline{z}$. For example, one simple assertion would be to select a small (near-zero) value for $\underline{z}$:

$$\text{Select } \hat{\underline{z}} \text{ by minimizing } \|\hat{\underline{z}}\| \text{ over those } \underline{z} \text{ satisfying } \underline{m} = C\underline{z}. \tag{11.37}$$

In practice, we do *both*. First, measurements are always subject to some degree of uncertainty, regardless whether existence fails or not, so (11.34) should be represented stochastically as

$$\underline{m} = C\underline{z} + \underline{w} \qquad \underline{w} \sim (0, R), \tag{11.38}$$

such that the noise $\underline{w}$ is characterized by an error covariance R. Second, regardless whether uniqueness fails or not, if a prior model is known then it should be asserted. As a result, (11.36), (11.37) are combined and generalized as

$$\text{Select } \hat{\underline{z}} \text{ by minimizing } \Big\{ \underbrace{\|\underline{m} - C\hat{\underline{z}}\|}_{\substack{\text{Model for}\\ \text{Measurement Error}}} + \underbrace{\gamma}_{\text{Tradeoff}} \underbrace{\|L(\underline{z} - \underline{\mu})\|}_{\substack{\text{Model for}\\ \text{Unknowns}}} \Big\} \tag{11.39}$$

for which the solution can be written [6] as a linear system,

$$\underbrace{(C^T R^{-1} C + \gamma L^T L)}_{A} \underbrace{(\hat{\underline{z}} - \underline{\mu})}_{\underline{x}} = \underbrace{C^T R^{-1}(\underline{m} - C\underline{\mu})}_{\underline{b}} \tag{11.40}$$

where γ represents a parameter to control the tradeoff between the influence of the measurements versus the influence of the prior model.

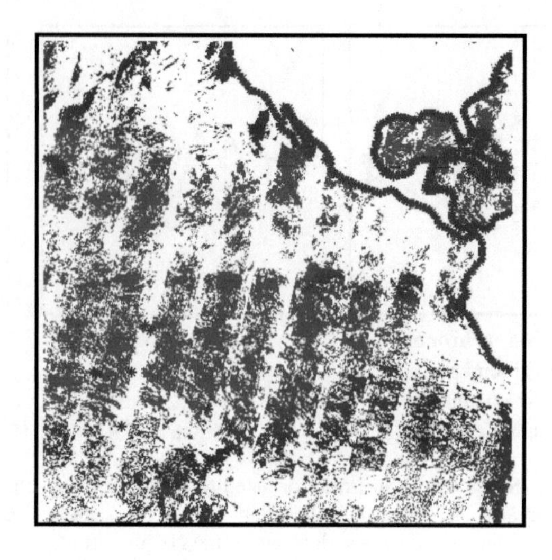

Subsampling in Space:	
Systematic:	Imaging swath width of 512km
Low Resolution:	Pixel spatial resolution of 1km
Irregular:	Imaging only over water, no signal over land
Random:	Infrared channels are cloud sensitive

Subsampling in Height:	Surface measurements only
Subsampling in Wavelength:	One visible and three infrared channels
Subsampling in Time:	Sun-synchronous orbit, thus 12-hour sampling

Fig. 11.15 In practice, real data, here from the Along-Track-Scanning-Radiometer (ATSR), will be subject to a wide variety of subsampling types.

ATSR data from the Rutherford Appleton Laboratory.

It is no accident that we encountered effectively equivalent linear systems (8.48) in solving spatial boundary-value problems, since in both cases we needed to combine a model (prior/differential equation) of the unknowns with additional constraints (measurements/boundary conditions). Indeed, the generalized stochastic measurement-boundary problem in (8.53) was formulated very much in the style of (11.39).

It is important to realize, however, that (11.40) cannot just be solved by matrix inversion $\underline{x} = A^{-1}\underline{b}$, particularly for large spatial problems:

1. **Sparsity:** Elements in matrix A describe the coupling (Ising model) or differences (discretized PDEs) between state elements. These couplings or differences are between neighbouring state elements only, therefore A is an exceptionally sparse matrix (mostly zeros), whereas the inverse of a sparse matrix is normally *dense*. As a result, in many circumstances it will be impossible even to store A^{-1}, let alone compute it.

2. **Conditioning:** Matrix singularity is a degenerate property: the tiniest perturbation of a singular matrix makes it nonsingular (invertible). Therefore essentially any square matrix A in (11.40) is invertible, however not all invertible matrices are equally easy to invert. The *condition number* κ or numerical stiffness of a problem measures the numerical difficulty in solving it, given by the ratio of the largest and smallest eigenvalues:

$$\kappa = \frac{|\mathrm{Re}(\lambda_{\max})|}{|\mathrm{Re}(\lambda_{\min})|}, \tag{11.41}$$

as we saw already in (8.27). Poor conditioning essentially stems from a need to make large changes ($\lambda_{\max}$) to the estimates $\hat{\underline{z}}$, but in only tiny steps ($\lambda_{\min}$), as illustrated in Figure 11.16.
3. **Stationarity:** Not only do we not want to find A^{-1}, in most cases we do not even want to write down A itself. As was discussed around (8.35) in Section 8.3, if the inverse problem of (11.39) is spatially stationary then we would much prefer to express A implicitly as a convolution

$$A\underline{x} \equiv a * x \tag{11.42}$$

To be sure, as we saw repeatedly in Figures 11.11 through 11.15, the measurement model will quite frequently be rather irregular, and not at all stationary. In such cases we would still not want to store A, rather we would normally use an implicit representation

$$A\underline{x} \equiv a(x) \tag{11.43}$$

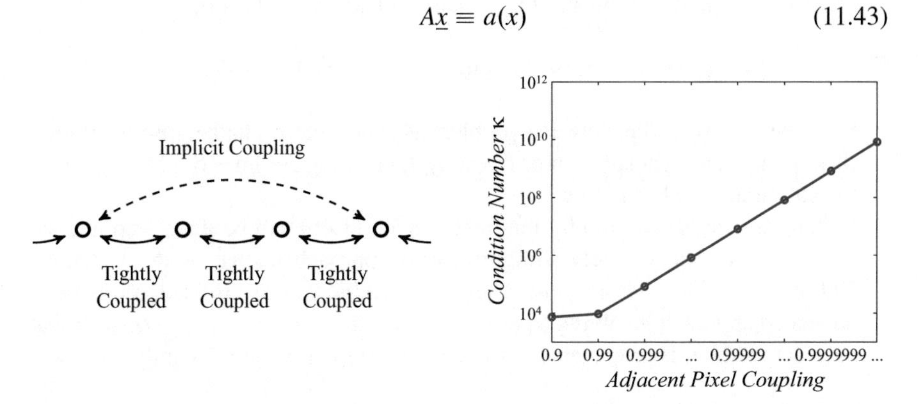

Fig. 11.16 Conditioning: the numerical difficulty of a problem is associated with κ, the *condition number*. κ is measured as the ratio of the largest to smallest eigenvalues, (11.41), since it is numerically difficult to simultaneously work with very large and very small numbers. For example, suppose we have a sequence of coupled state elements, left. The state value for any individual element is very flexible (large eigenvalue), however the state value *relative* to its neighbours may be tightly constrained (small eigenvalue), such that the numerical difficult increases as the correlation coefficient between adjacent pixels increases, right.

for some algorithm or function $a()$. In solving such problems, we need a linear systems solver, such as conjugate gradient [6], which only ever needs the *product* $A\underline{x}$, never the actual individual elements of A.

Finally, most certainly not all inverse problems can be solved in the linear systems form of (11.40). Indeed, the global-climate data assimilation problem back in Chapter 2 was both time-dynamic and nonlinear:

Over Time: Both in the steady-state problem of (8.48) and in the linear inverse problem of (11.39) we have a *single* unknown vector $\underline{z}$ to solve.
In contrast, in most socio-environmental problems we are interested in unknowns $\underline{z}(t)$ possessing dynamics and measured over time, as we already saw in (8.49), (8.52), (8.53):

$$\dot{\underline{z}} = A(t)\underline{z}(t) + noise \qquad \underline{m}(t) = C(t)\underline{z}(t) + noise, \tag{11.44}$$

The classic approach to such a time-dynamic inverse problem is the Kalman filter, for which large-scale data-assimilation variants continue to be proposed.

Discrete State: We saw discrete-state problems in Chapter 8, such as the Ising and other cellular automaton models. When the unknown $\underline{z}$ is discretely valued, we have a labelling or classification problem, the domain of Pattern Recognition.

Nonlinear Model: If we have a nonlinear forward problem, as in (11.1),

$$\underline{m}(t) = C\big(\underline{z}\big) + noise, \tag{11.45}$$

or, more likely, the nonlinear time-dynamic analogue of (11.44),

$$\dot{\underline{z}} = f\big(\underline{z}(t), t\big) + \ noise \qquad \underline{m}(t) = C\big(\underline{z}(t), t\big) + \ noise, \tag{11.46}$$

then the corresponding inverse problem is *much* more challenging. Certainly there is no analytical solution or linear system, along the lines of (11.40), rather the solutions are algorithmic.
If the problem is only mildly nonlinear, then (11.46) can be linearized at each time step, a comparatively straightforward approach known as the Extended Kalman filter. However in most cases an ensemble strategy is used, a relatively advanced approach involving a parallel set of estimates $\hat{\underline{z}}_1(t), \ldots, \hat{\underline{z}}_q(t)$, such that each set member is propagated through a time step of the nonlinear dynamics,

$$\hat{\underline{z}}_i\big(t - \delta_t\big) \quad \rightarrow \quad \hat{\underline{z}}_i\big(t\big), \tag{11.47}$$

and whether the member is then duplicated, removed, or kept as is depends on the degree of consistency with the measurements:

$$\left\| \underline{m}(t) - C\Big(\hat{\underline{z}}_i(t), t\Big) \right\|. \tag{11.48}$$

A final challenge of interest is that of *forecasting*: how can we use system measurements, say of global climate, to anticipate the future state of the system? In principle, given measurements to time T_m, we can solve an inverse problem to find the state $\hat{\underline{z}}(T_m)$, which is then fed into the system dynamic model to arrive at a state estimate $\hat{\underline{z}}(T_f)$ at future time $T_f > T_m$. There are three factors which limit our ability to infer such future states:

State Uncertainty: the estimated state $\hat{\underline{z}}(T_m)$ is subject to uncertainty due to noisy measurements, incomplete measurements, model approximations, and difficulties in exactly solving large nonlinear inverse problems.

Stochastic Systems: there are system influences that arrive, unanticipated, from outside of the modelled system envelope; these influences accumulate over time.

Chaotic Systems: even if the system state is known accurately at time T_m, nonlinear/turbulent/chaotic systems can exhibit exceptional sensitivities to initial conditions.

That is, there is an initial uncertainty inherited from the inverse problem, which unavoidably grows over time due to random forcing and chaotic elements in the dynamic system. Nevertheless, the forecasting range can be extended by acquiring more accurate or complete measurements, reducing the uncertainty in $\hat{\underline{z}}(T_m)$, and by developing more complete models, reducing unanticipated stochastic forcing.

Case Study 11A: Sensing — Synthetic Aperture Radar

From modest origins over a century ago, radar imaging has made astonishing advances. Radar's greatest strengths include its imperviousness to weather, penetrating cloud, rain, and fog, and the richness of the returned data, in the form of magnitude and phase information for multiple wavelengths and polarizations.

The basic principle of radar imaging is illustrated in Figure 11.17. Achieving a good resolution in *range* is relatively straightforward, requiring accurate timing to measure signal delay. However resolution in *azimuth* is much more challenging, since there are limits on how small the beam angle θ can be made, from (11.22), since reducing θ requires a larger antenna, and there are limits to antenna size. Because any reasonably-sized antenna implies a relatively wide beam angle, the radar return will necessarily be smeared over an arc.

For example, the ERS-1/2 satellites had a radar antenna of length 10 m operating in C-Band ($\lambda \simeq 5\,\text{cm}$), thus from (11.22) the beam angle is

$$\theta \simeq 1.22\frac{\lambda}{d} = 0.006\,\text{radians}. \tag{11.49}$$

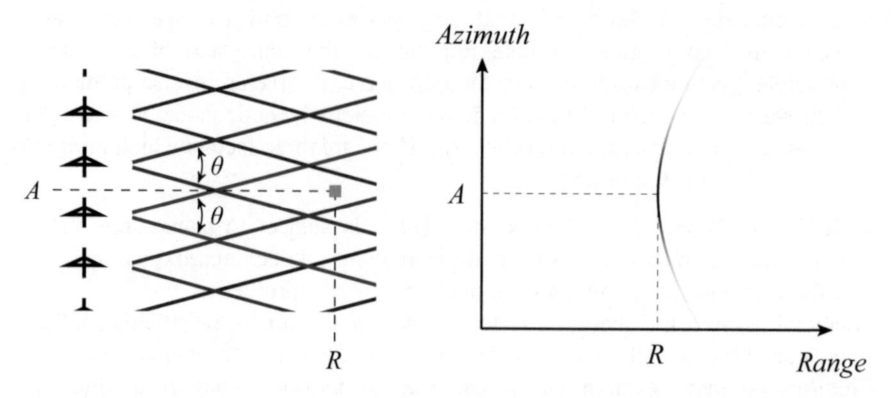

Fig. 11.17 Radar imaging: by flying along in azimuth, an aircraft can collect radar returns which, stacked row-by-row, produce an image. However for practical reasons an airborne radar antenna cannot be very long, therefore by (11.22) the beam width θ cannot be particularly narrow, meaning that a given radar reflector (green) appears over some extent of azimuth, right, and not just at one point. Each individual reflector will produce a characteristic arc, which leads to a rather blurry image.

That beam seems pretty narrow, until you consider the satellite's orbit of 780 km looking at an angle of 23°, leading to a spatial pixel size of

$$\text{Azimuth Resolution} = \frac{780\,\text{km}}{\cos 23°} \cdot \theta \simeq 5.2\,\text{km} \tag{11.50}$$

which isn't a particularly high resolution. This is, however, an inverse problem, for which the forward model is

$$\begin{matrix}\text{High Resolution} \\ \text{Surface Image}\end{matrix} \xrightarrow[\text{Imaging}]{\text{Wide Beam}} \begin{matrix}\text{Measured} \\ \text{Surface Image}\end{matrix} \tag{11.51}$$

We can *invert* the problem, essentially by looking for arcs, integrating in arc-shapes over the measured image. As sketched in Figure 11.18, by integrating over many radar returns, spread in azimuth, we have effectively created a synthetic antenna having a size d in (11.22) up to *kilometres* in length!, leading to a far narrower effective beam angle and a far superior resolution, essentially the super-resolution concept of Section 11.3. In the case of ERS-1/2, the 5200 m radar resolution is improved to a 30 m SAR pixel size, leading to spectacular images such as those shown in Figure 11.19.

So far our radar images have focused on the signal *magnitude*, which is a strong function of surface roughness and dielectric properties. The signal *phase*, in contrast, is nearly random, since the radar wavelength (5 cm) is so tiny relative to the spatial size (30 m) and height variations (10 m). With this in mind, it was particularly ingenious to consider acquiring *two* SAR images from nearby locations, as in Figure 11.20, and taking phase *differences*. The geometry of the problem leads

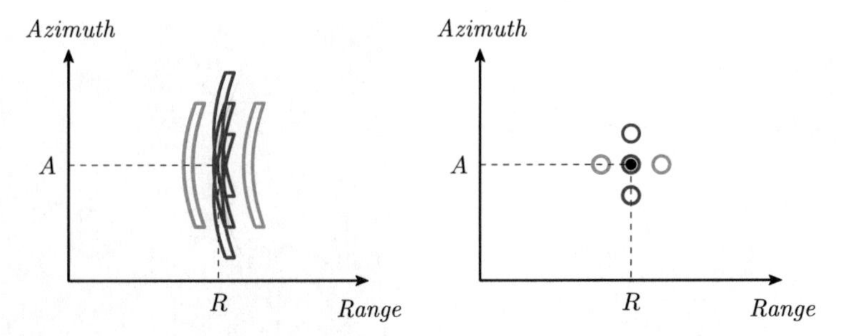

Fig. 11.18 Synthetic aperture radar (SAR): for a radar reflector at any point of range and azimuth, we can work out its corresponding radar arc from Figure 11.17, leading to many hypothesized arcs (left). We can deduce the presence or absence of a corresponding arc by integrating over the radar returns in Figure 11.17. Here the green and blue hypotheses do *not* have a corresponding arc in the data, only the red one is present, leading to a resolution, right, much higher than in the original radar image.

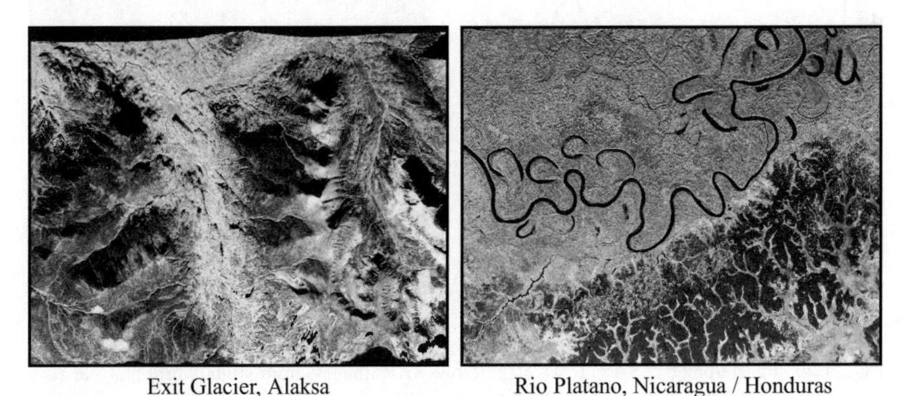

Exit Glacier, Alaksa | Rio Platano, Nicaragua / Honduras

Fig. 11.19 Airborne SAR images: radar does not capture colours the way our eyes see them; rather these images are false-coloured, with the presence of red/green/blue colours based on the radar response at different wavelengths.

Images from the AIRSAR mission, courtesy of NASA/JPL-Caltech.

to an interference pattern, such that 2π in phase corresponds to a topographic height offset of

$$H_{2\pi} \propto \lambda \frac{R}{B} \tag{11.52}$$

As shown in Figure 11.20, the fringes imply a *phase unwrapping* problem, a second inverse problem:

$$\begin{array}{c}\text{Surface Topography}\\ h\end{array} \xrightarrow[H_{2\pi}]{\text{InSAR Imaging}} \begin{array}{c}\text{Measured Phase}\\ m = \text{mod}\left(2\pi \dfrac{h}{H_{2\pi}}, 2\pi\right)\end{array} \tag{11.53}$$

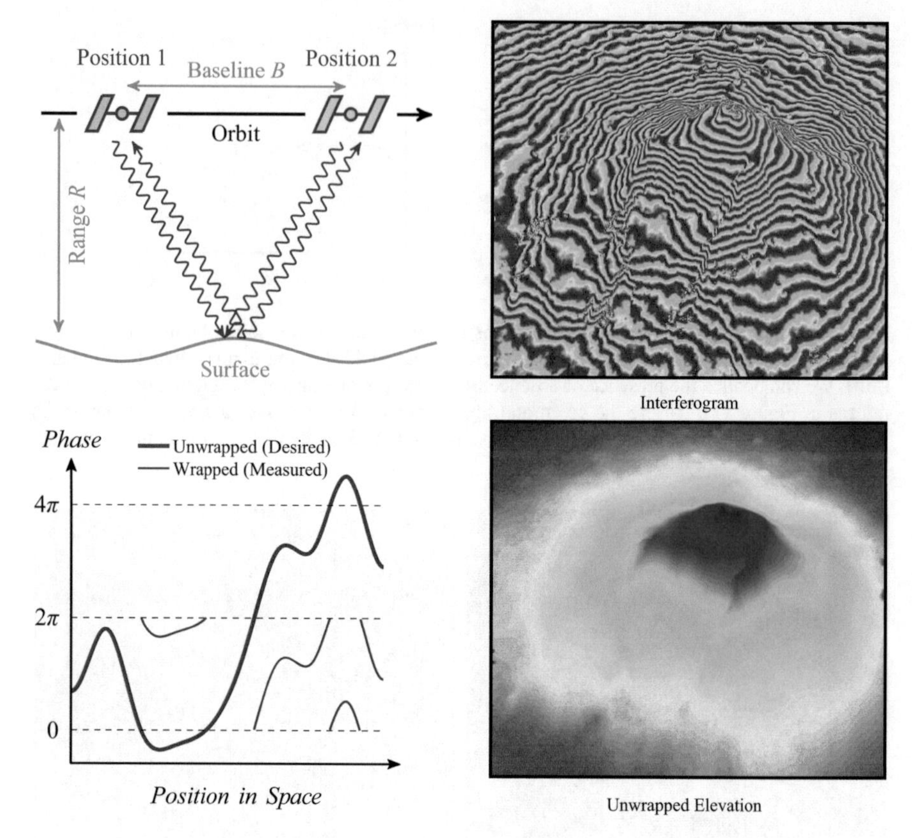

Fig. 11.20 Interferometric SAR: if we acquire *two* SAR images, from slightly different perspectives, then the phase *difference* is proportional to surface height. The measured phase forms an interferometric fringe pattern, which needs to be unwrapped to recover the surface topography, here of Mt. Etna in Italy. The view perspectives of the satellites is greatly exaggerated; in practice the baseline B is a tiny fraction of the range R.

ESA ERS 1/2 Interferometric Images from Carballo, Fieguth, *IEEE TGRS* (40) #8, 2002.

Thus, although a longer baseline B leads to a smaller offset $H_{2\pi}$, meaning finer resolution in range (surface height), larger B also leads to more measurement noise and closely spaced phase fringes, both of which make the inverse problem of (11.53) harder to solve.

A final step is *differential* InSAR, in which interferograms $\underline{m}_1, \underline{m}_2$ are measured before and after some event at intermediate time t. Whereas the phase fringes in $\underline{m}_1$ and $\underline{m}_2$ are dominated by the shape of the surface topography, the difference image $\underline{m}_1 - \underline{m}_2$ is *also* an interferogram, but one in which the phase fringes correspond only to the *change* in topography over time.

In this manner, exceptionally small surface deflections can be measured from satellite. Although satellite range resolution was able to measure height to 3 cm, the spatial resolution was very low; in contrast, differential InSAR offers an unprecedented spatial visualization of subsidence in geology (earthquakes, volcanoes), resource extraction (mining, oil, groundwater), and the layer thickening which we encountered in the Staufen example on page 271.

Case Study 11B: Inversion — Atmospheric Temperature

How do we measure the temperature of the atmosphere? Certainly weather balloons have been used for many years and satellite radiometers can measure infrared radiation brightness, however a balloon measures temperature at only a single location and infrared radiometers only of the earth's surface.

Let us begin with an unknown state $t(a)$, the temperature t of the atmosphere as a function of altitude a.

The *Forward Problem* is illustrated in Figure 11.21. It is well known that every molecule (such as oxygen O_2) is associated with a line (emission/absorption) spectrum. However the spectral line is not infinitesimally thin, rather it is a smeared line whose width is a function of pressure (the *Stark effect*), as shown. As a result, whether the atmosphere acts as a thick fog or is completely transparent at some frequency will be a function of frequency. Furthermore, the strength with which an Oxygen molecule radiates energy is a function of its temperature. Therefore a satellite, far above the atmosphere, observing the microwave energy relatively far from the O_2 spectral line at, say, 116 GHz, will be able to infer ground-level

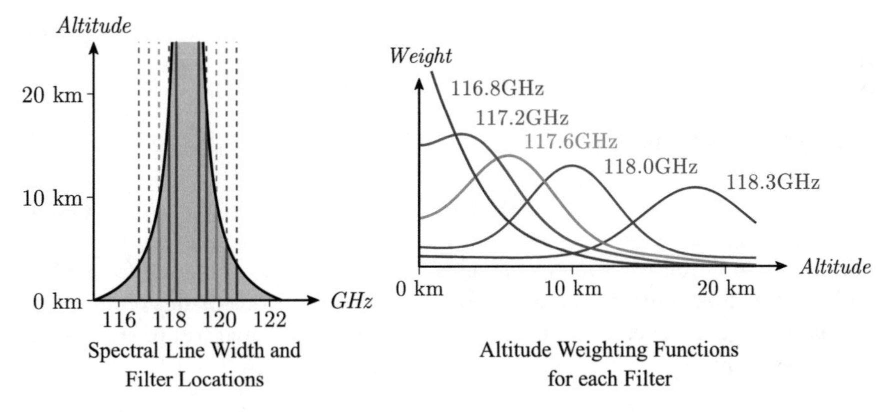

Fig. 11.21 Forward problem: the width of the oxygen 118.75 GHz spectral line (left, grey) broadens with increased pressure, thus it narrows with increased altitude. Five microwave filters are placed near the spectral line; *within* the spectral band (solid) the atmosphere interacts with the microwave signal, whereas *outside* of the spectral band (dashed) the radiation propagates freely. As a result, right, we obtain a set of curves, each describing how much a given frequency is influenced by temperature at a given altitude. These curves represent the forward model of the atmospheric temperature problem.

temperature, since at all altitudes *above* the ground the oxygen in the atmosphere is transparent at 116 GHz.

It would be nice to suppose that temperature higher in the atmosphere just corresponds to radiometric brightness closer to the centre of the spectral line, however the problem is a bit more complicated. The spectral power at 118 GHz, closer to the spectral line, is *not* just the atmospheric temperature at 15 km, nor is it the average temperature from 0 km to 15 km. Instead, from radiative transfer, a microwave photon emitted near the ground may be repeatedly absorbed and re-emitted by other molecules higher in the atmosphere. Because the atmosphere is transparent above 15 km at 118 GHz, certainly there is no dependence to air temperature above 15 km, however the dependence below 15 km needs to be found via a simulation of the radiative transfer process, leading to weighting functions $w(a,f)$ at altitude a and frequency f, such that the forward model for the observed spectral strength is

$$s(f) = \int t(a) \cdot w(a,f)\, da. \tag{11.54}$$

Normally our satellite would observe the spectrum at only a discrete number of frequencies $f_1, \ldots, f_q$, such as the filters shown in Figure 11.21.

To deduce the atmospheric temperature from the microwave brightness measurements we have to solve the corresponding *Inverse Problem*, normally by selecting discrete steps in altitude $a_1, \ldots, a_n$. We thus have state and measurement vectors

$$\underline{z} = \begin{bmatrix} t(a_1) \\ \vdots \\ t(a_n) \end{bmatrix} \qquad \underline{m} = \begin{bmatrix} s(f_1) \\ \vdots \\ s(f_q) \end{bmatrix} \tag{11.55}$$

leading to the linear forward model

$$\underline{m} = W\underline{z} + \underline{w}, \tag{11.56}$$

where matrix W follows from the weights in (11.54), and $\underline{w}$ is a noise term which will be a function of the instrument design. As an inverse problem, this problem will normally be quite underconstrained, so we need to assert some sort of smoothness in the temperature $\underline{z}$.

Further Reading

The references may be found at the end of each chapter. Also note that the textbook further reading page maintains updated references and links.

Wikipedia Links — Electromagnetics: Electromagnetic Radiation, Electromagnetic Spectrum, Scattering, Reflection (Physics), Radiative transfer, Planck's law

Wikipedia Links — Radar: Radar, Synthetic Aperture Radar, Interferometric Synthetic Aperture Radar

Wikipedia Links — Remote Sensing: Remote Sensing, Satellite, List of orbits

Wikipedia Links — Sensing Examples: Microwave Radiometer, Scatterometer, Landsat program, Weather satellite, Altimeter

Wikipedia Links — Inverse Problems: Inverse Problem, Least Squares, Data Assimilation, Regularization, Medical Imaging

I have written a text [6] which relates to inverse problems and image processing. Particularly Chapters 1 and 2 in [6] should be accessible to undergraduate students wanting to know more about inverse problems.

In general, the field of inverse problems is quite mathematical and advanced, so there are relatively few accessible texts. For readers wanting to read further in this area, I would recommend one of [1, 2, 4, 6].

To follow up on the SAR interferometry case study, the interested reader is encouraged to look at *InSAR Principles* from the European Space Agency, with a link from the textbook further reading page.

Sample Problems

Problem 11.1: Short Questions

(a) Define specular versus diffuse reflection.
(b) What does it mean for a satellite to have an *exact repeat* orbit?
(c) What is the problem of *phase unwrapping*?

Problem 11.2: Short Questions

(a) Define uniqueness
(b) Define existence
(c) What makes a linear inverse problem easier to solve than a non-linear one?
(d) What do we mean by a numerically stiff or ill-conditioned problem?

Problem 11.3: Analytical — Satellite Orbits

ERS1/2 were a "tandem pair" — two identical radar satellites flown at a precise orbital offset. Explain . . .

(a) The imaging principle behind such a pair — what does such a pair allow us to do, and how does it work?
(b) The applied principle behind such a pair — what are the earth systems phenomena that can be observed?

Problem 11.4: Analytical — Surface Reflection

(a) How is it that shampoo can be pretty much any colour (blue, green, orange, yellow, whatever) but the resulting froth of soap bubbles is always white?
(b) Everyone knows what a leaf (e.g., a maple leaf) looks like. Give an approximate description of the surface characteristics of a leaf.
(c) I have done a lot of painting (painting the walls in a house, not painting pictures). It is quite striking how wet paint is always shiny, but dry paint is not.
    (i) In terms of surface characteristics, what does "shiny" mean?
    (ii) Explain why it is that wet and dry surfaces normally have different reflection characteristics.

Problem 11.5: Analytical — Surface Reflection

(a) For each of specular and diffuse, give one example of a surface having such a reflection.
(b) Give two examples of surfaces having a reflection behaviour which is some sort of mix of specular and diffuse.
(c) Briefly, what are the physics principles underlying specular and diffuse reflections?

Problem 11.6: Analytical — Surface Reflection

I recently observed a Turkey Vulture (a large bird) flying overhead on a sunny day:

- At first, the front and back parts of the wing both looked black.
- As I watched it fly by, from another angle the front part of the wing stayed black, but the back part looked nearly white.

What can you say about the surface properties regarding the front and back parts of the wing?

Problem 11.7: Analytical — Diffuse Surfaces

On page 277 there appear to be two conflicting claims regarding diffuse surfaces:

- In the text, "a diffuse surface scatters light such that the surface appears equally bright at all angles."
- In (11.5), that $\rho(\psi_i, \psi_r) = \cos(\psi_r)$

That is, the former statement suggests no angular dependence, whereas the latter equation clearly shows an angular dependence.

Explain why the two claims are, in fact, equivalent.

Problem 11.8: Analytical — Resolution and Sensitivity

Your eye is essentially a remote sensing device.

Under reasonable assumptions, derive the maximum theoretical distance from which the word “STOP” can be read on a stop sign, assuming perfect eyesight.

State your assumptions and explicitly show your calculations.

Problem 11.9: Analytical — Uniqueness

Given a matrix

$$C = \begin{bmatrix} 3 & 1 \end{bmatrix}$$

(a) Argue why the inverse problem $m = C\underline{z}$ does not satisfy uniqueness.
(b) Give numerical values of m and $\underline{z}$ that demonstrate the failure of uniqueness.

Problem 11.10: Analytical — Existence

Given a matrix

$$C = \begin{bmatrix} 1 \\ 3 \\ 1 \end{bmatrix}$$

(a) Argue why the inverse problem $\underline{m} = Cz$ does not satisfy existence.
(b) Give numerical values of $\underline{m}$ and z that demonstrate the failure of existence.

Problem 11.11: Numerical/Computation — Phase Unwrapping

Interferometric data is available from textbook data site. The goal is using interferometric data is to *unwrap* it, per Figure 11.20. One of the challenges with unwrapping is that measurement noise causes the fringe edges to be hard to identify, and in steep terrain the fringes can be very close together.

Given a measured phase map ϕ_{xy} we wish to estimate the phase correction I_{xy}, where I_{xy} is a multiple of 2π, and $\phi + I$ is a smooth map of height.

In general, in comparing two adjacent phases ϕ_A, ϕ_B, if the phase values are similar then the phases are assumed to be in the same fringe, and if the phase values are somewhat different we assume there must be a fringe boundary in between:

$$\begin{aligned} |\phi_B - \phi_A| &< \pi & I_B &= I_A \\ \phi_B - \phi_A &< -\pi & I_B &= I_A + 2\pi \\ \phi_B - \phi_A &> \pi & I_B &= I_A - 2\pi \end{aligned}$$

So what we need is an origin, a place to start, and strategy for moving from one pixel to the next.

Attempt one or more strategies for unwrapping and report on the results:

1. Just unwrap each column separately, starting from the top.
2. Just unwrap each row separately, starting from the left.
3. Group pixels pairwise, ordered by reliability, first grouping those phase differences closest to 0 or to 2π.

Problem 11.12: Reading

Much has been written on data assimilation, particularly in the context of global climate models. Provide a brief overview, no more than one page, of the current state of data assimilation and the primary challenges in producing meaningful climate forecasts.

Possible places[7] to start:

- Hakim et al., "Overview of data assimilation methods," *PAGES news* (21) #2, 2013
- Kalnay, *Atmospheric Modelling, Data Assimilation …*, Cambridge, 2002
- Wunsch, *The Ocean Circulation Inverse Problem*, Cambridge, 1996
- Wunsch, *Inverse Methods, Inverse Problems, State Estimation, Data Assimilation, and All That* (online short course)

Problem 11.13: Policy and Sensing

Humans interact with and influence their environment in a wide variety of ways, many of which can be sensed. Examples include urban sprawl, land desertification, agriculture, and deforestation.

For the country in which you live, identify one of the dominant ecological issues, briefly describe the current governmental policies in place related to this issue, and describe what type of instrument or what form of sensing could be used to assess the issue.

References

1. R. Aster, B. Borchers, and C. Thurber. *Parameter Estimation and Inverse Problems*. Academic Press, 2005.
2. M. Bertero and P. Boccacci. *Introduction to Inverse Problems in Imaging*. Taylor & Francis, 1998.
3. A. Chandra and S. Ghosh. *Remote Sensing and Geographical Information System*. Alpha Science, 2015.
4. A. Tikhonov *et al.*. *Numerical Methods for the Solution of Ill-Posed Problems*. Kluwer Academic Publishers, 1995.
5. L. Kump *et al.*. *The Earth System*. Prentice Hall, 2010.
6. P. Fieguth. *Statistical Image Processing and Multidimensional Modeling*. Springer, 2010.
7. T. Lillesand, R. Kiefer, and J. Chipman. *Remote Sensing and Image Interpretation*. Wiley, 2015.
8. F. Mackenzie. *Our Changing Planet*. Prentice Hall, 2011.

[7]Links to most reading suggestions are available from the textbook reading questions page.

Chapter 12
Water

Much of the Western world takes purified tap-water for granted, assuming an indefinite supply at a price sufficiently low to allow us to use drinking water to flush toilets and irrigate golf courses.

However at the time of this writing, in 2016, water is increasingly making headlines, particularly as California faces a multi-year drought with Lake Mead, the source of water for seven southwest states and Mexico, at only 37 % of capacity.

In the case of Lake Mead, the challenge is political and legal. In 1922 the seven southwestern states (and later Mexico) drew up an agreement on how to share the water from the Colorado River: an allocation of approximately $9.3 \cdot 10^9\,\text{m}^3$ per year to the upper basin states, another $9.3 \cdot 10^9\,\text{m}^3$ to the lower basin states, and $1.9 \cdot 10^9\,\text{m}^3$ to Mexico. That is, $20.5 \cdot 10^9\,\text{m}^3$ per year of water has been promised, however the agreement was drafted at a time of high water flow, and the longer-term average flow of the Colorado River appears to be significantly less, closer to $17 \cdot 10^9\,\text{m}^3$ per year. With outflows exceeding inflows, the deplorable long-term decline of the Aral Sea comes to mind.

P. Fieguth, *An Introduction to Complex Systems*,
DOI 10.1007/978-3-319-44606-6_12

Although the agreement calls for water reductions as the lake level drops, the reduction is very modest, with a lower basin allocation from $9.3 \cdot 10^9\,\mathrm{m}^3$ reduced to $8.6 \cdot 10^9\,\mathrm{m}^3$ under extreme shortages. Since the "extreme shortage" allocation still exceeds some estimates of the *average* river flow, a crisis is all but guaranteed.

With hindsight, knowing that rainfall amounts are frequently power-law (Example 9.3), the optimistic allocation of Colorado River flow seems terribly naïve, and there will be no simple solutions to that particular predicament.

Far from being an isolated example, Lake Mead is only one of myriad contexts in which the spatial, power-law, and nonlinear systems explored in this text shed light on the geopolitical challenge of water. To be sure, water is just one of many worldwide complex systems topics. Indeed, the topics of energy, minerals, air, food, wealth, refugees, and carbon-dioxide could equally well have formed the basis of this chapter. Thus this chapter is not so much about water, as it is a framework for exploring a given topic of interest in the context of nonlinear/spatial/complex systems, and the interested reader is encouraged to re-imagine this chapter applied to any number of other topics.

Certainly water-related themes have appeared repeatedly throughout the text:

- At the *systems level* in global cycles, as discussed in Examples 3.7 and 8.4;
- In the illustrations of hystereses driven by *nonlinear systems*, in the lake-siltiness example of Scheffer in Example 6.6, and the global climate states of Figure 6.19 in Case Study 6;
- In the *power-law* distribution of water's appearance, such as rainfall (Example 9.3) and the California drought example at the beginning of this chapter.

Following on this latter point, the distribution of water on the planet certainly supports a power-law perspective, spread over four orders of magnitude:

Salty	98 %						
Fresh	2 %	→	Subsurface	99 %			
			Surface	1 %	→	Frozen	70 %
						Liquid	30 %

Water, being so fundamental to plant life, animal life, and chemistry, intersects with a wide variety of human, industrial, and extractive uses. That is, with respect to the water under, over, within, or around a country's land mass, there will be parallel individual, commercial, and political systems seeking to

Restrict access to it	Ownership, sovereignty
Travel on it	Military, commercial shipping
Pollute it	Industry, manufacturing
Harvest from within it	Fishing, aquaculture
Harvest from under it	Oil, gas
Control it	Electricity generation
Move it	Drinking water and irrigation

Given such a number of interacting systems, in turn further dependent on other spatial systems such as electricity, gas, and road infrastructures, based on the coupled–interdependence discussion in Section 10.4 it is almost certain that global water behaves as a highly complex system, with associated power-law and nonlinear state-transition challenges.

12.1 Ocean Acidification

Topics: Systems Theory, Nonlinear Systems, Inverse Problems

Ocean acidification is sometimes referred to as the lesser-known or evil twin of Global Warming, which we discussed in Chapter 2. Paralleling the asymmetric water input–output system of the Aral Sea, ocean acidification is due to carbon dioxide being absorbed into the ocean more quickly than carbon is removed, as illustrated in Figure 12.1.

Current estimates suggest that approximately one third of carbon dioxide released into the atmosphere is absorbed into the ocean, increasing the ocean concentration of carbonic acid. Over the past two hundred years the ocean pH has decreased by approximately 0.1, which sounds modest, but keeping in mind the logarithmic scale of pH units a 0.1 pH decline actually corresponds to a 30 % increase in acidity. Although the ocean possesses mechanisms to regulate pH level, the *rate* of carbon dioxide transfer into the ocean exceeds the response rate of regulation, so pH is expected to continue to drop.

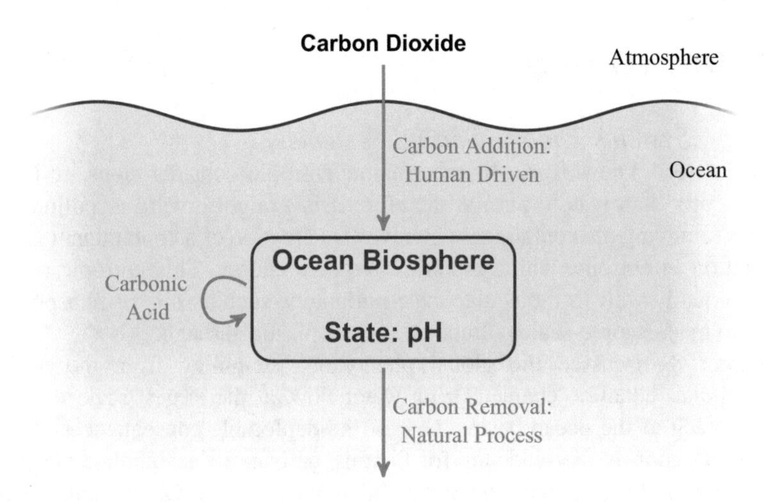

Fig. 12.1 Ocean acidification: the pH of the ocean is influenced by the rate at which CO_2 is absorbed from the atmosphere, relative to the rate at which CO_2 is sequestered to the deep ocean. The system is currently imbalanced, in that the input is controlled by human activities, whereas the output is not.

The key concern associated with acidification is its impact on calcifying organisms, such as corals or shellfish, since the rate at which calcium carbonate dissolves increases as pH drops. Since coral reefs and other calcifying creatures are so central to ocean life, the acidification issue is not a trivial question of lobster harvests, rather one with far reaching consequences on entire ocean ecosystems.

Acidification is a component of the global carbon cycle, driven by a system wide input–output mismatch, where our concern is a nonlinear response of the ocean ecosystem (e.g., coral bleaching and reef death), urgently necessitating global assessment and observation. Acidity measurement by research ships or moored buoys is technically straightforward, but prohibitively expensive to undertake globally; in contrast, satellite global observation of pH is challenging, since ocean acidity does not obviously affect electromagnetic radiation. There are multiple possibilities for indirect measurements, however:

- Inferring the partial pressure of CO_2 in seawater from measurements of sea-surface temperature (via infrared radiometry) and chlorophyll density (measuring biological activity, measured at visual wavelengths);
- Inferring pH from a combination of sea-surface temperature and ocean salinity (via microwave radiometry, since reflected radiation is a function of both temperature and salinity).

Ocean Acidification in *Wikipedia*
Q. Sun, D. Tang, S. Wang, "Remote-sensing observations relevant to ocean acidification," *IJRS* (33), 2012
S. Doney, "The Dangers of Ocean Acidification," *Scientific American*, 2006
R. Eisler, *Ocean Acidification: A Comprehensive Overview*, CRC Press, 2011
E. Kolbert, "The Acid Sea," *National Geographic*, 2011

12.2 Ocean Garbage

Topics: Systems Theory, Spatial Systems

As we well know from thermodynamic/entropic considerations, it is much easier to mix than it is to unmix, therefore it is straightforward to pollute water, whereas removing the contaminant involves the creation of a contaminant gradient, a reduction in entropy, which therefore requires energy. This entropic argument applies equally well to molecular-scale pollutants, such as oils or phosphates, as it does to macroscopic-scale pollutants, such as plastic garbage.

Chapter 8 discussed the global phenomena stemming from the nonlinear Navier-Stokes equation characterizing water flow, at the largest scale the natural gyres in each of the ocean basins. One of the deplorable consequences of plastic waste production is the tendency for floating garbage to accumulate and remain trapped within the gyre, with the largest being the so-called Great Pacific Garbage Patch.

There are widely varying estimates regarding the spatial extent, mass, and number of plastic pieces in the patch, however there is no doubt that the patch is enormous and poses significant ecological risks. Furthermore because plastic floats it remains near the surface, where most sea life is concentrated. The key concern is that plastic does not disappear or biodegrade; under the influence of sunlight it does break down, but into ever tinier plastic bits, right down to the molecular level, posing risks at all scales:

- *Large scale* plastics kill marine life, such as whales, seals, turtles by entrapment (discarded fishing nets) or by blocking the digestive system (water bottles or other large plastic pieces).
- *Mid scale* plastics, particularly pellets, kill marine and bird life since the pellets are confused with fish eggs, fed to chicks, which then die of starvation or organ rupture.
- *Fine scale* plastics may release pollutants, such as Bisphenol A, during the degradation process. Conversely, microscopic plastics may also absorb organic pollutants from the water, leading to toxic effects as the plastic pieces are ingested and concentrated in the food chain.

The ocean plastics problem is particularly challenging to address since the plastic pieces, being tiny and sub-surface, are essentially undetectable via satellite, consequently making it difficult to pinpoint the countries or industries responsible for the source of the garbage. There is growing interest in using SAR (synthetic aperture radar) to detect larger plastic pieces, or denser regions of aggregation, and InSAR (Interferometric SAR) to more accurately detail ocean currents, to better be able to model the dynamics of garbage accumulation.

Marine Debris, Great Pacific Garbage Patch in *Wikipedia*
L. Burns, *Tracking Trash: Flotsam, Jetsam, and the Science of Ocean Motion*, HMH Books, 2010
J. Decker, *Gyre: The Plastic Ocean*, Booth-Clibborn, 2014
C. Moore, C. Phillips, *Plastic Ocean*, Penguin, 2012
L. Parker, "Ocean Trash: 5.25 Trillion Pieces and Counting," *National Geographic*, 2015

12.3 Groundwater

Topics: Nonlinear & Spatial Systems, Power Laws, Inverse Problems

The catastrophe of the Aral Sea was already described at the opening of Chapter 4. At a high level, the Aral Sea system is straightforward:

Water Inflows: Through rivers, with diversion controlled by humans
Water Outflows: Through evaporation, determined by physics

The issue of groundwater depletion, also known as overdrafting and illustrated in Figure 12.2, very much parallels that of the Aral Sea:

Water Inflows: From rainfall, infiltrating through the ground, controlled by nature
Water Outflows: Up through drilled wells, controlled by humans

In those cases where humans control only one side of a system, such as the water system in groundwater or the carbon-dioxide in the ocean acidification example of

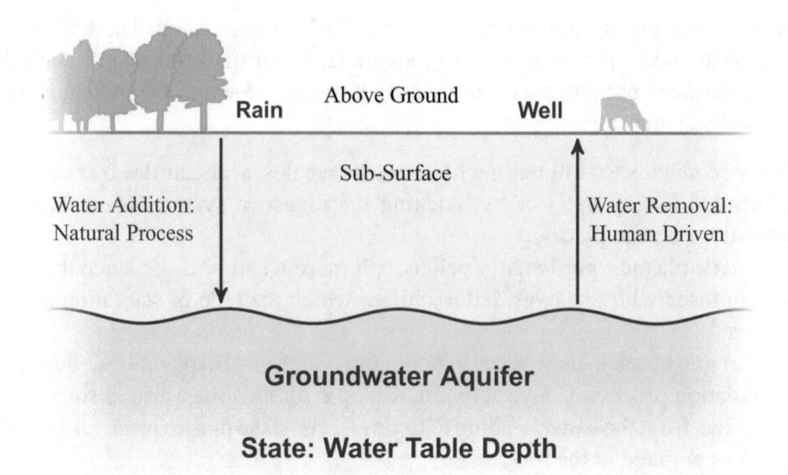

Fig. 12.2 Groundwater system: a groundwater aquifer, as a system, parallels the ocean acidification system of Figure 12.1: humans control the rate of water removal, however the physics of subsurface water flow control the rate of water addition.

Figure 12.1, we run into problems because of an inability to regulate the human behaviour, on the one side of the system, to match the natural flow on the other. As a result, there are serious underground aquifer depletion issues worldwide, some of the most famous being in India, California, and the Ogallala Aquifer in the central USA.

Assessing the rate of aquifer recharge is only one aspect of *Groundwater Hydrology*, the study of the sub-surface flow of water, a class of fascinating and challenging inverse problems. One of the key applications of groundwater hydrology is to contaminant flow, determining how fast/how far a surface contamination, for example from an industrial site or a heating-oil tank at a summer cottage, travels underground, ultimately contaminating streams or wells.

Groundwater hydrology is truly a remarkably difficult problem: The problem is very much power-law, in that water flow rates can vary by 12 or more orders of magnitude between gravel and granite, as outlined in Figure 12.3. Because pressure gradients are modest, and with limited permeability κ, the velocity of subsurface water flow is very low, typically on the order of cm per day. Therefore, rather than the Navier–Stokes description of above-surface water flow from (8.19) in Chapter 8, the low velocities prevent any sort of turbulence, and flow is described in terms of

$$\text{Velocity} \propto \kappa \frac{\Delta P}{\Delta x}, \tag{12.1}$$

similar to the diffusive flow of (8.13).

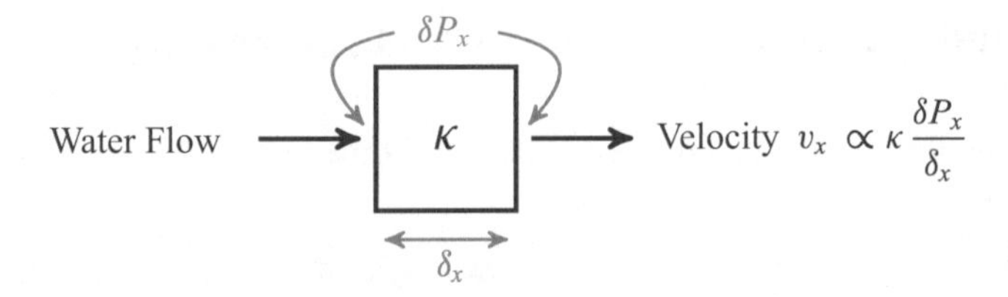

Examples of permeability κ in units of millidarcys:

Granite	Clay	Sand	Gravel
$10^{-4} - 10^0$	$10^0 - 10^2$	$10^4 - 10^6$	$10^5 - 10^8$

Fig. 12.3 Groundwater flow is characterized by a diffusion equation, driven by pressure gradients δP. The velocity of water flow is controlled by the material permeability κ, which can vary over twelve or more orders of magnitude.

Given a three-dimensional map of permeability κ, the forward model from κ to water flow is comparatively straightforward. In contrast, the inverse problem from measurements to κ is terribly challenging:

1. The problem is highly nonlinear: A tiny state perturbation, such as the introduction of a thin clay layer within a sand or gravel basin, can have large consequences on the velocity and direction of water flow.
2. The problem is unusually undersampled: Measurements cannot be acquired remotely, so *in-situ* measurements must be taken by drilling into the ground or following trace elements introduced into the groundwater, therefore measurements are very expensive and necessarily few in number.

Nevertheless, the highly related problem of oil reservoir characterization, which also requires the estimation of three-dimensional permeability, ensures a great deal of interest and motivation in such inverse problems.

Although remote satellite-based measurements may not help in estimating κ, remote measurements can tell us a great deal about a groundwater system:

Groundwater Inputs can be quantified by measuring rain (Doppler radar) and soil moisture (microwave radiometry).

Groundwater Outputs can be assessed on the basis of the extent of irrigated agricultural activity (visible wavelengths) and soil moisture.

The Aquifer State, in particular the degree to which there is overdrafting, can be directly observed via land subsidence (sinking), which can be measured using Differential InSAR.

Groundwater Recharge, Overdrafting, Ogallala Aquifer, Hydrogeology in *Wikipedia*
L. Brown, "Aquifer Depletion," *Encyclopedia of Earth*, 2013
M. Rosegrant, X. Cai, *World Water and Food to 2025*, IFPRI, 2002
K. Hiscock, V. Bense, *Hydrogeology: Principles and Practice*, Wiley, 2014
F. Schwartz, H. Zhang, *Fundamentals of Ground Water*, Wiley, 2002
D. Todd, L. Mays, *Groundwater Hydrology*, Wiley, 2004

Case Study 12: Satellite Remote Sensing of the Ocean

Fundamentally, a space-based satellite can really measure only three things of the ocean:

Altimetry: Send a signal and measure the length of time until it returns.
Scatterometry: Send a signal and measure the strength of its return.
Radiometry: Send no signal and just passively observe the surface brightness.

Clearly all of these can be measured as a function of location, time of day, and wavelength.

Since there are, however, a great many things which we would *like* to measure—temperature, salinity, ocean currents, tides, acidity, dissolved CO_2, biological activity—the issue is one of formulating the appropriate forward and prior models to allow the inference of the ocean state of interest:

Altimetry measures the height of the ocean surface, which is therefore a function of the gravitational equipotential, of tides, of ocean currents, and of ocean temperature.
Scatterometry measures the strength of a radar signal return. This returned echo is a strong function of surface roughness, therefore permitting an inference of waves and surface winds, and also a function of salinity. Interferometric SAR, also based on radar returns, can allow a reconstruction of ocean topography from which ocean currents can be deduced.
Radiometry measures reflected/emitted radiation, a function of temperature, salinity, and surface roughness.

Because of the thermal brightness of the earth, active altimetric or scatterometric measurements are possible only in the microwave/radar (long wavelength) part of the spectrum, where the earth is dark, since at other wavelengths the earth's brightness would simply overwhelm the transmitted signal.

From Section 11.3 we know that the measured footprint on the surface will be an increasing function of wavelength,

Microwave: Low resolution (25 km)
Infrared: Mid resolution (1–5 km)
Visible: High resolution (50 m)

however there is *also* a wavelength-dependence on *what* is being measured:

Microwave: Surface temperature and salinity, insensitive to clouds and atmosphere,
Infrared: Surface temperature, sensitive to clouds and atmosphere,
Visible: Not a function of temperature, sensitive to clouds.

Finally, the definition of the ocean "surface" becomes problematic, since the depth of electromagnetic penetration is similarly wavelength dependent:

Microwave: Senses surface to a depth of 1–2 mm,
Infrared: Senses the surface to a depth of 10–20 μm,
Visible: Penetrates into the water column, half of the light reaching 1 m,

thus ironically the surface of Mars is better imaged than the 71 % of the Earth which is covered in water.

The concept of surface is a significant challenge in observing the ocean, since satellite radiometers measure only the upper *skin*, moored instruments measure the near-surface *bulk*, and altimeters are sensitive to the integrated ocean temperature over all depths. In response to sunlight the skin will clearly warm much more quickly than the bulk, and will also cool more quickly at night. Finally the skin-bulk temperature difference will depend greatly on mixing/agitation, which is a strong function of wind and waves.

In summary, nearly every aspect of the water cycle can be measured: infrared measurements to infer sea-surface temperature, which determines evaporation; infrared measurements to infer atmospheric humidity; infrared and visible measurements to infer clouds; Doppler radar to observe the presence and strength of precipitation; and microwave radiometry to observe the resulting soil moisture. Nevertheless the majority of the ocean is very poorly sampled, and it is only through significant research effort that the interior of the ocean can be inferred from surface measurements via inverse methods.

Further Reading

The references may be found at the end of each chapter. Also note that the textbook further reading page maintains updated references and links.

A great many books have been written on water issues, the water cycle, and the world's oceans. Certainly the topic of water is becoming increasingly prevalent, due to droughts, concerns of freshwater availability in developing countries, water being used as a political tool, and a growing perception that the oceans, once thought impossibly large, are indeed not so vast that humans cannot have significant deleterious effects.

Two books looking at the topic of water, comprehensively, are those of De Villiers [7] and Solomon [6].

Two further books focusing more specifically on the oceans, and really offering an outstanding overview of the history of the oceans and the major issues facing them, are the two books [3, 4] by Roberts, *An Unnatural History of the Sea* and *The Ocean of Life*.

In terms of the complex systems theme of this text, two texts which examine water from a related perspective includes the book by Scheffer [5], which has been mentioned frequently in this text, and the book on weather and climate [2] by Lovejoy and Schertzer.

Sample Problems

Problem 12.1: Reading

The beginning of this chapter claimed that any of a wide variety of topics, such as

energy, minerals, air, food, wealth, refugees, carbon-dioxide

could be examined through the perspective of two or more of the systems-related topics discussed in this text. Select one of the above topics, or a topic of your choice of significant environmental, social, and/or geopolitical significance, and describe how this topic relates to or is influenced by two or more of the chapters in this text, at a level of detail similar to that of the Ocean Acidification, Ocean Garbage, and Groundwater examples developed in this chapter.

References

1. S. Grace. *Dam Nation: How Water Shaped the West and Will Determine Its Future*. Globe Pequot Press, 2013.
2. S. Lovejoy and D. Schertzer. *The Weather and Climate: Emergent Laws and Multifractal Cascades*. Cambridge, 2013.
3. C. Roberts. *An Unnatural History of the Sea*. Harper Collins, 2009.
4. C. Roberts. *The Ocean of Life*. Viking, 2012.
5. M. Scheffer. *Critical transitions in nature and society*. Princeton University Press, 2009.
6. S. Solomon. *Water: The Epic Struggle for Wealth, Power, and Civilization*. Harper Collins, 2009.
7. M. De Villiers. *Water: The Fate of Our Most Precious Resource*. McClelland & Stewart, 2003.
8. A. Weisman. *The World Without Us*. Picador, 2007.

Chapter 13
Concluding Thoughts

The topics of this book are meant to go well beyond mathematical equations, to allow the reader to obtain some insights into broader socio-enviro-political events, of which a few are presented here as illustration.

At the time of this writing, in 2016, there have been significant concerns regarding droughts (weather power-law); in some places these concerns may be a matter of lifestyle and convenience, in others a matter of crisis and starvation.

The recent news and discussion surrounding the Trans-Pacific Partnership (TPP) reflects issues around coupled economic systems (complex interconnected systems) which are, based on some political perspectives, not connected *enough*, or for other people, already *too* connected and increasingly fragile.

The Arab Spring of 2010/2011 clearly illustrated an example of an unanticipated, emergent phenomenon stemming from an interdependent complex system. An emergence is, to be sure, not always positive or desirable; the current Syrian refugee crisis (to be sure, influenced by many factors, including a great deal of political and military meddling) certainly makes it very clear how quickly a formerly comparatively stable system/apparently resilient country, such as Syria, such can trip (nonlinear state transition) into an undesirable state.

Indeed, the topic of resilience has been a recurring theme throughout this book:

Resilience I:	Linear Systems:	Based on return dynamics
Resilience II:	Nonlinear Systems:	Based on basin geometries
Resilience III:	Complex Systems:	Based on criticality distance

Finally, the deplorable disaster of the Fukushima reactor complex in 2011 is one which captures many of the complex systems perspectives discussed in this text:

Power Laws: The earthquake and resulting tidal wave precipitated the initial damage. Both earthquakes and tidal waves are governed by power-law distributions, for which it is exceptionally challenging to plan, since it is very difficult to know the size or likelihood of extreme events.

P. Fieguth, *An Introduction to Complex Systems*,
DOI 10.1007/978-3-319-44606-6_13

Complex Systems: Whether the further disaster could have been avoided, with hindsight, may be difficult to say. However the confusion at the time was certainly compounded by interdependent complex systems, involving an interplay of electrical and communications infrastructures with human error and cultural influences, leading to cascading failures.

Nonlinear Systems: The actual reactor core melt and resulting hydrogen explosion were due to nonlinear factors, taking the reactor well outside its intended operating range, introducing a massive hysteresis (destruction).

Spatial Systems: The spread and distribution of radiation is dominated by wind and ocean currents, which can be simulated in global climate models. The radiation itself cannot be sensed remotely, however winds can be measured by radar scatterometry, and ocean currents by altimetry and InSAR techniques.

Inference: Finally, in-situ measurements of radiation can be taken and combined with ocean-atmospheric models, leading to an inverse problem which can allow estimates of radiation exposure to be produced.

The intent here is not to paint a depressing picture of complex power-law systems which defy predictability, leaving us to the whims of fate. Rather, the intent is that we require a degree of humility in working with complex systems, recognizing that the precautionary principle, a greater margin of safety, may be required to avoid the hubris of human over-confidence in pushing systems too close to the edge of catastrophic state transitions, unrecoverable hysteresis loops, unanticipated power-law extrema, or complex cascading failures.

Further Reading

The references may be found at the end of each chapter. Also note that the textbook further reading page maintains updated references and links.

The goal of this text has been to provide a broad, interdisciplinary perspective on complex and nonlinear dynamic systems.

My hope is that, having read this text, most readers would be interested in following up in greater depth on some of the specific topics covered within this text, and suggestions for further reading in many of these topics have, of course, been offered in their respective chapters.

For readers interested in other similarly-broad systems-level discussions of ecological and social issues, I would suggest the six books listed below in the references [1–6].

References

1. F. Capra and P. Luisi. *The Systems View of Life: A Unifying Vision*. Cambridge, 2014.
2. A. Ghosh. *Dynamic Systems for Everyone: Understanding How Our World Works*. Springer, 2015.
3. M. Mitchell. *Complexity - A Guided Tour*. Oxford, 2009.
4. M. Scheffer. *Critical transitions in nature and society*. Princeton University Press, 2009.
5. A. Weisman. *The World Without Us*. Picador, 2007.
6. R. Wright. *A Short History of Progress*. House of Anansi Press, 2004.

Appendices

Appendix A
Matrix Algebra

A matrix is really nothing more than a collection of numbers, a way of grouping together longer mathematical expressions into a simpler representation. For example, one of the most fundamental mathematical representations is a linear system of equations,

$$\begin{aligned} 3x_1 + 2x_2 &= 8 \\ 2x_1 - 2x_2 &= 2 \end{aligned} \tag{A.1}$$

where we have two equations and two unknowns, x_1 and x_2. The linear system of (A.1) can be equivalently represented in matrix–vector form, as

$$\underbrace{\begin{bmatrix} 3 & 2 \\ 2 & -2 \end{bmatrix}}_{A} \underbrace{\begin{bmatrix} x_1 \\ x_2 \end{bmatrix}}_{\underline{x}} = \underbrace{\begin{bmatrix} 8 \\ 2 \end{bmatrix}}_{\underline{b}} . \tag{A.2}$$

Here A is a 2×2 square matrix, however there are, as illustrated in Figure A.1, many possibilities for A, depending on the linear system being represented:

- A could be an $n \times n$ square matrix;
- A could be an $m \times n$ rectangular matrix;
- A could be dense, meaning that most matrix entries are non-zero;
- A could be sparse, in that only few entries are non-zero;
- A could be diagonal, such that all matrix elements are zero, except the elements A_{ii} along the diagonal;

where $A_{i,j}$ refers to the scalar element in row i and column j of matrix A.

P. Fieguth, *An Introduction to Complex Systems*,
DOI 10.1007/978-3-319-44606-6

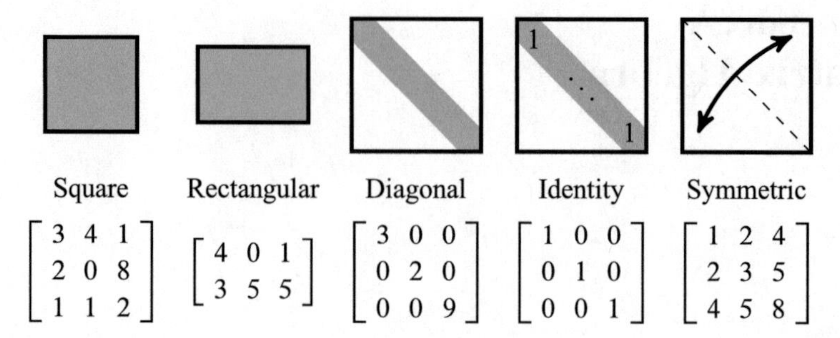

$$\begin{bmatrix}3&4&1\\2&0&8\\1&1&2\end{bmatrix} \quad \begin{bmatrix}4&0&1\\3&5&5\end{bmatrix} \quad \begin{bmatrix}3&0&0\\0&2&0\\0&0&9\end{bmatrix} \quad \begin{bmatrix}1&0&0\\0&1&0\\0&0&1\end{bmatrix} \quad \begin{bmatrix}1&2&4\\2&3&5\\4&5&8\end{bmatrix}$$

Fig. A.1 Five very basic matrix types, in cartoon form (top) and given simple examples (bottom).

The representation in (A.2) makes sense only if we understand the principle of matrix–vector and matrix–matrix multiplication, in which we take dot products of rows on the left with one or more columns on the right, thus

$$\begin{bmatrix}3&2\\2&-2\end{bmatrix}\begin{bmatrix}1\\2\end{bmatrix}=\begin{bmatrix}3\cdot1+2\cdot2\\2\cdot1+-2\cdot2\end{bmatrix}=\begin{bmatrix}7\\-2\end{bmatrix} \tag{A.3}$$

or, in general,

$$C = A\cdot B \qquad \longrightarrow \qquad C_{i,j}=\sum_p A_{i,p}\cdot B_{p,j} \tag{A.4}$$

To take a dot product, the number of elements in the left-row and right-column must be equal, therefore in general the matrix dimensions must obey

$$[k \times m] \cdot [m \times n] = [k \times n] \tag{A.5}$$

for matrix dimension parameters k, m, n. As a result, if we can multiply $A\cdot B$ we may *not* be able to multiply $B\cdot A$; furthermore even if both products are valid, in general

$$A\cdot B \neq B\cdot A. \tag{A.6}$$

Matrix addition and subtraction are simpler than multiplication, and proceed element by element:

$$C=\alpha A+\beta B \qquad \longrightarrow \qquad C_{i,j}=\alpha A_{i,j}+\beta B_{i,j} \tag{A.7}$$

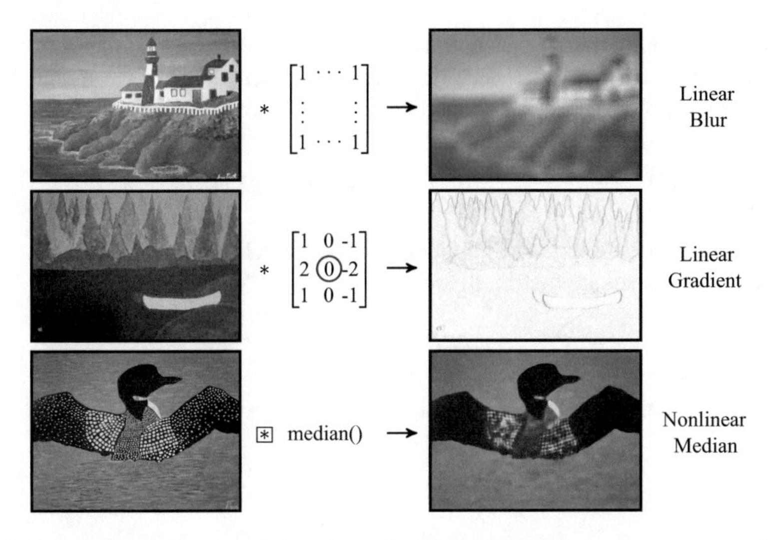

Fig. A.2 Convolution: the convolution shown in (A.10) allows the compact representation of stationary linear transformations, such as the blurring or edge-detection operations common in image processing. If the transformation is nonlinear, such as a median filter, the transformation may still be local and stationary. We need to know the location of the origin in the convolutional kernel a of (A.10); if there is any ambiguity that element will be circled, as shown here in the middle row.

Given a matrix A, the most fundamental operation on A is its *transpose*, essentially a mirror image,

$$\text{Transpose}(A) = A^T \qquad \longrightarrow \qquad A^T_{i,j} = A_{j,i} \tag{A.8}$$

such that a matrix is said to be *symmetric* if $A^T = A$.

The matrix operation $\underline{y} = A\underline{x}$ is essentially a linear transformation

$$\underline{x} \xrightarrow{A} \underline{y} \tag{A.9}$$

However if $\underline{x}$ represents a large map (for example, of ocean surface temperature), and if the transformation implied by A is *stationary*, meaning that the nature of the transformation is unchanged from one location to another, then the transformation of (A.9) is more compactly expressed as a convolution

$$y = x * a \qquad \longrightarrow \qquad y_{i,j} = \sum_{g=-m}^{m} \sum_{h=-m}^{m} a_{g,h} x_{i-g,j-h} \tag{A.10}$$

such that transformation A is implicitly represented via a local $2m + 1 \times 2m + 1$ kernel a, where $*$ represents the convolution operation. Convolution is illustrated in Figure A.2 for the two-dimensional case, but is in principle applicable to 3D volumetric (x, y, z) or 4D space–time (x, y, z, t) problems. Now the convolution (A.10), as written, applies strictly to linear transformations; however the local–kernel *concept* of (A.10) certainly generalizes to the nonlinear case

$$y = x \circledast \phi() \quad \longrightarrow \quad y_{i,j} = \phi\Big(\big\{x_{i-g,j-h}, |g| \le m, |h| \le m\big\}\Big) \tag{A.11}$$

for some nonlinear function $\phi()$ applied to a local patch of values in x.

Next, if A is a square matrix, then

$$\text{Inverse}(A) = A^{-1} \quad \longrightarrow \quad A \cdot A^{-1} = A^{-1} \cdot A = I. \tag{A.12}$$

The matrix inverse of A is essentially the reversed transformation, such that

$$\text{If} \quad \underline{x} \xrightarrow{A} \underline{y} \quad \text{then} \quad \underline{y} \xrightarrow{A^{-1}} \underline{x}. \tag{A.13}$$

For all rectangular matrices[1] and many square matrices there is no inverse, and a matrix which cannot be inverted is said to be *singular*. The singularity of a matrix A can be assessed by its *determinant* $\det(A)$:

$$\begin{aligned} A \text{ singular} \quad &\longleftrightarrow \quad \det(A) = 0 \\ A \text{ invertible} \quad &\longleftrightarrow \quad \det(A) \ne 0 \end{aligned} \tag{A.14}$$

where

$$\begin{gathered} \det([a]) = a \qquad \det\left(\begin{bmatrix} a & b \\ c & d \end{bmatrix}\right) = ad - bc \\ \det\left(\begin{bmatrix} a & b & c \\ d & e & f \\ g & h & i \end{bmatrix}\right) = a(ei - fh) - b(di - fg) + c(dh - eg) \end{gathered} \tag{A.15}$$

Certainly the determinant can be defined for larger matrices, but it is notationally cluttered and offers little insight. Intuitively, $\det(A)$ sort of measures the "size" of A or the overall degree of amplification associated with transformation A. It therefore stands to reason, from (A.13), that for A^{-1} to represent the inverse transformation of A it must undo the amplification of A, from which it becomes clear that

$$\det\big(A^{-1}\big) = \frac{1}{\det(A)} \tag{A.16}$$

[1] There is the related notion of a *pseudo-inverse*, which does define a sort of inverse for rectangular matrices.

If $\det(A) = 0$ then the effect of transformation A is to multiply by zero in some direction; therefore in the same way that we cannot divide by zero, similarly we cannot invert a matrix having a determinant of zero.

If a linear system, such as in (A.2), is characterized by an invertible matrix A, then the system can be uniquely solved as

$$A\underline{x} = \underline{b} \qquad \longrightarrow \qquad \underline{x} = A^{-1}\underline{b}. \tag{A.17}$$

In general it is challenging to compute A^{-1} by hand, however this text is not concerned with the numerical methods behind such matrix operations, and it is assumed that the reader will use numerical software such as Matlab or Octave.

The *eigendecomposition* of a square[2] matrix A is one of the most powerful tools in linear algebra. The eigendecomposition essentially reveals those key vectors, the eigenvectors, which do not change direction when transformed (multiplied) by A. In particular, $\underline{v}$ is an eigenvector of A if

$$A\underline{v} = \lambda\underline{v}, \tag{A.18}$$

such that $\underline{v}$ has not changed direction, only length, where the change in length is captured by the *eigenvalue* λ. Formally, the eigenvalues of A are found as the roots of the *characteristic polynomial* of A

$$\det(A - \lambda I) = 0. \tag{A.19}$$

In general, for an $n \times n$ matrix A there will be n eigenvalues $\lambda_1, \dots, \lambda_n$, which relate to A as

$$\det(A) = \prod_{i=1}^{n} \lambda_i \qquad \mathrm{trace}(A) = \sum_{i=1}^{n} A_{ii} = \sum_{i=1}^{n} \lambda_i, \tag{A.20}$$

from which it is clear that any singular matrix A, meaning that $\det(A) = 0$, must be associated with at least one eigenvalue equal to zero.

For a given eigenvalue, the associated eigenvector is found from (A.18) by solving the linear system

$$\left(A - I\lambda_i\right)\underline{v}_i = 0 \tag{A.21}$$

although, as with matrix inversion, in practice we will rely on numerical software to compute the eigendecomposition for us.

[2]There is the related concept of the singular value decomposition, which is essentially the generalization of eigendecompositions to rectangular matrices, a powerful concept but beyond the scope of this text.

Since it is only the direction of the eigenvector which matters, we will always assume eigenvectors to be normalized to unit length, that is[3]

$$\sum_i v_i^2 = 1. \tag{A.22}$$

It is frequently convenient to collect the eigenvalues/eigenvectors into matrices as

$$A\underline{v}_i = \lambda_i \underline{v}_i, \quad 1 \le i \le n \quad \longrightarrow \quad V = \begin{bmatrix} | & & | \\ \underline{v}_1 & \cdots & \underline{v}_n \\ | & & | \end{bmatrix} \quad \Lambda = \begin{bmatrix} \lambda_1 & & \\ & \ddots & \\ & & \lambda_n \end{bmatrix} \tag{A.23}$$

where V is invertible, in which case

$$A = V\Lambda V^{-1} \qquad A^{-1} = V\Lambda^{-1}V^{-1} \qquad A^n = V\Lambda^n V^{-1}, \tag{A.24}$$

where the inverse Λ^{-1} or powers Λ^n are trivial to compute because Λ is diagonal.

The simplest sorts of systems are *decoupled* ones, where the elements of the system do not interact. To be sure, in general systems are *coupled*, however one of the key uses of the eigendecomposition is to undo such coupling, to simplify analysis. If $A_{i,j}$ describes the interaction of elements z_i and z_j, then a decoupled problem is one characterized by a diagonal matrix. The systems concept of decoupling is thus equivalent to the mathematical operation of matrix diagonalization.

Given a dynamic system $\dot{z}(t) = Az(t)$, suppose we apply a non-singular transformation T to $\underline{z}$

$$\underline{y} = T\underline{z} \qquad \dot{\underline{y}} = T\dot{\underline{z}} \tag{A.25}$$

such that the dynamic system transforms as

$$\dot{z} = Az \quad \rightarrow \quad T^{-1}\dot{\underline{y}} = AT^{-1}\underline{y} \quad \rightarrow \quad \dot{\underline{y}} = TAT^{-1}\underline{y}. \tag{A.26}$$

From (A.26) it is clear that we seek T such that the product TAT^{-1} is a diagonal matrix. Matrix A is said to be *diagonalizable* if such a transformation T exists; furthermore, the diagonalizing transformation T is given by the eigenvectors of A:

$$T = V^T \qquad T^{-1} = V \tag{A.27}$$

[3] Note that we need to be careful with indexing; $\underline{v}_i$ means the ith vector in a sequence, whereas v_i means the ith element of vector $\underline{v}$.

where, remarkably, the inverse of T is simply its transpose, a special property of what are known as *orthogonal* matrices.

Not all matrices are diagonalizable, thus not all systems can be decoupled. The best we can do in such situations is to maximally decompose a system into its non-interacting parts. Non-diagonalizability occurs when the characteristic polynomial of (A.19) has repeated roots (eigenvalues) which share a common eigenvector. Repeated eigenvalues with a shared eigenvector imply a higher-order behaviour which cannot be decoupled into separated, low-order pieces; such a system can, at best, be reduced to a block-diagonal representation known as a *Jordan form*:

$$\begin{bmatrix} \boxed{\lambda_1} & & & & & \\ & \boxed{\lambda_2} & & & & \\ & & \boxed{\lambda_2} & & & \\ & & & \lambda_3 & 1 & \\ & & & & \lambda_3 & 1 \\ & & & & & \lambda_3 \end{bmatrix} \tag{A.28}$$

In this case the root λ_2 (red) has a so-called multiplicity of two, but is still diagonalizable (because of unique eigenvectors). In contrast λ_3, which has a multiplicity of three, has a single common eigenvector, giving rise to the blue third-order block.

The details of the Jordan form are not so important to us, particularly since non-diagonalizable matrices are degenerate, in that even an infinitesimal perturbation of the matrix values will change the Jordan form. The emphasis here is on understanding the conceptual connections between matrix algebra and systems coupling.

Further Reading

There are many books on linear algebra, and the interested student will find a wide selection at Library of Congress call number QA251 at any library.

Three initial suggestions to get you started would include the books by Poole [3], Lay et al. [2], and Damiano et al. [1].

References

1. D. Damiano and J. Little. *A Course in Linear Algebra*. Dover, 2011.
2. D. Lay, S. Lay, and J. McDonald. *Linear Algebra and Its Applications*. Pearson, 2014.
3. D. Poole. *Linear Algebra: A Modern Introduction*. Brooks Cole, 2014.

Appendix B
Random Variables and Statistics

There are a great many quantities which have some degree of uncertainty: the roll of a die, a randomly chosen playing card, or tomorrow's weather. In all cases, a value x having some uncertainty is referred to as a random variable. To be sure, x being random does *not* mean that nothing is known about x; to the contrary, we may very well have detailed information or *prior knowledge* about x, such that some values of x are known to be more likely than others.

The most complete description of x is given by its probability distribution. If x describes *discrete* events, such as the six sides of a die, then the probability distribution is just a list of events and their respective probabilities:

$$\text{Die Probability} \quad P(x) = \begin{cases} 1/6 & x \in \{1, \ldots, 6\} \\ 0 & \text{Otherwise} \end{cases} \tag{B.1}$$

Since x must take on some value, the sum of all of the probabilities must equal one.

In contrast, if x describes a *continuous* distribution, such as adult human height, we can no longer talk about the *probability* of x, since the probability that an adult has a height of *exactly* 164.5246897501 cm is zero. Rather the probability *distribution* $p(x)$ is defined as

$$\text{Probability}(x < \alpha) = \int_{-\infty}^{\alpha} p(x)\, dx \tag{B.2}$$

Perhaps a slightly simpler description is that

$$\text{Probability that } x \text{ lies within a distance } \beta \text{ of } \alpha = \int_{\alpha-\beta}^{\alpha+\beta} p(x)\, dx \tag{B.3}$$

P. Fieguth, *An Introduction to Complex Systems*,
DOI 10.1007/978-3-319-44606-6

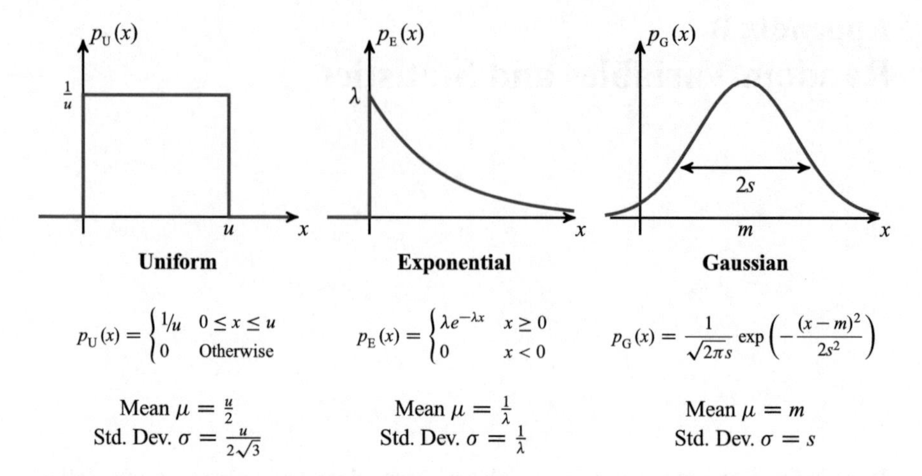

Fig. B.1 Probability distribution functions: the figure shows mathematical definitions and representative illustrations for three distributions. In all cases x is a random variable, whereas u, λ, μ, σ are deterministic parameters controlling the width (u, λ, s) or location (m) of the distribution.

As before, since x must take on some value, it is now the *integral* of the probability distribution which must equal one:

$$\int_{-\infty}^{+\infty} p(x)\, dx = 1. \tag{B.4}$$

Three of the most common probability distribution functions are illustrated in Figure B.1.

A probability distribution is often an inconvenient description of a random variable to be useful, normally for one of two reasons:

1. The equation of the probability distribution for x is much too complicated, or
2. We don't actually know the probability distribution of x.

Instead, in most cases we will want to work with much simpler notions of distribution:

- The *mean* or average is a simple measure of central tendency,

$$\text{Mean}(x) = \mu_x = \int x \cdot p(x)\, dx \tag{B.5}$$

- The *variance* is a measurement of variability or spread, calculated as the typical squared deviations of random variable x from its mean:

$$\text{Variance}(x) = \sigma_x^2 = \int \left(x - \mu_x\right)^2 \cdot p(x)\, dx \tag{B.6}$$

Because the variance is in squared units, it is frequently much simpler to think about the spread of the distribution in terms of the *standard deviation* σ_x, the square-root of the variance, since σ_x is measured in the same units as x.

Thus, per the notation of Appendix C, it will be much more compact to refer to $x \sim (\mu, \sigma^2)$ to mean that random variable x has mean μ and variance σ^2. The rules for manipulating means and variances are very simple; given independent (completely unrelated) random variables x and y,

$$x \sim (\mu_x, \sigma_x^2) \qquad y \sim (\mu_y, \sigma_y^2) \tag{B.7}$$

then for any constant α,

$$\begin{aligned} \text{Mean}(\alpha x) &= \alpha \mu_x & \text{Mean}(x+y) &= \mu_x + \mu_y \\ \text{Variance}(\alpha x) &= \alpha^2 \sigma_x^2 & \text{Variance}(x+y) &= \sigma_x^2 + \sigma_y^2 \end{aligned} \tag{B.8}$$

Two further notions of distribution, which we will not need very often, are included here for completeness:

- The *skewness*

$$\text{Skewness}(x) = \gamma_x = \int \left(\frac{x - \mu_x}{\sigma_x} \right)^3 \cdot p(x)\, dx \tag{B.9}$$

  assesses asymmetry in the distribution, whether the distribution has a long extension or tail on only one side, for example. Thus for the examples in Figure B.1, the uniform and Gaussian distributions both have zero skewness, however the exponential distribution is clearly skewed, with $\gamma > 0$.
- The *kurtosis*

$$\text{Kurtosis}(x) = K_x = \int \left(\frac{x - \mu_x}{\sigma_x} \right)^4 \cdot p(x)\, dx \tag{B.10}$$

  assesses the heaviness of the tails of the distribution, a topic of great significance in Chapter 9, since the heaviness of the tail measures the likelihood of extreme outliers. The kurtosis corresponding to a few familiar distributions is as follows:

Uniform Distribution	No tails at all	$K = 1.8$
Gaussian Distribution	Thin tails	$K = 3$
Power-law Distribution	Heavy tails	$K \gg 3$

Certainly higher moments (integrals of higher powers of x) are possible, however the interpretation of such moments becomes increasingly unclear. Indeed, in general even the skewness and kurtosis can be difficult to interpret, so in many cases it is only the mean and the variance which are calculated.

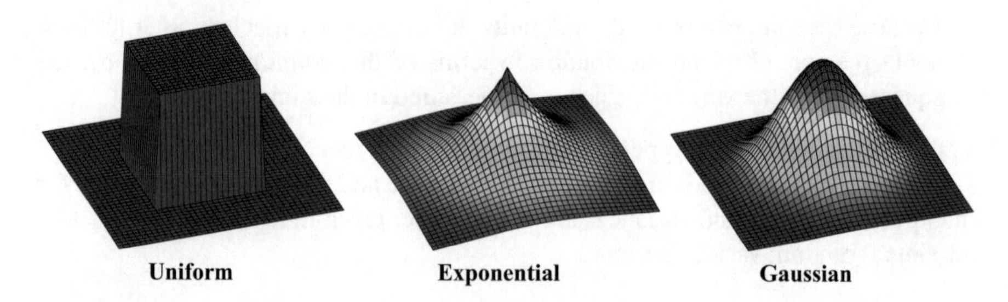

Fig. B.2 Probability distribution functions: the one-dimensional distributions defined in Figure B.1 can all be generalized to multi-dimensional distributions corresponding to random vectors. In all cases, the two-dimensional function shows the probability distribution (height) as a function of two random variables.

Frequently we have some information about a random variable, leading to a *conditional statistic* $(x \mid y)$, read as x *given* y, meaning that random vector x is now constrained by the information present in y. The information in y can take essentially any form: a measurement of x, a constraint on x, or information on another random variable that may or may not be related to x:

(T)	Random variable describing a temperature
$(T \mid T \geq 0°\text{C})$	T, given an inequality constraint
$(T \mid m = 20°\text{C} \pm 0.5°\text{C})$	T, given a noisy measurement
$(T \mid \text{Waterloo, Canada})$	T, given a geographic context
$(T \mid \text{Nov.21/2016, 3:45pm})$	T, given a temporal context
$(T \mid \text{Undergraduate GPA})$	T, given irrelevant information

The conditional statistic $(x \mid y)$ is still a random variable, and therefore we can talk about its probability density $p(x \mid y)$, its mean $\mu_{x|y}$, its variance $\sigma^2_{x|y}$ etc. If the information in y is completely irrelevant to or unrelated to x, then the conditioning has no effect:

$$p(x \mid y) = p(x) \qquad \mu_{x|y} = \mu_x \qquad \sigma^2_{x|y} = \sigma^2_x \quad \text{etc.} \tag{B.11}$$

In many cases we don't just have one random element, but several. If stacked into a vector

$$\underline{x} = \begin{bmatrix} x_1 \\ \vdots \\ x_n \end{bmatrix} \tag{B.12}$$

we refer to $\underline{x}$ as a *random vector*. In principle we can generalize the probability densities of Figure B.1 to the random vector case, as sketched in Figure B.2. Although the plots offer a nice intuitive sense of the dependence of the probability distribution on multiple variables, in practice the probability distribution equations will be even more difficult to work with than before for individual random variables.

Instead, as before, we prefer the simpler concepts of mean and variance, but now generalized to a random vector. The vector mean is straightforward, being the means of the individual random variables:

$$\text{Mean}(\underline{x}) = \text{Mean}\left(\begin{bmatrix} x_1 \\ \vdots \\ x_n \end{bmatrix}\right) = \begin{bmatrix} \mu_1 \\ \vdots \\ \mu_n \end{bmatrix} = \underline{\mu}_x \tag{B.13}$$

The variance becomes more complex, since we don't just have the variability of individual random variables, but also their interdependence in terms of the variability of their *product*. Thus the *co*-variance is an $n \times n$ matrix, describing the interaction of all n variables with one another:

$$\text{Cov}(\underline{x}) = \begin{bmatrix} \sigma_1^2 & \sigma_1\rho_{1,2}\sigma_2 & \sigma_1\rho_{1,3}\sigma_3 & \cdots \\ \sigma_2\rho_{2,1}\sigma_1 & \sigma_2^2 & \sigma_2\rho_{2,3}\sigma_3 & \cdots \\ \sigma_3\rho_{3,1}\sigma_1 & \sigma_3\rho_{3,2}\sigma_2 & \sigma_3^2 & \cdots \\ \vdots & \vdots & \vdots & \ddots \end{bmatrix} = \Sigma_x \tag{B.14}$$

where σ_i is the standard deviation of x_i, and $\rho_{i,j}$ is the *correlation coefficient* between x_i and x_j, measuring the strength of the linear relationship between the two random variables:

Given	Then x_i and x_j are
$\rho_{i,j} = -1$	On a straight line with negative slope
$-1 < \rho_{i,j} < 0$	Negatively related
$\rho_{i,j} = 0$	Uncorrelated
$0 < \rho_{i,j} < 1$	Positively related
$\rho_{i,j} = 1$	On a straight line with positive slope

Conveniently, given any linear transformation T of a random vector $\underline{x}$, the mean and the covariance transform as

$$\text{Mean}(T\underline{x}) = T\underline{\mu}_x \qquad \text{Cov}(T\underline{x}) = T\Sigma_x T^T \tag{B.15}$$

The covariance measures the correlations, the strength of the *linear* relationships between all pairs of random variables in a vector, however the *actual* relationship between a pair of variables may certainly not be linear. The stronger test, whether there is *any* relationship between two random variables, is that of *independence*:

- x and y are *uncorrelated* if there is no *linear* relationship between them;
- x and y are *independent* if there is no relationship of *any* kind between them, meaning that the conditional statistics obey

$$p(x \mid y) = x \qquad \text{or} \qquad p(y \mid x) = y. \tag{B.16}$$

Independent variables are therefore also uncorrelated; however uncorrelated variables may or may not be independent.

The covariance matrix of (B.14) has a close connection with the algebra of eigendecompositions. All real-symmetric matrices are diagonalizable, therefore *any* covariance matrix will be diagonalizable. From (A.26), given a covariance Σ_x we know that the covariance can be diagonalized via its eigendecomposition:

$$\text{Eigendecomposition} \quad \Sigma_x \underline{v}_i = \lambda_i \underline{v}_i \quad \longrightarrow \quad T \Sigma_x T^{-1} = \Lambda \tag{B.17}$$

From (A.27) we know that T is an orthogonal matrix, therefore $T^{-1} = T^T$ and (B.17) becomes

$$T \Sigma_x T^T = \Lambda \tag{B.18}$$

From (B.15) we recognize (B.18) to be the covariance stemming from a linear transformation T applied to random vector $\underline{x}$; that is,

$$\underline{y} = T\underline{x} \quad \text{Cov}(\underline{x}) = \Sigma_x \quad \longrightarrow \quad \text{Cov}(\underline{y}) = \text{Cov}(T\underline{x}) = T \Sigma_x T^T = \Lambda. \tag{B.19}$$

So the covariance of the transformation $T\underline{x}$ is *diagonal*, meaning that all of the transformed elements are uncorrelated with one another, thus the statistics of $\underline{x}$ have been decoupled by transformation T.

Since the diagonal elements in Λ are both the variances of the elements in $\underline{y}$ and also the eigenvalues of Σ_x, we interpret the eigenvalues as variances. One of the more powerful applications of this interpretation is to dimensionality reduction, in particular a method known as *principal components*, whereby one can most efficiently preserve the most significant statistical variability in $\underline{x}$ by keeping the largest eigenvalues.

That is, suppose we are given an n-element random vector $\underline{x}$ characterized by covariance Σ_x, where we wish to compress $\underline{x}$ by transforming it to a smaller vector $\underline{y}$ consisting of $m < n$ elements. We can find the eigendecomposition of Σ_x and order the eigenvalues

$$\lambda_1 \geq \lambda_2 \geq \ldots \geq \lambda_n \geq 0. \tag{B.20}$$

The transformation T in (B.19) consists of the eigenvectors of Σ_x. Suppose we define a transformation T_m consisting of the m eigenvectors, one per row, associated with the m largest eigenvalues, then

$$\underline{y} = T_m \underline{x} = \begin{bmatrix} \underline{v}_1^T \\ \vdots \\ \underline{v}_m^T \end{bmatrix} \underline{x}, \tag{B.21}$$

that is, $\underline{y}$ contains the principle components of $\underline{x}$, the optimum compression of $\underline{x}$ maximally preserving the statistical variability.

All of the preceding discussion is premised on somehow knowing the statistics of a random variable or vector. In practice, however, we need to take given sample data and *deduce* the statistics, then referred to as *sample* statistics. The general problem of inferring a model from data, known as system identification or machine learning, is well outside the scope of this text, however estimating a sample mean or covariance is relatively straightforward.

Given q samples $x_1, \ldots, x_q$ of a random variable x,

$$\begin{aligned} \text{Sample Mean}\big(\{x_i\}\big) &= \frac{1}{q}\sum_i x_i \\ \text{Sample Variance}\big(\{x_i\}\big) &= \frac{1}{q-1}\sum_i \big(x_i - m_x\big)^2 \end{aligned} \tag{B.22}$$

The slightly more complex equations in the random vector case appear as

$$\begin{aligned} \text{Sample Mean}\big(\{\underline{x}_i\}\big) &= \frac{1}{q}\sum_i \underline{x}_i \\ \text{Sample Covariance}\big(\{\underline{x}_i\}\big) &= \frac{1}{q-1}\sum_i \big(\underline{x}_i - \underline{m}_x\big)\big(\underline{x}_i - \underline{m}_x\big)^T \end{aligned} \tag{B.23}$$

Given a random variable x with true mean μ and standard deviation σ, the standard deviation of the *error* between the true and sample means is given by

$$\text{Standard Deviation}\big(\mu_x - m_x\big) = \frac{\sigma}{\sqrt{q}} \tag{B.24}$$

Therefore to improve the accuracy of the sample mean requires either a reduction in sample variability (σ) or an increase in the number (q) of samples.

Further Reading

As in the preceding appendix on linear algebra, there are a great many books written on all aspects of probability and statistics. In general, you would begin your search at Library of Congress call number QA273.

Three initial suggestions to get you started would include the books by Haigh [2], DeGroot et al. [1], and Walpole et al. [3].

References

1. M. DeGroot and M. Schervish. *Probability and Statistics*. Pearson, 2010.
2. J. Haigh. *Probability: A Very Short Introduction*. Oxford, 2012.
3. R. Walpole, R. Myers, S. Myers, and K. Ye. *Probability & Statistics for Engineers & Scientists*. Pearson, 2006.

Appendix C
Notation Overview

To keep the text as accessible as possible, I have tried to emphasize a relatively simple and (hopefully) consistent mathematical notation.

First and foremost, we will regularly need to distinguish between scalars, vectors, and matrices:

x	A *lower case* variable: a scalar, a single value
$\underline{X}$	An *underlined* variable: a column vector
X	An *upper case* variable: a matrix

The above rules continue to apply when we are talking about parts of vectors or matrices:

x_i	A scalar, the ith element of vector $\underline{x}$
$\underline{x}_i$	A vector, the ith vector in some sequence
x_{ij}	A scalar, an entry in matrix X

and, although not occurring frequently, to constants as well:

0	The scalar value of zero
$\underline{0}$	A vector of zeros

For dynamic systems,

$$\dot{\underline{z}}(t) = f\big(\underline{z}(t), \theta\big) \qquad \text{or} \qquad \dot{\underline{z}}(t) = A(\theta) \cdot \underline{z}(t)$$

which play a major role in the text, we have the following dedicated notation:

$\underline{z}$	The state of the dynamic system
t	Time
n	Discrete–time index, if the dynamic system is discretized over time
$f()$	A nonlinear or linear representation of the dynamics
A	A necessarily linear representation of the dynamics
θ	A parameter influencing the behaviour of the system
u	An optional input to the system, possibly deliberate or accidental

P. Fieguth, *An Introduction to Complex Systems*,
DOI 10.1007/978-3-319-44606-6

with associated derived notation which describes the behaviour of the system:

$\bar{\underline{z}}$	A fixed point
$\mathcal{B}$	A basin of attraction
λ	An eigenvalue of the linear or linearized system
Λ	A diagonal matrix collecting all of the eigenvalues of a system
$\underline{v}$	An eigenvector of the linear or linearized system
V	A matrix collecting all of the eigenvectors of a system
Δ	The determinant of the system matrix, the product of the eigenvalues
τ	The trace of the system matrix, the sum of the eigenvalues
ϵ	A small perturbation
δ	A discretization step, possibly in time or in space

Although the text does not lean heavily on statistical theory, we will frequently need to be able to describe some sense of uncertainty or distribution:

μ	A mean, some measure of the centre of a distribution
σ	A standard deviation, with corresponding variance σ^2
$\sim$	Is distributed as, thus
	$x \sim (\mu, \sigma^2)$ — x has mean μ and variance σ^2
	$x \sim \mathcal{N}(\mu, \sigma^2)$ — x obeys a Gaussian distribution
	$x \sim \mathcal{U}(a, b)$ — x is uniformly distributed from a to b
	$x \sim \mathcal{E}(\lambda)$ — x is exponential with parameter λ
ρ	A correlation coefficient, a measure of statistical relationship
$p()$	A probability density, a statistical measure of relative likelihood
γ	The exponent for a power law distribution
$\underline{w}$	A stochastic element, a random vector, random noise

Finally, in terms of inference and system observation, we have

C	The forward model, the function measuring an underlying state
ρ	The reflectivity of a surface as a function of angle or wavelength
λ	Wavelength

Index

P. Fieguth, *An Introduction to Complex Systems*,
DOI 10.1007/978-3-319-44606-6